T0291797

CAMBRIDGE LIBRARY COLLECTION

Books of enduring scholarly value

Earth Sciences

In the nineteenth century, geology emerged as a distinct academic discipline. It pointed the way towards the theory of evolution, as scientists including Gideon Mantell, Adam Sedgwick, Charles Lyell and Roderick Murchison began to use the evidence of minerals, rock formations and fossils to demonstrate that the earth was older by millions of years than the conventional, Bible-based wisdom had supposed. They argued convincingly that the climate, flora and fauna of the distant past could be deduced from geological evidence. Volcanic activity, the formation of mountains, and the action of glaciers and rivers, tides and ocean currents also became better understood. This series includes landmark publications by pioneers of the modern earth sciences, who advanced the scientific understanding of our planet and the processes by which it is constantly re-shaped.

Travels Through the Alps of Savoy and Other Parts of the Pennine Chain

The renowned geologist James D. Forbes (1809–68) presents an account of his systematic exploration of alpine mountain regions and glaciers in this important 1843 publication. Forbes' graphic descriptions of alpine scenery and his mountaineering feats are combined with detailed records of his scientific research and experiments. The study cemented Forbes' reputation in the field, which was later to be commemorated by the naming of the Aiguille Forbes in the Alps and of Mount Forbes in both Canada and New Zealand. The aim in writing the book, says Forbes, was to 'illustrate the physical geography of a particular district in one of the most frequented regions of the Alps'. In doing this, he draws upon the important work of the 'bold mountaineer' Horace Bénédict de Saussure, whose feats and achievements Forbes clearly admired. The book is still viewed today as one of the most influential and important publications on mountaineering.

Cambridge University Press has long been a pioneer in the reissuing of out-of-print titles from its own backlist, producing digital reprints of books that are still sought after by scholars and students but could not be reprinted economically using traditional technology. The Cambridge Library Collection extends this activity to a wider range of books which are still of importance to researchers and professionals, either for the source material they contain, or as landmarks in the history of their academic discipline.

Drawing from the world-renowned collections in the Cambridge University Library and other partner libraries, and guided by the advice of experts in each subject area, Cambridge University Press is using state-of-the-art scanning machines in its own Printing House to capture the content of each book selected for inclusion. The files are processed to give a consistently clear, crisp image, and the books finished to the high quality standard for which the Press is recognised around the world. The latest print-on-demand technology ensures that the books will remain available indefinitely, and that orders for single or multiple copies can quickly be supplied.

The Cambridge Library Collection brings back to life books of enduring scholarly value (including out-of-copyright works originally issued by other publishers) across a wide range of disciplines in the humanities and social sciences and in science and technology.

Travels Through the Alps of Savoy and Other Parts of the Pennine Chain

With Observations on the Phenomena of Glaciers

James D. Forbes

CAMBRIDGE
UNIVERSITY PRESS

CAMBRIDGE UNIVERSITY PRESS

Cambridge, New York, Melbourne, Madrid, Cape Town,
Singapore, São Paolo, Delhi, Mexico City

Published in the United States of America by Cambridge University Press, New York

www.cambridge.org
Information on this title: www.cambridge.org/9781108037662

© in this compilation Cambridge University Press 2012

This edition first published 1843
This digitally printed version 2012

ISBN 978-1-108-03766-2 Paperback

This book reproduces the text of the original edition. The content and language reflect
the beliefs, practices and terminology of their time, and have not been updated.

Cambridge University Press wishes to make clear that the book, unless originally published
by Cambridge, is not being republished by, in association or collaboration with, or
with the endorsement or approval of, the original publisher or its successors in title.

The original edition of this book contains a number of colour plates,
which have been reproduced in black and white. Colour versions of these
images can be found online at www.cambridge.org/9781108037662

Pl. 1.

Draw from Nature by Professor Forbes L. Edin. del.

GLACIER TABLE, ON THE MER DE GLACE.

Day & Haghe lith.rs to the Queen.

TRAVELS

THROUGH

THE ALPS OF SAVOY

AND

OTHER PARTS OF THE PENNINE CHAIN

WITH OBSERVATIONS ON THE PHENOMENA OF

GLACIERS.

BY

JAMES D. FORBES, F.R.S., Sec. R.S. Ed., F.G.S.,

CORRESPONDING MEMBER OF THE INSTITUTE OF FRANCE,
AND PROFESSOR OF NATURAL PHILOSOPHY IN THE UNIVERSITY OF EDINBURGH.

Sage mir was du an diesen kalten und starren Liebhabereyen gefunden hast.

GOETHE.

EDINBURGH:
ADAM AND CHARLES BLACK,
AND LONGMAN, BROWN, GREEN, AND LONGMANS, LONDON.

M.DCCC.XLIII.

EDINBURGH : PRINTED BY T. CONSTABLE,
PRINTER TO HER MAJESTY.

DEDICATION.

TO M. BERNARD STUDER,

DOCTOR IN PHILOSOPHY,

CORRESPONDING MEMBER OF THE ROYAL ACADEMY OF SCIENCES OF TURIN, OF THE NATIONAL INSTITUTE OF WASHINGTON, OF THE GEOLOGICAL SOCIETY OF FRANCE, ETC. ETC., AND PROFESSOR OF GEOLOGY AND PHYSICS IN THE UNIVERSITY OF BERNE.

MY DEAR SIR,

In former times, Dedications were usually the fulfilment of a stipulation, by which Patronage was to be purchased by Eulogy. But since Patronage has ceased to secure success to mediocrity, and complimentary phrases have become too trite to be gratifying, a Dedication has become a rare appendage to a book.

Nevertheless, it has always appeared to me an opportunity to be valued by a literary man, of expressing publicly his respect for the talents, and his esteem for the character of another, in terms requiring no rhetorical embellishments, because, in that case, the language of Truth and of Eulogy is the same.

That you were my companion through several of the most interesting scenes described in this Volume, would alone be a good reason for requesting permission to dedicate it to you, especially as its appearance is not wholly unconnected with conversations which then passed between us.

But when I add, that your intimate acquaintance with the Alps and their structure, derived from many years of unwearied research, gives you an especial right to judge of a Work relating to their Geography and Natural History— a farther reason for this Dedication will be understood by those who are aware that the best-informed are usually the most candid judges of the merits of others.

Independently of this, I am happy in being able to claim your sympathy and friendship as the best reason of all,—a sympathy derived from common pursuits, and a friendship which, though not yet old, may certainly be affirmed to be not untried.

I remain,

My Dear Sir,

Yours very sincerely,

James D. Forbes.

Edinburgh, 1st July 1843.

CONTENTS.

LIST OF ILLUSTRATIONS.

LITHOGRAPHS.

MAPS.

TOPOGRAPHICAL SKETCHES.

THE PENNINE CHAIN
OF ALPS

based on Keller's Map of 1842
with corrections

French Leagues.
English Miles.

Eng.d by W.& A.K. Johnston

TRAVELS

THROUGH

THE ALPS OF SAVOY,

&c. &c.

CHAPTER I.

THE ALPS, AND ALPINE TRAVELLERS.

WAYS OF TRAVELLING AND BOOKS OF TRAVELS—THE ALPS AN UNEXHAUSTED FIELD—DE SAUSSURE—HIS SUCCESSORS—THE AUTHOR'S EXPERIENCE— TRAVELLING IN SWITZERLAND—ACTION AND SPECULATION IN TRAVEL- LING—PLAN PROPOSED—THE PENNINE CHAIN OF ALPS.

Men travel from a great variety of motives, and they publish their travels perhaps from a still greater. The manner of travelling, and the forms of publication are equally diverse, and mark strongly the features of the age. The folio of the sixteenth and seventeenth cen- turies, and the quarto of the eighteenth, and even of our own time, have melted into the modern duodecimo : and something like a cor- responding change may be traced in their contents. " Pilgrimages" are out of date, and the traveller's portfolio on his return is as light in comparison as his portmanteau at starting : both are necessarily pro- portioned to the rapidity of his movements. The modern facilities for locomotion extend not only to England, France, Germany, and what in former days was called the *grand tour*, but gentlemen now walk across Siberia with as little discomposure as ladies ride on horseback to Flo- rence. Even the Atlantic is but a highway for loungers on the

A

American continent, and the overland route to India is chronicled like that from London to Bath.   The Desert has its post-houses, and Athens has its omnibuses.

One consequence of this surprising change has been, upon a great scale, like that which the existence of railways has produced in any particular district.   Persons who travel for the purpose of seeing, and relating what they have seen, are in such haste to escape from more familiar and accessible objects, that the world gradually accepts it as a principle, that what is worth describing must be distant by at least the breadth of an ocean, or half a continent, from the home of the traveller.   The result is, that those who write books of travels with other objects than to make money, or to indulge a harmless vanity, have usually sought remote countries for the subject of their writings. Thus by an insatiate thirst for novelty, and for communicating what is most new or strange, rather than what is worth knowing, we find that the proper dignity of an intelligent book of travels has been often overlooked.   The question may yet remain, whether it is not a greater service to the community to show how much remains to be seen and studied in countries, comparatively speaking, accessible to all, than to write detailed descriptions of regions presenting few natural objects of importance, or of remote tribes, unvisited perhaps only because uninteresting or dangerous.

To write a book of Travels in the Alps, will no doubt appear to many persons a very unpromising as well as superfluous undertaking, it being taken for granted that what is so easily accessible must be perfectly well known ; and the absence of any recent book of the kind, intended for more than a temporary object, and speedy oblivion, might tend to confirm the idea that no such work is required.   It has, however, been the result of journeys continued throughout several summers, in countries commonly called the best known in Europe, that I am persuaded that even in these there is yet much to be seen, much explained, and much of which a general account may prove as interesting as that of visits to more distant, though scarcely more unknown lands.   An excellent work might be written —and it would be a large one—on the less known or undescribed parts of the most frequented districts of Europe, which would show what a narrow line it is—no broader sometimes than the coloured border on a common map, separating one province or kingdom from another—which divides the known from the unknown; the highway along which roll daily the luxurious travelling equipages of Russian wealth or English fashion, and the remote valley scarcely outlined in our atlases, with a population speaking no acknowledged European language, and to whom the sight of a foreign pedestrian occurs perhaps but once or twice a-year. Nor this alone. Even where all men go, none may have stopped; what all men see, none may have observed. As in many parts of experimental science unexpected discoveries are made in a work-shop or manufactory, so the book of nature, whose pages are open to all, is read but by a few; and the notoriety of a fact, or a supposed fact, is often exactly the cause why no explanation of it is sought, or its questionable authenticity tested.

It is not too much to say that the natural history of a great part of the chain of Alps, the most instructive and grandest theatre of natural operations in Europe, is in this predicament. Thousands of travellers, many of them amongst the most enlightened men of their day, frequent them: But where is the fruit? Whilst Parry, and Franklin, and Foster, and Sabine, and Ross, and Darwin brave the severities of arctic and antarctic climates, to reap the knowledge of the various phenomena of earth and atmosphere, climate and animals, the geology, meteorology, and botany of countries comparatively uninteresting to us, are we perfectly informed of all these particulars even in our own quarter of the globe? Undoubtedly not. Where are we to look for travels like De Saussure's, and why are comprehensive works, adapted for the general reader and student of nature, to be replaced entirely by studied monographs connected with some single science in some single district?

The belief that the narrative form is at once the most agreeable and the most natural, both to author and reader, when truths progressively attained, and founded on numerous observations of detached facts, are the subjects, has finally determined me to fulfil an early and nearly abandoned project of writing a book of travels. The present volume is the result. It may be considered as an attempt to show, upon a small scale, what it is believed might advantageously be pursued upon a larger one. The aim of the work is confessedly to illustrate the physical geography of a particular district in one of the most frequented regions of the Alps; and more especially to arrive at results of a definite kind respecting the natural history of glaciers, those great masses of ice which so generally attract the casual, though only the casual, notice of travellers.

It is a duty which every one who writes owes to the public and to himself, to be informed, generally at least, of the labours of his predecessors, that he may not, even involuntarily, assume to himself credit for that which belongs to another, nor invite attention to that which is already well known.

The duty is not an easy one. Topographical literature, more than almost any other, is diffused over bulky and unindexed compilations, or more irrevocably lost in fugitive pamphlets. I well know, from some former experience, the labour of an attempt to analyse all the writings connected with even a small district, and, generally speaking, its little value as regards substantive information: and I soon saw that such an attempt in the present case would be wholly incompatible with the proposed extent of this work, and with the time which I could withdraw from other duties for writing it. I hope that it will appear, notwithstanding, that I have not been inattentive to what my predecessors have done, and that I have endeavoured throughout, in matters of original observation, to render to them their due. I do not, indeed, pretend to have read the *whole* works of Simler, Scheuchzer, and Gruner, the older alpine historians; but I have carefully examined them in many parts, especially those which bear upon the doctrine of glaciers.

The writings of De Saussure have been the subject of perpetual reference—not only at home, but amongst the very scenes which he has described, and where it is easy to retrace the exactness of his assertions, and the faithful yet sober colouring of his descriptions. Himself a man of independence and station at Geneva, early imbued with a taste for exploring mountain scenery; well instructed in the then existing state of natural history and the allied branches of physics, he was exactly in the proper position for advancing a knowledge of his own country, and of those natural laws which may best be studied amongst its mountains. His journeys were not " *tours de force*," miracles of rapidity and boldness, from which, if any thing were gained, it must have been by a sort of intuition. On the contrary, even his more adventurous expeditions were commenced with a calm foresight, peculiar to himself, of the ends to be gained, and the best methods of attaining them. He did not court dangers; he did not affect to despise even inconveniences. His fortune permitted him to travel and observe in a manner which is as rare at the present day as formerly. He was frequently accompanied by ten or twelve men, and four or six mules carrying baggage, provisions, instruments, beds, and a tent; and perhaps to this precaution may be partly attributed the long period of life through which he was able to extend his laborious researches, trying to most constitutions, and from which, he states, that even he did not fail to suffer at last. He acquired by his convenient position, (for he always resided at Geneva,) a familiarity with many of the scenes which he described by repeated visits, each one clearing up the doubts of the last. For many years he made an annual journey, and a great part of the Alps was traversed by him, although unnoticed in his published journeys. De Saussure had a particular caution and anxiety about the editorial part of his writings;—it is probable that he only selected the most complete for publication.* It undoubtedly requires a very long apprenticeship to the art of travelling to learn how to group facts,—to observe with intelligence, and to record observations on the spot with sufficient clearness and detail.† De Saussure had seen some other countries

---

* See the Advertisement to the first volume of his Travels, dated 1779.

† The practice which I have long adopted with advantage, is this:—to carry a memorandum-book with Harwood's prepared paper and metallic pencil, in which notes, and observations, and slight sketches of every description, are made on the spot, and in the exact order in

which he was able to compare with the structure of the Alps, although he does not appear to have travelled much beyond Switzerland and France, excepting one journey to Sicily. It is not easy to ascertain from his published journeys, and still less from the meagre biography which exists of him by Senebier, the exact time which he spent upon his travels each summer : As far as I can gather, however, it appears not to have been very long ; and we are struck with the circumstance that many remarkable parts of the Alps, within easy reach of Geneva, are wholly undescribed, and that he would appear very seldom to have taken up his residence for a considerable time at one station.

That De Saussure was a bold mountaineer is plain from his well known ascent of Mont Blanc, at a time when such difficulties, little understood, seemed far more formidable than at present, when the chief obstacles to such a feat, arise from its very familiarity, and the ostentatious and expensive precautions which, not unwisely perhaps, have been interposed to its accomplishment. But the most interesting and most adventurous feat which De Saussure performed, was his residence of seventeen days on the Col du Géant, a height of above 11,000 feet ; of which I shall give a separate account in the course of this work.

De Saussure's style is generally easy and interesting, without any pretension to elaboration, and in this respect his work contrasts most happily with that of Bourrit, published about the same time; who, though by no means an uninteresting writer, conveys the simplest facts through a medium of such unmixed bombast as to disgust the reader, rather than arouse his sympathies for admiration or for awe. In both, however, it must be admitted that here and there a natural

which they occur. These notes are almost ineffaceable, and are preserved for reference. They are then extended, as far as possible, every evening, with pen and ink, in a suitable book, in the form of a journal—from which, finally, they may be extracted and modified for any ultimate purpose. The speedy *extension* of memoranda has several great advantages ; it secures a deliberate revision of observations, whether of instruments or of nature, whilst the circumstances are fresh in the mind, whilst farther explanation may be sought, and very often whilst ambiguities or contradictions admit of removal by a fresh appeal to facts. By this precaution, too, the not inconsiderable risk of losing all the fruits of some weeks of labour, by the loss of a pocket-book, is avoided.

passage of calm eloquence may be found descriptive of natural beauty, and of the sentiments, irrepressible in most minds, of Natural Religion, which familiarity with great mountain scenery peculiarly calls forth.

De Saussure has aimed at variety in his work—and beyond a doubt successfully. Topography, natural history, and personal adventure are happily combined ; and many persons who would have been repelled by a professed work on the geology of the Alps, have read with avidity one which offers so much else to their attention. Even at the present time De Saussure's Travels can hardly be called obsolete, because no other work has replaced them ; and though the geology of his day is in some degree exploded, the texture of the work is sufficient to retain for it a permanent interest. The arrangement is generally topographical; if any place has been repeatedly visited, the description refers to the general result of the observations. If only once visited, the narrative form is adopted. Occasional chapters are devoted to the explanation of subjects which have occupied much of his attention, without reference to particular localities. Such an arrangement has many advantages. The plates in De Saussure's work, though more faithful than those of Bourrit, are not happy, and the maps give but an unfavourable impression of the state of topography or of art at that time in Switzerland.

There is scarcely one of the more modern writers with whom I am acquainted, whose writings can be classed with those of the great historian of the Alps. The reputation of De Saussure seems to have deterred others, however well qualified, from resuming and continuing a work which, whether we regard the state of knowledge when it was written, or the vast extent of Alpine country scarcely noticed or unmentioned in its pages, ought rather to have been considered as a commencement and a model, than as the completion of an undertaking so vast and so varied. Far be it from me to underrate what has been accomplished since his time for the natural history of Switzerland by most able and zealous observers in special departments of science, to which the excellent Journals, and the valuable Academical Transactions published there, especially those of Geneva, and of the Swiss Society, bear ample testimony. Far be it from me to overlook the

monographs by Necker and Studer, Escher and De Charpentier,
Lardy and Zumstein in the country itself, and those of many eminent
foreigners connected with geology; and of Venetz, De Charpentier,
Agassiz, and Rendu on the subject of glaciers, which have recalled attention to what De Saussure had only outlined, and whilst showing the
incompleteness of his generalisations, revived in us at the same time
admiration of his genius, his fidelity, and his varied knowledge. Two
works only of late years seem at all to emulate De Saussure in style
or matter—the "Naturhistorische Alpenreise" of Hugi, and the
"Etudes dans les Alpes" of Necker.

The former is a singular, we might call it a fantastical work. With
a praiseworthy desire to benefit experimental philosophy, as well as
the sciences of observation, by his very unusual and intrepid journeys,
the Professor of Soleure, describes in detail his instruments and the
results he obtains with them, which are often, however, so much
at variance with those of indubitable authority, as to render us somewhat diffident in the adoption of them. We cannot but remark, too,
that the ostentatious style of travelling which he preferred, often with
twelve or fifteen companions and guides largely paid, was necessarily
confined to very short and interrupted excursions, which in most cases
were brought to a premature close by bad weather, when he was compelled to break up his band, and relinquish his objects. Amidst much
which appears so paradoxical in Hugi's writings, as to pass with many
for fabulous, we perceive a bold and determined spirit daring to follow
nature, and in the lively, sometimes really eloquent, descriptions of
scenery, we discover, too, the heart that can feel nature. Amidst a
mass of dry details there is sufficient narrative to render the volume of
Hugi agreeable reading, and this is another testimony to the value of
the style of writing of which we speak, which presents even scientific
truths in a form calculated to interest persons at large.

Very different is the volume of M. Necker, Honorary Professor of
Mineralogy at Geneva, and grandson of De Saussure. It forms only
a portion of an extended work, calculated to embrace the geology of a
large district of the Alps. Its arrangement is, however, rather systematic than local, and, therefore, it wants some of the liveliness which

characterises the writings of the author of the " Voyages dans les Alpes." Containing as it does abundant references to the localities of the Alps, to which the author's enquiries are especially directed, it is enriched with the fruits of his observations and long residence in other countries. Still by its nature and arrangement it is a book of geology and not of travels.*

With respect to the present work, it is considerably more special than I could have wished it to be, had my sole object been to give a specimen of a continuation of travels in the Alps in the manner of De Saussure. Whilst general geology may be considered as the basis of his work, or the investigation which guided the course of his travels, the theory of glaciers, and of the departments of geology and topography more immediately connected with them, forms the groundwork of mine. This circumstance has led me in the journeys, which are to be described in this volume, through more wild and remote scenery than any other inducement, except perhaps a passion for the chase, is likely to carry a traveller. Some account of these scenes may have sufficient interest for the general reader to induce him to excuse the more scientific details. I can at least plead, in excuse for an attempt which I feel to border on presumption,—the endeavour to follow the great historian of the Alps in his own country, and to meet him on his own ground,—that it is upon no sudden impulse that I came forward with the hasty notes of a few months' tour to lay them before the public. It is now a good many years since I proposed to myself to travel, not as an amusement, but as a serious occupation, and with De Saussure before me as a model. I have reason to be glad that circumstances, by postponing its execution, led me to appreciate more fully the

---

* For a copious list of works published on Switzerland, see EBEL's *Guide du Voyageur*, tom. i. Many of these contain valuable information ; and even from common tour and guide-books, useful facts may often be gained. Those of Ebel himself, of Latrobe and Simond, of Fröbel and Engelhardt, of Brockedon and Bakewell, may be mentioned ; but they do not properly come under the class of works referred to in the text.

difficulties of the plan, and to come to its fulfilment after some experience, with moderated expectations of ultimate success. The habit
of observation, I have already observed, is of slow growth—to use
opportunities, we must prepare to seize them. I had the advantage of
receiving my first impressions of Switzerland in early youth, and I
have carefully refreshed and strengthened them by successive visits to
almost every district of the Alps between Provence and Austria. I
have crossed the principal chain of Alps twenty-seven times, generally
on foot, by twenty-three different passes,* and have of course intersected the lateral chains in very many directions. In my journeys I
have most frequently been alone, although occasionally I have had the
advantage of eminent naturalists and esteemed friends for my companions, from whose superior knowledge I have been happy to gain in-

---

* The following are the names and directions of these Passes, commencing with the Ma-
ritime Alps :—

Col d'Argentière,	From Barcellonette,	To Coni.
Col de Vallante,	Abries,	Castel Delfin.
Col de la Traversette,	Abries,	Saluces.
Col de la Croix,	Abries,	Lucerna.
Mt. Genévre,	Briancon,	Fenestrelles.
Mt. Cenis,	Lans-le-Bourg,	Susa.
Petit St. Bernard,	Moutiers,	Aosta.
Col du Bonhomme,	St. Gervais,	Courmayeur.
Col du Géant,	Chamouni,	Courmayeur.
Col de Ferret,	Martigny,	Courmayeur.
Grand St. Bernard,	Martigny,	Aosta.
Col de Fenêtres,	Bagnes,	Valpelline.
Col d'Arolla,	Evolena,	Valpelline.
Col du Mont Cervin,	Zermatt,	Chatillon.
Monte Moro,	Saas,	Macugnaga.
Simplon,	Brieg,	Duomo d'Ossola.
St. Gothard,	Altorf,	Bellinzona.
Splügen,	Coire,	Chiavenna.
Stelvio,	Innsbruck,	Bormio.
Brenner,	Innsbruck,	Botzen.
Velber-Tauern,	Mittersill,	Windisch Matrei.
Malnitzer Taureen,	Gastein,	Ober Vellach.
Prebühel,	Eisenerz,	Bruck am Mur.

formation as a learner, and by whose urbanity and kindness the roughest way has been smoothed, and the longest day beguiled.

I have likewise undertaken similar journeys in other mountainous countries with a view to compare the results. I have spent a part of ten summers on the Continent, and six of these in the Alps and adjacent country. I have thus repeated my visits to the same spot, and without almost any exception, I have found more to enjoy, to admire, and to learn on the renewal of my acquaintance with it. Most of the places described in this volume have been visited twice, and several of them in four different years. As the mere novelty of travelling wears off, its deeper charms impress themselves more indelibly—the habits of observation and of thought are strengthened—the short term of human life itself seems to expand in proportion to the variety and greatness of the objects contemplated; and if the solitary pedestrian in foreign parts feels his heart often glow with thoughts which bear him untiring company, incommunicable, and with which the stranger cannot intermeddle, he may yet have an honest gratification in attempting to convey to others some part of his enjoyment in the conquest of obstacles, and in the pursuit of truth.

Switzerland is undoubtedly one of the most agreeable, as well as most interesting countries in the world to travel in. Its parts rise to all the elevation which is necessary in order to convey to the imagination the fullest sense of the sublime in such objects, whilst their dimensions—gigantic, no doubt, compared to the mountains of the British Islands—do not present the unwieldy extent of the Andes or Himalaya. There is no *transverse* valley in the Alps—that is, one leading directly from the plains to the highest ridge—up which an active man cannot walk in two days, and the actual passage of the chain may usually be effected in one. Now, any great increase upon such a scale necessarily wearies the traveller with monotony, even though it be the monotony of grandeur, whilst it tasks his physical powers by keeping them too long upon the stretch. The circuit of Mont Blanc or Monte Rosa is quite as long and fatiguing as most persons will consider necessary to give them a vivid conception of an immense hill; and if we accurately examine the slow progress which the uneducated eye makes to a correct estimate of magnitudes and distances in the Alps, we find that, practically, their scale is sufficiently great to afford to at least nine-tenths of travellers the most majestic conceptions with which such objects can at all inspire them.

Add to this, that the actual height of the zone of perpetual snow is as great as that of any mountains in the world, with one or two exceptions; for the highest land on the surface of the globe is near the equator, where the corresponding high temperature raises the limit at which perpetual snow *commences* to nearly the extreme height of European mountains. The eye, which must always have some actual or conventional standard of reference, if it cannot judge by the level of the sea, takes the level of the plain as a starting-point, or, if there be no plain, the level of perpetual snow is a natural index of elevation, which, connected as it is with height, solitude, and vastness, impresses the mind with the highest sense of grandeur in natural scenery. It has often been observed, that Chimborazo is less elevated above the table land from which it rises than Mont Blanc is above the valley of Chamouni; and taking the level of perpetual snow in the Alps at 8500 feet, Mont Blanc is snow-clad throughout its higher 7000 feet. Now, a peak in the Himalaya range, in order to show as much, would need to rise to above 22,000 feet,—a height which few of them exceed.

The climate of the Alps, as well as their scale, is highly favourable to observation and to personal exertion; and it must not be reckoned a small advantage, that shelter, if not accommodation, is to be found within a moderate distance of the most retired and wildest scenery. Obstacles to travelling, whether from rude curiosity or violence, on the part of the inhabitants, are undoubtedly smaller in Switzerland than in any other country in Europe. The traveller who makes a sojourn of some length in the remoter parts of even the most frequented countries, is as often subjected to the suspicions of the authorities as of the people. The mere fact of his traversing mountains where no one habitually passes, is a sufficient crime in the eyes of the vigilant police; and if to this he add a turn for sketching, or the use of a hammer or barometer, or any such instrument, he is likely to raise a host of popular prejudices, whose extent can often only be guessed from the extraordinary conjectures which he occasionally finds to be current respecting his character and pursuits. Having, at different times, had my own share of these troubles, I appreciate highly the happy independence of a pedestrian in Switzerland, where, partly from the peculiar character of the people, partly from their form of government, and partly from their familiarity with strangers of every country, race, occupation, and fancy, no one need fear being set down either for a magician or a political agent—the two offensive categories in which he is often elsewhere included; and even the philosopher, with all his whims and his chattels, his labours and hardships which seem to end in nothing, is allowed, after a short cross-examination, and a significant shrug from the questioner, to pursue quietly an avocation, which is considered at least as harmless as walking in a motley suit would be, or twenty other vagaries. I own, that although the character of the Swiss or Savoyard peasant can rarely excite much enthusiasm or admiration amongst those who know them well, I always feel a satisfaction and a freedom from restraint when I approach these mountains and their exhilarating atmosphere, which dispel anxiety, and invite to sustained exertion.

What a field, indeed, for those whom professional and other cares, and even the habits of the society which they frequent, leave, during a great portion of life, but a few hours together, never a whole day, which can be called their own, to find themselves transplanted to a new position—time at command—no interruptions—no calls, invitations, or engagements—no letters to write or receive but those which give pleasure—surrounded by nature in its grandest forms, delighting the eye, yet affording far keener pleasure to the intellect, by the interest of the problems which it presents for solution! The attention, undistracted, dwells on the objects around without hindrance or satiety. The sense of perfect health—the rapid and refreshing sleep which attends most persons escaped from the hot-bed languour of towns to the freshness of the Alps, stimulate the powers of thought; and thought is without fatigue when each passing event gives a varied tone to it—

when each step furnishes a new subject for its exercise—when all nature is our laboratory, and we read the axioms of *her* philosophy indelibly engraven on the eternal hills.

Mere change of scene and active exercise produce fatigue at last, unless the mind have some wholesome employment as well as the body; and most of those who have made the trial will probably regard as amongst the happiest periods of their lives those in which a favourite study has been pursued in the retirement of mountain scenery. Mornings of active exercise, from sunrise till afternoon, and evenings of quiet thought and speculation, with here and there a day interposed of easy society with intelligent travellers, or employed in reducing and digesting the knowledge previously acquired by observation, give the sense of living twice over. The body and the mind are alike invigorated and refreshed; weariness from fatigue, and weariness from inactivity, are forgotten, together with the other evils of our more artificial existence. The student in his closet exhausts his powers by one kind of toil, whilst the fox-hunter and deer-stalker exhausts them by another; both call it pleasure; but the one is all too exclusively speculative, the other too exclusively active. Let speculation and action minister to one another; then, like a well-compacted body, the members act in harmony,—the double exercise prevents fatigue. Happy the traveller who, content to leave to others the glory of counting the thousands of leagues of earth and ocean they have left behind them, established in some mountain shelter with his books, starts on his first day's walk amongst the Alps in the tranquil morning of a long July day, brushing the early dew before him, and, armed with his staff, makes for the hill-top, (begirt with ice or rock, as the case may be,) whence he sees the field of his summer's campaign spread out before him, its wonders, its beauties, and its difficulties, to be explained, to be admired, and to be overcome.

" Ignotis errare locis, ignota videre
Flumina gaudebat ; STUDIO MINUENTE LABOREM."

It only remains to be added here, that the country which it is proposed to describe in the present volume, includes exactly that part of the Alpine chain called by the ancients the "Pennine Alps," a term of doubtful origin, but which it is convenient to retain, as having no modern synonyme. It extends from the Col du Bonhomme on the west side of Mont Blanc, to Monte Rosa inclusive, thus comprising the highest ground in Europe, and the two most colossal mountain groups. The map facing the first page of this work shows its limits, and will be found useful in tracing the routes to be described. It has been compiled with care, though on a small scale. The basis is the last edition of Keller's map. The valley of Erin and its neighbourhood are corrected from Fröbel; those of St. Nicolas and Saas from De Charpentier and Engelhardt; the *southern* side of Mont Rosa from Von Welden and the new Sardinian Survey; and the passes and glaciers between St. Bernard and Mont Cervin from my own observations, so far as they go.

The Pennine chain is particularly distinguished by the number and extent of its glaciers; and as the study of these formed the chief object of my journey in 1842, upon which the material of this volume is based, it presented itself as the most natural field for my enquiries. The *mer de glace* of Chamouni, from its very easy access, and its great extent and variety of surface, seemed to me the most eligible post, and I am inclined to think that it is, on the whole, the best fitted in Switzerland or Savoy for investigations like those which I had in view. Within a stone's throw of the ice, at the Montanvert, is to be found sufficient shelter, fitted for a permanent residence of some weeks or months, which is of the very first importance in the prosecution of a task requiring much perseverance, detail, the use of a multiplicity of instruments, the performance of calculations, and the making of drawings. I know from experience how little of this can be accomplished in a temporary residence, such as a tent or hut, without tables, chairs, or a fire; and however amusing such privations are for a time, and however pleasant it may be to laugh over them in good company, such expeditions tend rather to amusement than edification. I preferred, therefore, in general, the least expensive and least ostentatious methods of pursuing my enquiries, and I felt the necessity of carrying them out alone. I employed neither draughtsman, surveyor, or naturalist; every thing that it was possible to do I executed with my own hands, noted the result on the spot, and extended it as speedily as possible afterwards. My only assistant was a very intelligent and very worthy guide of Chamouni, Auguste Balmat by name, to whom I shall have frequent occasion to refer in these pages; and I am indebted to the friendship of the Curé of Chamouni, M. Lanvers, for having recommended him to me, as well as for many other acts of substantial kindness, for which I shall ever remain his debtor.*

Although, as has been said, I was acquainted, from former visits, with many of the places to be described, yet all the detailed observations which will be given were conducted during the course of last season, (1842.) The information collected in that time will at least, I hope, be thought creditable to my industry, and it may be an encouragement to persons who might be withheld (as no doubt many have been) from similar undertakings, by an erroneous estimate of the scale of assistance and expenditure required, which may truly be termed the trappings and paraphernalia of science, to know what may be effected with patience and previous study, in a moderate space of time, and in a very simple way.

I spent the latter part of June 1842 at the Montanvert, (Chamouni,) the first half of July on the southern side of Mont Blanc and in Piedmont. I then returned to the Montanvert by the Col du Géant, and continued my experiments on the *Mer de glace* until the 9th August. I then passed a month on a journey (partly in company with M. Studer) to Monte Rosa and the adjacent country, when I returned for the second time to Chamouni, and spent the remainder of September on or near the glacier.

---

* It happened very rarely indeed that I required any other assistance than that of Balmat.

Moraine near the Montanvert, Chamouni.

CHAPTER II.

SOME ACCOUNT OF GLACIERS GENERALLY.

THE SNOW LINE—THE WASTE OF ICE AND ITS SUPPLY IN GLACIERS—CAUSES OF WASTE—MOTION—FALLEN BLOCKS—MORAINES, MEDIAL AND LATERAL —GLACIER TABLES AND CONES—FORMATION OF HOLES IN ICE—VEINED STRUCTURE OF THE ICE IN GLACIERS—THE GLACIERS OF THE AAR AND RHONE—THE NEVE—CAUSE OF GLACIER MOTION—DE SAUSSURE'S GRAVITATION THEORY—DE CHARPENTIER'S THEORY OF DILATATION—OBJECTIONS TO EACH.

> " Where so wide,
> In old or later time, its marble floor
> Did ever temple boast as this, which here
> Spreads its bright level many a league around."
>
> DYER'S *Fleece.*

IT has already been said, that no small part of the present work refers to the nature and phenomena of glaciers. It may be well therefore, before proceeding to details, to explain a little the state of our present knowledge respecting these great ice-masses, which are objects of a kind to interest even those who know them only from description, whilst those who have actually witnessed their wonderfully striking and grand characteristics can hardly need an inducement to enter into some inquiry respecting their nature and origin.*

I have already alluded to the fact, that high mountains in every part of the world are covered with snow. It is enough for our present purpose, that *the fact is*, that the atmosphere becomes colder as we ascend in it, until that cold reaches a great and hitherto unmeasured intensity. Consequently, by merely ascending the slope of a hill, we pass through successive gradations of seasons. Whilst the plains are covered with the verdure of summer, eternal winter reigns upon the summits, and thus the stupendous ranges of the Himalaya or the Andes present, in one condensed picture, all the climates of the earth, from the tropics to the poles.

Since then, the long summer's day, of six months' duration in the arctic regions, is insufficient to melt the accumulated ice, it is not surprising, that at a certain height above the earth's surface, snow always lies,—a height greatest at the equator, amounting there to 16,000 feet above the sea, which, in the Swiss Alps, has diminished to 8700 feet, and which in very high latitudes reaches to the level of the ocean, so that there the natural covering of the earth is snow, and the very soil is frozen to an increasing depth. The mere continuance of snow on any spot does not suppose that snow never melts there. Were that the case, a progressive and unceasing accumulation would be the result; the position of the *snow line*, or what is often erroneously called *the line of perpetual congelation*, is determined solely by this circumstance, that during one complete revolution of the seasons, or in the course of a year, *the snow which falls is just melted and no more.*

Now, a snow-clad mountain is *not* a glacier. Whence the real difference, or how it comes that in some climates glaciers are produced in situations and circumstances apparently similar to those which yet do

---

* For some further details than would be consistent with the due length of this preliminary chapter, I would refer the reader to an article in the *Edinburgh Review*, for April 1842, on the subject of Glaciers, which has been admirably translated into French in the *Annales de Chimie*, for October and November 1842. From that article, (written by myself,) I have extracted one or two passages in this and the next chapter.

not produce them in others, is a question which we do not mean now to handle.   But let us first see what is understood by a glacier in the more familiar sense of the word.*   The common form of a glacier is a river of ice filling a valley, and pouring down its mass into other valleys yet lower.   It is not a frozen ocean, but a frozen torrent.   Its origin or fountain is in the ramifications of the higher valleys and gorges, which descend amongst the mountains perpetually snow-clad. But what gives to a glacier its most peculiar and characteristic feature is, that it does not belong exclusively or necessarily to the snowy region already mentioned.   The snow disappears from its surface in summer as regularly as from that of the rocks which sustain its mass.   It is the prolongation or outlet of the winter-world above; its gelid mass is protruded into the midst of warm and pine-clad slopes and green sward, and sometimes reaches even to the borders of cultivation.   The very huts of the peasantry are sometimes invaded by this moving ice, and many persons now living have seen the full ears of corn touching the glacier, or gathered ripe cherries from the tree with one foot standing on the ice.

Thus much, then, is plain, that the existence of the glacier in comparatively warm and sheltered situations, exposed to every influence which can insure and accelerate its liquefaction, can only be accounted for by supposing that the ice is pressed onwards by some secret spring, that its daily waste is renewed by its daily descent, and that the termination of the glacier, which presents a seeming barrier or crystal wall immoveable, and having usually the same appearance and position, is, in fact, perpetually changing—a stationary form, of which the substance wastes—a thing permanent in the act of dissolution.

The result of the heat of the valley in thawing the ice, is a stream of ice-cold turbid water, which issues from beneath its extremity, and which, by gradually undermining, works out a lofty cavern, from beneath which it rolls.   This water is derived from various sources: in the

---

* *Glacier* French, *Gletcher* German, *Ghiacciaia* Italian.   But the glaciers have also provincial names, as *Fern* in the Tyrolese Alps, *Käss* in Carinthia, *Vedretto* in part of Italy, *Biegno* in the Vallais, *Ruize* in Piedmont, *Serneille* in the Pyrenées.

first place, from the natural springs which, it may be conceived, rise from the earth beneath the ice, just as they would do in any other valley. This source remains, in a great measure, even in winter, when the glacier stream, though diminished, does not vanish. Secondly, from the heat of the earth in contact with the ice, which probably melts annually a very small thickness of its mass. This, too, will not depend upon the season. Thirdly, the fall of rain upon the whole area which the glacier valley drains—which acts, in the first place, by melting the superficial ice and snow ;—and the rain water being thus reduced to the freezing point, washes through the cracks and fissures of the ice by innumerable streamlets, which unite beneath its mass, and swell the general stream. Fourthly, the waste of the glacier itself due to the action both of sun and rain—a most important item, and which constitutes the main volume of most glacier streams, except in the depth of winter. It is on this account that the Rhine and other great rivers, derived from Alpine sources, have their greatest floods in July, and and not in spring or autumn, as would be the case if they were alimented by rain-water only. On the same account, the mountain torrents may be seen to swell visibly, and roar more loudly, as the hotter part of the day advances, to diminish towards evening, and in the morning to be smallest.

The lower end of a glacier is usually very steep and inaccessible. This arises, in some cases, from the figure of the ground, over which the glacier tumbles in an icy cascade often a thousand feet high. Its middle course is more level, and its highest portion, again, steeper: thus the final ice-fall of the glacier Des Bois at Chamouni is inclined 20°, the mean portion between 4° and 5°, and the higher part at least 8° or 10°.

The mean or middle portion of the glacier is a gently sloping icy torrent, from half a mile to three miles wide, more or less undulating on its surface, and this undulating surface more or less broken up by *crevasses*,* which, generally nearly vertical in their direction, have a

---

* The translation of the French word *crevasse* into the English *crevice*, is so evidently inapplicable to these vast fissured chasms, that we shall constantly adopt the French spelling.

width of from a few inches to many feet; and a length which sometimes
extends almost from side to side of the glacier. In all this, there is
little or no resemblance to water tranquilly frozen. The surface is not
only uneven, but rough ; and the texture of the ice wants the homo-
geneity of that formed on the surface of lakes. The hollows, which
appear but trifling when viewed from a height and compared with the
expanse of ice, are individually so great as to render the passage
amongst them toilsome in the extreme, even independent of the crevasses ; and the traveller who has to walk for several hours along a
glacier, will often prefer scrambling over stones and rocks on the side,
to the harassing inequalities which appeared at first so trivial. In a
day of hot sunshine or of mild rain, the origin of the hummocky ridges
is apparent : the intervening hollows have every one of them their rill,
which, by a complicated system of surface-draining, discharge the
water, copiously melted by the solar influence, the contact of warm air
and the washing of the rain. These rills combine and unite into larger
streams, which assume sometimes the velocity and volume of a common
mill-race. They run in icy channels, excavated by themselves, and
unlike the water escaping from *beneath* the glacier, being of exquisite
purity, they are both beautiful and refreshing. They seldom, however,
pursue their uninterrupted course very far, but reaching some crevasse,
or cavity in the glacier, mechanically formed during its motion, they
are precipitated in bold cascades into its icy bowels—there, in all probability, to augment the flood which issues from its lower termination.
Nothing is more striking than the contrast which day and night produce in the superficial drainage of the glacier. No sooner is the sun
set, than the rapid chill of evening, reducing the temperature of the air
to the freezing point or lower, the nocturnal radiation at the same
time violently cooling the surface—the glacier life seems to lie torpid
—the sparkling rills shrink and come to nothing—their gushing mur-
murs and the roar of their water-falls gradually subside—and by the
time that the ruddy tints have quitted the higher hill-tops, a deathlike
silence reigns amidst these untenanted wilds.

Winter is a long night amongst the glaciers. The sun's rays have
scarcely power to melt a little of the snowy coating which defends the

proper surface of the ice ; the superficial waste is next to nothing ; and
the glacier torrent is reduced to its narrowest dimensions.

The glacier in this part of its course is more or less covered with
blocks of stone which move along with it, or rather are borne down
upon its surface. The motion of the glacier we have already inferred
from the subsistence of the ice in valleys where the daily waste is immense, and where yet the glacier maintains its position ; but its progress is also well marked by the displacement of great blocks of stone
upon its surface, which, from their size or figure, cannot be mistaken,
and which may be watched from year to year descending the icy stream
whose deliberate speed they mark, as a floating leaf does that of a current of water. These detached rocks fall from the cliffs which usually
bound both sides of a glacier in its middle portion, and from which the
alternate effects of frost and thaw rapidly and surely separate them. They
may be seen to fall almost every summer's day, in consequence of the loos-
ening of the icy bands which hold together fragments previously wrenched
asunder by the irresistible expansion of freezing water. A single promontory may yield a great stream of those blocks in the course of years;
were the ice stationary, they would accumulate on its surface at the
base of the promontory, but as the ice advances, its charge is carried
along with it, and the glacier becomes burdened on both sides with a
band of blocks, which by their geological character bear the impress
of their origin, and thus not unfrequently bring down to the reach of
the mineralogist specimens which otherwise would be quite unattainable, and whose native place may be surely inferred by observing the
direction of the ice stream which is charged with them. Such, for
instance, are fragments of the *Gabbro* of Saas, which has not yet been
found *in situ*, but which is discharged by the glacier of Allalein, in the
Vallais, near Monte Rosa.

What a curious internal historical evidence, then, does a glacier bear
to the progress of events which have modified its surface ! It is an
endless scroll, a stream of time, upon whose stainless ground is engraven the succession of events, whose dates far transcend the memory of
living man. Assuming, roughly, the length of a glacier to be twenty
miles, and its annual progression 500 feet, the block which is *now* dis-

charged from its surface on the terminal moraine, may have started from its rocky origin in the reign of Charles I. ! The glacier history of 200 years is revealed in the interval, and a block larger than the greatest of the Egyptian obelisks, which has just commenced its march, will see out the course of six generations of men ere its pilgrimage too be accomplished, and it is laid low and motionless in the common grave of its predecessors.

The stony borders now described are called *Moraines* in French, *Guffer* or *Gufferlinien* in German. The glacier retains a portion of them on its own surface, and throws up a part upon the bank or shore which confines it. If the shore be precipitous, it will be conceived that the blocks cannot be stranded, and therefore either remain on the surface of the ice, or fall into the occasional vacuities left between the ice and its wall, and there are ground and chafed, acting, of course in a notable manner upon the rock, and producing rounded surfaces, the angles being worn off, and grooves and scratches parallel to the direction of motion of the ice. All this is an immediate and necessary consequence of the fact of the glacier moving and heaving blocks along its edges. When the rocky slope or shore of the glacier is less steep, since, owing to the heat reflected and communicated from the ground, the ice almost invariably sinks towards the sides, a portion of the load of blocks falls over, and is accumulated in a ridge as from an over filled waggon. But the more striking cause of this accumulation is the oscillation of dimension of the glacier at different seasons, and in different years. If the glacier from any cause whatever becomes enlarged, and, like a swollen torrent, occupies its bed to an unusual depth, the *Moraines* are uplifted with it, and when the return of summer or warmer seasons reduces the ice to its former bulk, the blocks are deposited at the higher level. Such moraines are to be seen in the neighbourhood of most modern glaciers, and they are important to be observed, because the existence of similar mounds in places remote from existing glaciers, has been inferred to demonstrate their former presence. The sketch at the head of this chapter represents a moraine about a hundred feet above the present level of the Mer de glace of Chamouni.

It often happens that two glaciers, having separate sources, unite in a common valley, exactly as two rivers would do. Each, of course, has

its edging moraine or *list*, and therefore where the glaciers unite, the
two inner moraines must unite also. This does not, however, alter
their character; as in the case of the Rhone and Saone uniting their
streams at Lyons, each preserves the characteristic colour of its water
for a long way down, unmixed with its neighbour river—so, much more,
does the compact and firm glacier. The *debris* proper to each unite
upon the surface, and mark by a band of stones, often for miles, the
actual separation of the two ice streams, which otherwise would (at
that distance) have become undistinguishable. These united bands,
which are equal in number to the junctions of tributary glaciers which
combine to form a great one, are called *medial moraines*, whilst those
formerly described are called *lateral moraines*. The former have only
been distinctly explained of late years, by Agassiz* and De Charpentier,† whilst the latter have been long perfectly understood. There is
nothing more surprising to be found in the writings of De Saussure
than the most unsatisfactory explanation which he gives of medial
moraines.‡

As these facts are important to be
distinctly apprehended, some slight
figures may tend to illustrate them.
Thus figure 1 represents a plan of an
ideal glacier composed of five streams,
A, B, C, D, E, each of which has its
lateral moraines, and the union of
these represented by the dotted lines,
1, 2, 3, 4, forms the superficial trains
of rocks which are carried along on
the surface of the ice. A mere prominent rock or islet in the ice, as that
between D and E, may yield also its
small contribution of blocks. The
section in figure 2 represents a glacier
having a steep wall, *a*, where consequently the debris are ingulfed

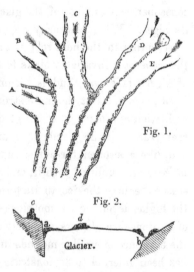

Fig. 1.

Fig. 2.

Glacier.

---

* Études sur les Glaciers, 1840.   † Essai sur les Glaciers, 1841.   ‡ Voyages, § 537.

Echellets.

Frelaporta.

Drawn from Nature by Professor Forbes, T. Picken. lith   Day & Haghe Lith.rs to the Queen.

MER DE GLACE OF CHAMOUNI, FROM LES CHARMOZ.

between the wall and the ice, producing friction, and an inclined shore, *b*, on which the lateral moraine has been deposited. There is also shown, at *c*, the position of an ancient moraine, deposited at a time when the glacier was elevated enough to have submerged the promontory *a*; one of the *medial moraines* is shown at *d*: The ice rises to a greater height under it than at any other part, owing to a circumstance to be mentioned immediately. An exact idea of the general phenomena of moraines will be obtained from the large map of the Mer de glace of Chamouni accompanying this work, and from Plate II., which gives a view of it.

The presence of these blocks upon the surface of the glacier, and of the fine sand and debris which is produced by their trituration, gives rise to a peculiar and striking class of phenomena, easily explained, yet at first sight most astonishing. The surface of a glacier is usually divided by numerous rents or crevasses, stretching as we have seen, often nearly from side to side, and into these rents blocks are continually falling. Still, the fact is, that the moraines remain upon the surface, and unless after a very long or very uneven course, they are not dissipated or ingulfed. On the contrary, the largest stones are set on a conspicuous pre-eminence,—the heaviest moraine, far from indenting the surface of the ice, or sinking amongst its substance, rides upon an icy ridge as an excrescence, which gives to it the character of a colossal back-bone of the glacier, or sometimes appears like a noble causeway, fit indeed for giants, stretching away for leagues over monotonous ice, with a breadth of some hundreds of feet, and raised from fifty to eighty feet above the general level. Almost every stone, however, rests on ice; the mound is not a mound of debris, as it might at first sight appear to be. Nor is this all. Some block of greater size than its neighbours, covering a considerable surface of the ice, becomes detached from them, and seems shot up upon an icy pedestal, in the way represented in the Frontispiece, from a real and very striking example which occurred in 1842 on the Mer de glace of Chamouni. This apparent tendency of the ice to rise wherever it is covered by a stone of any size, results from the fact, that its surface is depressed every where else by the melting action of the sun and

rain ; the block, like an umbrella, protects it from both ; its elevation
measures the level of the glacier at a former period, and as the depression of surface is very rapid—amounting even to a foot per week,
during the warm months of summer—the ice, like the fields, puts
forth its mushrooms, which expand under the influence of the warm
showers, until the cap, becoming too heavy for the stalk, or the centre
of gravity of the block ceasing to be supported, the slab begins to slide,
and, falling on the surface of the glacier, it defends a new space of
ice, and forthwith begins to mount afresh. These appearances are
called Glacier Tables. Their origin was perfectly explained by De
Saussure.

Where sand derived from the moraines has been washed by superficial water-runs into the deep cavities which are occasionally formed in
the glacier, the accumulation is at length sufficient to check the progress of the waste of ice, and what was a hole filled with sand, becomes
a pyramid projecting above its surface, and coated with the protecting
layer.* These produce glacier cones, which are amongst the most singular and apparently unaccountable of this class of phenomena. They
are sometimes astonishingly regular, 20 or 30 feet in height, and 80 or
100 in circumference ;—but this is one of the rarer appearances.

From what has been said, it will appear that a glacier has a remarkable tendency to reverse its *contour*, or to present at one time the mould
or cast of what it was at another ; any part of the surface prominently
exposed is sure to be speedily reduced, and the hollows, whether holes
or cracks or water-runs, by being silted up are protected from farther
decay. The valleys are literally exalted, and the hills levelled. It is
owing to this beautiful compensation that the glacier maintains a
tolerable evenness of surface.

A converse action, however, may be noticed. It is always on a
small scale, and there are two causes. The first occurs from the collection of small objects of a dark colour and in no great quantity on the
surface of the ice, which absorbing the solar heat, transmit it quickly to
the ice beneath, and such particles of black sand, or even leaves which

---

* See AGASSIZ, *Études*, chap. x.

are wafted by the wind from vast distances upon the glaciers, are found sinking into cavities, whilst blocks, larger than a cottage, and weighing millions of pounds, rise above the surface. The other fact is the deepening of cavities in the ice, once formed and filled with water, but containing no considerable quantity of *detritus*. These basins, or *baignoirs*, as they are usually called, appear to be formed in the following manner, first explained by Count Rumford. Water, just freezing, is *lighter* than water at a temperature somewhat higher; the water at 32°, therefore, floats on the surface of the other. Imagine a small cavity in ice, filled with water just thawed. The sun's rays first heat the surface of the water, which becoming *denser* descends, and is replaced by water at 32°. But the water which subsided with a temperature, suppose of 36°, soon communicates its heat to the sides of the icy receptacle, and being cooled to 32° it rises in its turn. The heat of the denser water is thus spent in melting the ice of the bottom of the cavity, which is thus deepened by the continual current.

The ice of which the glacier, in the stage which we have described, is composed, is unlike that produced by freezing still water in a lake or pond. Although remarkably pure and free from all intermixture of earthy matter, and even the smallest fragments of rock (except very near where it touches the soil) it is far from homogeneous, or uniformly transparent. It has been described as composed of layers of perfect ice and of frozen snow intermixed, but this does not express the fact as observed in the middle and lower glacier. The ice is indeed porous and full of air bubbles, and it is very probable that these bubbles result from the freezing of snow imbibed with water; but as it exists in the glacier it is not granular. Laminæ, or thin plates of compact transparent blue ice, alternate in most parts of every glacier, with laminæ of ice not less hard and perfect, but filled with countless air bubbles, which give it a frothy semi-opaque look. This peculiar structure, which gives to glacier ice its extreme brittleness, (which makes the formation of steps with a common hatchet a very easy task compared to what it would be in common ice,) may be compared to what geologists call the *slaty cleavage* of many rocks, rather than to stratification, properly so called. The distinction is important, and amounts

to this: that strata are deposited in succession, and owe their form
and separation to that circumstance only; whereas, slaty cleavage, or
structural planes, occur in rocks, and in many bodies, wholly irrespective of stratification or deposition, and may be communicated to a mass
after complete or partial consolidation.

The alternation of bands, then, in a glacier, is marked by blue and
greenish-blue or white curves, which are seen to traverse the ice
throughout its thickness whenever a section is made. It is, therefore,
no external accident, it is the intimate structure of a glacier, and the
only one which it possesses, and may be expected to throw light upon
the circumstances of the formation and motion of these masses. I
became acquainted with this fact by observing these bands on the
lower glacier of the Aar, when I visited it for the first time in company
with M. Agassiz and Mr. Heath in August 1841. It appeared so plain,
that I was surprised to find that M. Agassiz, who had passed a part of
two preceding summers on the same glacier, should have overlooked it.
At first he maintained that it was a superficial striping of the ice, owing
to the washing of sand along its surface; but when I showed that it
descended to a depth of twenty feet or more in the crevasses, he stated
that it must certainly have appeared since the previous year. I
speedily, however, verified its occurrence in other glaciers, where it had
not been remarked any more than on the glacier of the Aar, and from
that time the attention of glacial theorists has been generally directed
to this curious, important, and quite general phenomenon. M. Guyot,
an ingenious professor of Neufchâtel, stated, after I had left Switzerland, that he had observed a similar appearance in one glacier (that of
the Gries) some years before, which he described along with a number
of other facts connected with glaciers, in a Memoir read before a provincial meeting of naturalists at Porrentruy in France. This Memoir
remained unprinted, and the insulated fact observed on the glacier of
the Gries was forgotten, until I drew attention to its importance and
generality in 1841. It is singular, that not only in the writings of De
Saussure and the older naturalists (so far as yet appears) can there be
traced no notice of this veined structure which pervades glaciers, but in
the modern literature of the subject, Hugi's Travels, published in 1830,

and the writings of Agassiz, Godefroy, De Charpentier, and Rendu, devoted exclusively to glaciers, and published in 1840 and 1841, there is an equal silence as to the real nature of glacier structure, which we can scarcely account for, if so obviously important a fact, however difficult to explain, had been known to any of these authors. It will be seen, from the descriptions we shall have to give in another place, as well as from Plates I. and V., that this appearance is in many glaciers a striking one. It has, I know, been distinctly remarked by several ingenious persons, both in this country and abroad, who yet, from not having been engaged in the special study of glaciers, or from having attached to it no particular importance, or perhaps from a very natural supposition that it must be already described, have published no account of it. Amongst others, Colonel Sabine and M. Elie de Beaumont have mentioned it to me as a circumstance which they recollected to have attracted their attention whilst on the ice; and Sir David Brewster has shown me a memorandum to the same effect made in 1814.

As observed on the glacier of the Aar it exhibited an appearance of almost vertical layers nearly parallel to the length of the glacier, inclin-

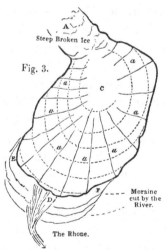

ing outwards a little, like the rays of a fan, as it approached either shore. It was difficult to make out its form at the lower termination. A visit which I subsequently paid to the glacier of the Rhone, satisfied me that these apparent layers bent round the lower extremity of the glacier, dipping forwards as the surface was depressed, and at last becoming nearly or quite horizontal. This circumstance was explained by me at the time to M. Agassiz, (who was not present when I visited the Rhone glacier,) and also in a paper read in December 1841 to the Royal Society of Edinburgh, which forms Appendix, No. I., to this volume. It was illustrated by the sketch of a ground plan of the glacier of the Rhone, fig. 3, which shows, by the dotted lines, how

these structural veins
followed the circum-
ference of the ice ; but
the circumstance of the
varying dip and cup-
like form which they
assume, will be better
understood from fig. 4,
in which an attempt is
made to represent by
ideal sections a glacier

Fig. 4. Ideal sectional view of a glacier.

of this kind. My later observations will be detailed in another part
of the present work.

The phenomena we have described belong to the *middle* and *lower*
part of glaciers : let us now trace them to their origin amongst the perpetual snows, of which it is impossible to doubt that the glaciers are in
some sense the outlets,—that is, that they are fed or maintained by the
snow which otherwise would accumulate in the higher valleys. But let
us at once and for good dispel the natural error which induces many
persons who have never seen a glacier, to suppose that in its middle or
lower part it is fed or increased by the snows which fall annually upon
its surface, or are wedged in at its sides. Let it be distinctly understood that the snow as regularly disappears and melts from the surface
of the glacier as it does from the surface of the ground in its neighbourhood. Here and there a patch of the last winter's snow in a shady
nook, or a deep crevasse, enables us in a moment to draw the distinction
between ice and mere hardened snow ;—the one is blue or bluish-green,
and transparent, though filled here and there with air bubbles and
cavities ; the other remains throughout the whole year of a dull white,
without an approximation to the character of ice, or the least tendency
to enter into a complete union with it at the point of contact : the two
things remain as distinct as the geological contact of sandstone with
slate. However, therefore, the middle and lower glacier is maintained,
it is most assuredly not by the assimilation into its substance of the
fallen snow of winter, either superficially or laterally.

The case, however, differs in the higher ice-world. We find the snow disappearing more and more tardily from the surface of the ice as we ascend, and at length we reach a point where it never disappears at all. This is, of course, the *snow line* upon the glacier. It is somewhat lower than the snow line upon the ground, but it is fundamentally the same thing. Here a well marked change occurs. There is often a passage nearly insensible from perfect snow to perfect ice : at other times the level of the superficial snow is well marked, and the ice occurs beneath it.* No doubt the transition is effected in this way :— The summer's thaw percolates the snow to a great depth with water ; the frost of the succeeding winter penetrates far enough to freeze it at least to the thickness of one year's fall, or by being repeated in two or more years, consolidates it more effectually. Thus M. Elie de Beaumont most ingeniously accounts for the alleged non-existence of glaciers between the tropics,† by the fact that the seasons there have no considerable variations of temperature, and the thaw and frost do not separately penetrate far enough to convert the snow into ice.

The part of a glacier covered with perpetual snow is what I understand to be meant by the term *névé* in the writings of the modern glacialists, although that term is vaguely defined. It will appear, however, to offer a very distinct and important line of demarcation in this view. It is where the surface of the glacier begins to be annually renewed by the unmelted accumulation of each winter. It is called *firn* in German Switzerland. These accumulations of snow produce a true stratification, which has been recognised by De Saussure, Zumstein, Hugi, and all later writers. I agree with De Charpentier‡ in thinking, that this stratification is entirely obliterated as the névé passes into complete ice. Other writers, and particularly M. Agassiz,||

---

* See the description of the Glacier of Macuguaga in the latter part of this volume, and the references to *Veined Structure* in the index.

† *Annales des Sciences Geologiques, par Rivière,* 1842. I find a notice of Glaciers in Little Thibet, lat. 36° N., in VIGNE's *Travels in Kashmir,* ii., 285.

‡ *Essai,* p. 3.

|| " Un autre caractère propre à la glace des glaciers et qui tient à son mode de formation, c'est qu'elle est stratifiée," &c.—*Etudes sur les Glaciers,* p. 40.

have attempted to trace these layers throughout the lower glacier, and
maintain that the whole glacier is stratified horizontally—an oversight
which appears to have arisen from the appearance of the varied structure or the terminal front of the glacier being nearly in parallel horizontal lines, (see fig. 5, p. 29,) which were imagined to be a continuation of the stratification of the névé, the intermediate structure having
been overlooked.

The granulated structure of the névé is accompanied with the dull
white of snow passing into a greenish tinge, but rarely, if ever, does it
exhibit the transparency and hue of the proper glacier. The deeper
parts are more perfectly congealed, and the bands of ice which often
alternate with the hardened snow, are probably due to the effect of a
thaw succeeding the winter coating, or any extraordinary fall. On
exposed summits, where the action of the sun and the elements is
greater, the snow does not lie so long in a powdery state, and the exposed surface becomes completely frozen. Hence the highest part of
Mont Blanc, the Jungfrau, and other summits, is covered with hard
ice, though always of a whitish colour. The floating masses called
ice-bergs in the polar seas are, for the most part, of the nature of *névé*,
mere consolidated snow. The occurrence of true ice is comparatively
rare, and is justly dreaded by ships.

The crevasses in the névé differ from those in the glacier by their
greater width and irregularity, and by the green colour of the light
transmitted by their walls, and the appearance of horizontal stratification. The substance is far more easily fractured than ice, and also more
readily thawed and water-worn; hence the caverns in the névé are
extensive and fantastical, often extending to a great distance under a
deceptive covering of even snow, which may lure the unwary traveller
to destruction. Sometimes, through a narrow slit or hole opening to
the surface of the névé, he may see spacious caverns of wide dimensions, over which he has been ignorantly treading, filled with piles
of detached ice-blocks, tossed in chaotic heaps, whilst watery stalactites
—icicles of ten or twenty feet in length—hang from the roof, and
give to these singular vaults all the grotesque varieties of outline
which are so much admired in calcareous caverns, but which here

show to a far greater advantage, in consequence of their exquisite transparency and lustre, and from being illuminated, instead of by a few candles, by the magical light of a tender green which issues from the very walls of the crystal chambers.

Considering, then, the glaciers as the outlets of the vast reservoirs of snow of the higher Alps,—as icy streams moving downwards, and continually supplying their own waste in the lower valleys, into which they intrude themselves like unwelcome guests, in the midst of vegetation, and to the very threshold of habitations,—it is a question of the highest interest to explain the cause of this movement of the ice. The inquiry may not result in any immediately useful application, but its interest is the same as that which belongs to the theories of physical astronomy, or to the cause of any other natural effect which commends itself to our attention by its grandeur, its regularity, and its resistless power. The glacier moves on, like the river, with a steady flow, although no eye sees its motion ; but from day to day, and from year to year, the secret silent cause produces the certain slow effect ;—the avalanche feeds it, and swells its flowing tide, the mightiest masses which lightning or the elements roll from the mountain side upon its surface, are borne along without pause ; when the glacier, advancing beyond its usual limit, presses forward into the lower valleys, it turns up the soil, and wrinkles, far in advance, the greensward of the meadows, with its tremendous ploughshare ; it brings amongst the fields the blasts of winter, and overthrows trees and houses like stubble in its ruthless progress ; no combination of power and skill can stay its march, and who can define the limit of its aggression ? Its proud waves are however stayed, and by causes as mysterious as those of its enlargement, it retreats year by year within its former limits ; but where the garden and the meadow were, it has left a desolate spread of ruin, like the fall of a mountain, which never again may be tilled, and over which for at least half a century not even a goat shall pick the scanty herbage.

The theory which appears at first sight most readily to account for

C

the leading facts, is that maintained by De Saussure, that the valleys in which glaciers lie being always inclined, their weight is sufficient to urge them down the slope, pressed on by the accumulations of the winter snows above, and having their sliding progress assisted by the fusion of the ice in contact with the ground resulting from the natural heat of the earth.*

This cause of motion has been rejected as insufficient by M. de Charpentier, who has supported another which, though like the last, suggested originally by an older author, Scheuchzer, as De Saussure's was by Gruner, having received a scientific form and detail in his hands we will call "Charpentier's Theory of Dilatation," as the other may be called "Saussure's Gravitation Theory," or the sliding theory.

De Charpentier's theory is this. The snow is penetrated by water and gradually consolidated. It remains however, even in the state of ice, always permeable to water by means of innumerable fissures which traverse the mass; these are filled with fluid water during the heat of the day, which the cold of the night freezes † in these fissures, produc-

---

* I wish to quote De Saussure's own statement of his views, which is very distinct,— " Ces masses glacées entraînées par la pente du fond sur lequel elles reposent, dégagées par les eaux de la liaison qu' elles pourraient contracter avec ce même fond, soulevées même quelquefois par les eaux* doivent peu-à-peu glisser et descendre en suivant la pente des vallées ou des croupes quelles couvrent. C'est ce glissement lent, mais continu, des glaces sur leurs bases inclinées, qui les entraîne jusque dans les basses vallées, et qui entretient continuellement des amas de glaces dans les vallons assez chauds pour produire de grands arbres, et même de riches moissons."—*Voyages*, § 535. For De Saussure's very clear views respecting the action of the heat of the earth, see § § 532, 533, 534, 535, 739, &c.

† The following quotations make it quite plain, that it is to the difference of the temperature of the day and night alone, that the freezing of the water in the capillary fissures is attributed :

" Il resulte .... que pendant les jours d'été les glaciers s'imbibent d'eau, et que celle-ci s'y congèle pendant les nuits."—CHARPENTIER, *Essai*, p. 11.

" Dans la plupart des nuits durant l'été les glaciers augmentent de volume par la congélation de l'eau qu'ils ont absorbée pendant le jour'.'—*Ib.*, p. 14.

" Cette alternative de gelée et dégel, comme je viens de le dire, a lieu pendant la belle saison, surtout à l'époque des jours les plus chauds suivis de nuits fraîches.—P. 15. See also p. 23. Compare AGASSIZ, *Etudes*, pp. 165, 211.

* This singular expression seems to point to a cause of motion like that developed in a curious paper on Glaciers, published by Mr. Robert Mallet at Dublin in 1838.

ing by the expansion which freezing water undergoes in that process, an immense force, by which the glacier tends to move itself in the direction of least resistance—in other words, down the valley: This action is repeated every night during summer, in winter the glacier being assumed to be perfectly stationary.*

In the *Edinburgh Review* for April 1842, I have stated some leading objections to both of these theories, to which I refer the reader. I will content myself with specifying one against each, which seems conclusive.

1. If the glacier *slide* down its bed, why is not its motion continually accelerated—*i. e.*, why does it not result in an avalanche? And is it conceivable that a vast and irregular mass like a glacier, having a mean slope of only 8° and often less than 5°, can *slide*, according to the common laws of gravity and friction, over a bed of uneven rock, and through a channel so sinuous and irregular, that a glacier is often embayed in a valley whence it can only escape by an aperture of half its actual width? On all mechanical principles, we answer, that it is impossible. We may add, that many small glaciers are seen to rest upon slopes of from 20° to 30°, without taking an accelerated motion; and this is conformable to the known laws of friction. It is known, for instance, to architects, that hewn stones, finely dressed with plane surfaces, will not slide over one another until the slope exceeds 30°.

2. The dilatation theory is founded on a mistake as to a physical fact. I am sorry to put it in this way, but it is unavoidable; and the

---

* " Une troisième objection contre le mouvement des glaciers par leur propre poids, se tire de leur immobilité pendant l'hiver. Car c'est un fait reconnu et attesté par tous ceux qui demeurent dans leur voisinage, tels que les habitans de Chamounix, de Zermatt, de Saas, de Grindelwald, &c., que les glaciers restent *parfaitement stationaires* dans cette saison, et ne commencent à se mouvoir qu'à la fonte des neiges."—CHARP., p. 36.

" Le mouvement des glaciers suppose des alternances fréquentes de chaud et de froid. * * * Il en resulte que l'hiver est pour les glaciers l'époque de repos."—AGASSIZ, p. 175.

"Pendant l'hiver toute sa masse (*c. a. d.*, du glacier) est dans un état de rigidité permanente qui la maintient dans *une immobilité complète* jusqu'à l'époque du retour des variations de la température."—AGASSIZ, p. 212.

respectable author of the only intelligible or precise account of the theory will, I hope, excuse me for pointing it out.

"The maximum temperature which a glacier can have," observes M. de Charpentier,* "is 0° centigrade, or 32° Fahr, and the water in its fissures is kept liquid only by *the small quantity of heat* which reaches it by the surface water and by the surrounding air. Take away this sole cause of heat—*i. e.*, let the surface be frozen, and the water in the ice must congeal." Now, this is a pure fallacy; for the fact of the Latent Heat of water is entirely overlooked. The latent heat of water expresses the fact, that where that fluid is reduced to 32°, it does not immediately solidify, but that the abstraction, not of "a small quantity," but a very large quantity indeed, is necessary to convert the water at 32° into ice at 32°. Not a great deal less heat must be abstracted than the difference of the heat of boiling water and that at common temperatures. The fallacy, then, consists in this: Admitting all the premises, the ice at 32° (it is allowed that in summer, during the period of infiltration, it cannot be lower,) is traversed by fissures extending to a great depth, (for otherwise the dilatation would be only superficial,) filled with surface water at 32°. Night approaches, and the surface freezes, and water ceases to be conveyed to the interior. Then, says the theorist, the water already in the crevices and fissures of the ice, and in contact with ice, instantly freezes. Not at all; for where is it to deposit the heat of fluidity, without which it cannot, under any circumstances, assume the solid form? The ice surrounding it cannot take it; for, being already at 32°, it would melt it. It can only, therefore, be slowly conveyed away through the ice to the surface, on the supposition that the cold is sufficiently intense and prolonged to reduce the upper part of the ice considerably below 32°. The progress of cold and congelation in a glacier will therefore be, in general, similar to that in earth, which, it is well-known, can be frozen to the depth of but a few inches in one night, however intense the cold. Such a degree and quantity of freezing as can be attributed to the cold

* *Essai*, pp. 9 and 104.

of a summer's night must therefore be absolutely inefficient on the mass of the glacier.

I will not stop to consider the attempt made by M. de Charpentier to show, that the friction of any length of a glacier upon its bed may be overcome as easily as the shortest, from a consideration of the forces producing dilatation; but it is as indefensible on mechanical grounds as the preceding theory is on physical ones, (*Essai*, p. 106.) I quote from M. de Charpentier, not because his defence of the theory of dilatation is more assailable than that of others, but because his work is the only one in which an attempt is made to explain its physical principles with precision.

I cannot admit, then, that either the sliding or dilatation theory can be true in the form which has hitherto been given to them. When I first began to study the subject minutely, under the auspices of M. Agassiz, in 1841, its difficulty and complication took me by surprise, and I soon saw, that to arrive at any theory which, consistent with the rigour of physical science at the present time, would be worthy of the name, a very different method of investigation must be employed from that which was then in use by any person engaged in studying the glaciers.

To a person accustomed to the rigour of reasonings about mechanical problems, the very first *data* for a solution were evidently wanting— namely, the amount of motion of a glacier in its different parts at different times. A few measures had indeed been made from time to time by M. M. Hugi and Agassiz, of the advance of a great block on the glacier of the Aar from one year to another, but with such contradictory results as corresponded to the rudeness of the methods employed; for in some years the motion appeared to be *three times* as great as in others. I then pointed out to M. Agassiz, how, by the use of fixed telescopes, the minutest motions of the glacier might be determined,—a suggestion which he has, I believe, since put in practice. It seems very singular that ingenious men, with every facility for establishing facts for themselves, should have relied on conclusions vaguely gathered from uncertain data, or the hazarded assertions of the peasantry about matters in which they take not the slightest interest. The supposed immobility of the glaciers in winter,—the supposed greater velocity of the sides than the centre of the ice, were amongst the assumptions traditionally handed down, upon no sufficient authority, and, I believe that I may safely affirm, that not one observation of the rate of motion of a glacier, either on the average, or at any particular season of the year, existed when I commenced my experiments in 1842. Far from being ready to admit, as my sanguine companions wished me to do in 1841, that the theory of glaciers was complete, and the cause of their motion certain, after patiently hearing all that they had to say, and reserving my opinion, I drew the conclusion that no theory which I had then heard of could account for the few facts admitted on all hands, and that the very structure and motions of glaciers remained still to be deduced from observation.

The preceding sketch of the phenomena of glaciers is, I am aware, very imperfect. It would, however, make this chapter too long, and encroach upon the special topics of this work, to enlarge further; but several explanations and references to other authors will be made immediately in the chapters where they may be naturally introduced.

Pierre à Bot, near Neufchâtel.

CHAPTER III.

ON THE GEOLOGICAL AGENCY OF GLACIERS.

REASONS FOR SUPPOSING GLACIERS TO HAVE CAUSED THE TRANSPORTATION OF
PRIMITIVE BLOCKS IN SWITZERLAND——PLAYFAIR——VENETZ——DE CHAR-
PENTIER——AGASSIZ——ACTION OF GLACIERS UPON ROCKS——THE PIERRE À
BOT——THE BLOCKS OF MONTHEY——ABRADED SURFACES NEAR THE PISSE-
VACHE——OBJECTIONS TO THE THEORY OF ANCIENT GLACIERS CONSIDERED.

" Zuletzt wollten zwey oder drei stille Gäste sogar einen Zeitraum grimmiger Kälte zu
Hülfe rufen, und aus den höchsten Gebirgszugen, auf weit in's Land hingesenkten Glet-
schern, gleichsam Rutschwege für schwere Ursteinmassen bereitet, und diese auf glatter

Bahn fern und ferner hinausgeschoben im Geiste sehen. Sie sollten sich, bei eintretender Epoche des Aufthauens, niedersenken und für ewig in fremden Boden liegen bleiben."

WILHELM MEISTER's *Wanderjahre.* Edit. 1829.

TRANSLATION.

Finally, two or three hitherto silent guests called to their aid a period of intense cold, with glaciers descending from the highest mountain ranges, far into the low country, upon which, as on an inclined plane, heavy primitive blocks were slid farther and farther onwards: So that, at the period of thawing of the ice, they sank down and remained permanently on the foreign soil.

It has been stated in the last chapter that glaciers are useful geological emissaries, which bring down from the inaccessible mountain chains, where they originate, specimens of rock which otherwise would be unattainable. The glaciers have a *carrying* power which exceeds that of any other agent, vital or mechanical. Hence, geologists having observed the benefit which existing glaciers conferred on their cabinets, naturally enough considered whether the enigmatical dispersion of blocks of foreign materials upon wide surfaces of country, in the most singular positions, might not be due to the former existence of extensive glaciers in those regions.

The occurrence of vast masses of primitive rocks, apparently without any great wear and tear of travelling, upon secondary or alluvial surfaces, at great distances from their origin, has been one of the numerous *opprobria* of geology. It is peculiarly so, because a thousand circumstances demonstrate that the deposition of these masses has taken place at the very last period of the earth's history. No considerable changes of surface have occurred since. These blocks are superficial, naked, deposited upon bare rock, which has received no coating of soil since, and are often placed in positions of such ticklish equilibrium that any considerable convulsion of nature, whether by earthquake or *débacle*, must inevitably have displaced them. A geologist might, therefore, fairly be asked,—" If you cannot account for these very latest and plainest phenomena of change and transport on the earth's surface, whose various revolutions you pretend to explain, how shall we follow you when you tell us of the metamorphoses of slates and the throes of granite?" And certainly geologists were put to their wits' end by such questioning, for no hypothesis seems too absurd to have found a place amidst their conjectures on the subject. Explosions, without apparent origin or cause, which projected the primitive blocks in a shower carrying them to a distance of a hundred miles or more,—currents of water which, derived from some unknown source, took their way on either side of the axis of a great chain of mountains, and with so stupendous a velocity as to carry with them blocks containing hundreds of thousands of cubic feet, and not only that, but transported them across lakes and up hills, and finally deposited them unshivered, and even with sharp angles and edges,—such were amongst the speculations proposed to account for these phenomena. A more plausible theory was that of ice rafts, by which (as on the ice bergs of the Polar seas, which are masses detached from the great glaciers of the north) blocks of stone were to be transported across lakes and wafted to the sides of distant mountains; but the immense changes which must in many cases have been admitted in the *contour* of the country, to permit the existence of such lakes, besides many peculiarities in the distribution of the blocks, at least in Switzerland, renders this ingenious theory not universally applicable.

The first person, so far as I know, who perceived the possible importance of glaciers as geological agents, was my respected predecessor, Professor Playfair. This indication, which forms part of the very note on the Transportation of Stones, in the *Illustrations of the Huttonian Theory*, is neither vague nor indirect. It is put forward as the most probable explanation of all cases of transport where immense power was obviously required:—

" For the moving of large masses of rock," says Professor Playfair, " the most powerful agents without doubt which nature employs are the glaciers, those lakes or rivers of ice which are formed in the highest valleys of the Alps, and other mountains of the first order. These great masses are in perpetual motion, undermined by the influx

of heat from the earth, and impelled down the declivities on which
they rest, by their own enormous weight, together with that of the
innumerable fragments of rock with which they are loaded. These
fragments they gradually transport to their utmost boundaries, where
a formidable wall ascertains the magnitude, and attests the force, of
the great engine by which it was erected. The immense quantity and
size of the rocks thus transported, have been remarked with astonishment by every observer, and explain sufficiently how fragments of rock
may be put in motion even where there is but little declivity, and
where the actual surface of the ground is considerably uneven. In this
manner, before the valleys were cut out in the form they now are, and
where the mountains were still more elevated, huge fragments of rock
may have been carried to a great distance; and it is not wonderful if
these same masses, greatly diminished in size, and reduced to gravel or
sand, have reached the shores, or even the bottom of the sea. *Next in
force to the glaciers*, the torrents are the most powerful instruments
employed in the transportation of stones."*

Now, as the passage immediately preceding that which we have
quoted contains a statement of the problematical facts mentioned above,
respecting the distribution of the travelled blocks over the plains of
Switzerland and on the Jura, we cannot but give to Professor Playfair
the credit of having clearly pointed out the probability of the former
greater extension of glaciers as the most powerful known agents of
transport. This was in the year 1802, before the author had had the
opportunity of personally estimating the applicability of the theory to
phenomena. The following passage from the notes of his journey in
1816, shows that his views in this respect had undergone no change in
the interval, and were only confirmed by an inspection of the erratic
blocks on the Jura, which he unhesitatingly ascribes to the former existence of glaciers which once *crossed* the lake of Geneva and the plain of
Switzerland. " A current of water," he says, " however powerful, could

---

* *Huttonian Theory*, Art. 349.

never have carried it (the Pierre à Bot, near Neufchâtel,) up an accli-
vity, but would have deposited it in the first valley it came to, and
would in a much less distance have rounded its angles, and given to it
the shape so characteristic of stones subjected to the action of water.
A glacier which fills up valleys in its course, and which conveys rocks
on its surface free from attrition, is the only agent we now see capable
of transporting them to such a distance, without destroying that sharpness of the angles so distinctive of these masses."*

Like many other anticipations of new theories, these pointed and just
observations of Professor Playfair lay dormant until the opinion which
he advanced had been separately originated and discussed. M. Venetz,
an intelligent engineer of the canton of Valais, speculating upon the
irregular periods of increase and decrease of glaciers, collected partly
from history and partly from tradition a variety of curious and distinct
facts bearing upon these oscillations of the great glaciers of the Alps.
He united them with judgment and impartiality in a memoir which
was read in 1821 to the Swiss Natural History Society, and published
in the second part of the first volume of their *Transactions*. In this
paper, M. Venetz classifies separately the facts which prove an increase,
and those showing a decrease of glaciers in modern times. The former
are certainly the most remarkable—showing that passes the most inaccessible, traversed now perhaps but once in twenty years, were frequently passed on foot, sometimes on horseback, between the eleventh
and fifteenth centuries. Thus the Protestants of the Haut Valais
took their children across what is now the Great Glacier of Aletsch to
Grindelwald for baptism ; and at the same period horses passed the
Monte Moro from Saas into Italy; and the peasantry of Zermatt, at
the foot of the Monte Rosa, went annually in procession through the
Eringer Thal to Sion, by a pass which few inhabitants of either valley
would now venture to attempt. We regard these facts, not as forming
any proof of the former great extension which carried the glaciers even
over to the Jura, but as evidencing one only of many oscillations

---

* PLAYFAIR's *Works*, I., p. xxix.

which the glacier boundaries have undergone; and as important in showing that a *very notable* enlargement of these boundaries was consistent with the limits of atmospheric temperature, which we know that the European climate has not materially overpassed within historic times. It may not, therefore, require so violent a depression of temperature as we might at first sight suppose, to account for any extension of the glaciers which the facts may require us to admit. The causes of these oscillations are as yet very obscure. I purposely refrain (for the sake of conciseness) from analyzing the theories which have been given, because I find them all unsatisfactory.

M. Venetz has, in his memoir, further pointed out certain ancient moraines, belonging to modern glaciers, which indicate their previously greater extension; an evidence which had formerly been accepted by De Saussure, especially in the case of the Glacier du Bois at Chamouni,* and that of the Rhone.† The remark is important, because it requires us to investigate the character of a moraine, so as to recognize it wherever it may be found.

It does not appear that M. Venetz has published any other memoir on the subject of glaciers; but it is quite certain that he was the first person publicly to maintain in Switzerland the doctrine of the former extension of the glaciers to the Jura, as the transporting agents of the erratics. I was introduced to M. Venetz in 1832, as the man who had originated a speculation, which, though it had not then perhaps another advocate, was acknowledged to be novel, ingenious, and bold; and the reputation which the author of it had acquired, as the intrepid and skilful engineer of the works on the glacier of Gétroz, (the cause of inundations which threatened the town of Martigny with destruction,) gave it a consequence which might not otherwise have been conceded to it.

In the second edition of Göthe's *Wilhelm Meister*, he has introduced a discussion as to the cause of the transport of erratic blocks, which I have placed at the head of this chapter, and in which the glacier theory is not forgotten, and was most likely borrowed from Playfair.

---

* *Voyages*, § 623.   † *Voyages*, § 1722.

The farther history of the geological theory need not be detailed. It received in Switzerland the powerful support of De Charpentier, and was yet farther pushed by Agassiz, who attempted to extend it with some variations to every part of the temperate zone, and to explain the distribution of the Scandinavian blocks, and those of Great Britain, by a similar action. We will confine ourselves, however, for the present, to a brief consideration of the erratic phenomena as they present themselves in Switzerland, and without attempting to demonstrate the absurdity of other suppositions, give some reasons for considering the former existence of glaciers 100 miles long or more, as a less extravagant hypothesis than almost any one will at first sight be disposed to regard it.

There are two principal grounds upon which it is maintained that the former presence of a glacier can be proved. In the first place from the TRANSPORTATION OF BLOCKS; and, secondly, from the FORM AND POLISH which glaciers give to the rocks which they chafe during their descent. The most weighty objections urged against the theory are (1,) the difficulty of admitting a former condition of climate cold enough to permit so vast an extension of glaciers as would be required; and, (2,) that under any circumstances of climate it is difficult or impossible to conceive that glaciers could have existed in the particular situations conjectured, on account of the little declivity which the surface could have had, and which it is assumed is inconsistent with their progression.

We shall consider these points briefly in order.

The transportation of blocks by existing glaciers has been already spoken of as one of their most marked prerogatives. The quantity is often so great as almost entirely to conceal the mass of the ice under the prodigious load which during a long descent is accumulated upon it. Thus, the lower parts of the Glacier de Miage, near Mont Blanc, and the Glacier of Zmutt, near Monte Rosa, are completely darkened by the quantity of rocks which they transport. And although in some cases the disappearance of these moraines, which it would seem ought to have formed in the course of ages a vast accumulation at the foot of the glacier, may require some farther explanation; in others, there is no want of evidence of their geological power, filling up entire valleys, and forming lakes, as in the case of the Glacier de Miage, just mentioned, and that of Allalein in the valley of Saas.*

The *dimensions* of the transported masses, of which we shall presently speak, offer no difficulty on this theory ; masses of nearly or quite equal size may be seen on existing glaciers, nor does there appear to be any limit to their magnitude except the cohesion of the granite or other rock of which they are formed. I have seen a mass actually on the ice of the glacier of Viesch in the Vallais, nearly 100 feet long (judging by the eye) and 40 or 50 feet high. There is also a block of green slate in the valley of Saas, pushed forward by the glacier of Schwartzberg, which contains, according to Venetz, 244,000 cubic feet. It was deposited about twenty years ago, and the glacier has now retreated at least half a mile, leaving the intervening space covered with true erratics, and which in that condition is called by the German writers, *Gletscherboden.*

Again, a very remarkable action of existing glaciers is to chafe and polish the rocks over which they are pushed or dragged, whether by their weight or by any other cause. The fact is certain that, at least at their sides, there is a continued contact between the supporting rock or wall of the glacier, and the glacier itself. Its stupendous unwieldy mass is dragged over the rocky surface, it first denudes it of every blade of grass and every fragment of soil, and then proceeds to wear down the solid granite, or slate, or limestone, and to leave most undeniable proofs of its action upon these rocks. It is very strange that this most evident and seemingly natural action should have been so long overlooked, and finally contested ; it is to De Charpentier that we owe its clear assertion and the proof, in the following passage, published in 1835.—" We know that the glaciers rub, wear, and polish the rocks with which they are in contact. Struggling to dilate, they follow all the sinuosities, and press and mould themselves into all the hollows and excavations they can reach, polishing even overhanging surfaces,

---

* For an account of all these phenomena in this work, see the index under the proper names.

which a current of water, hurrying stones along with it, could not effect."* An attentive survey of the glaciers cannot leave the slightest doubt of this action on the mind of any unprejudiced person.

There can be no doubt from observation, that a glacier carries along with its inferior surface a mass of pulverized gravel and slime, which, pressed by an enormous superincumbent weight of ice, *must* grind and smooth the surface of its rocky bed. The peculiar character of glacier water is itself a testimony to this fact. Its turbid appearance, constantly the same from year to year, and from age to age, is due to the impalpably fine *flour* of rocks ground in this ponderous mill betwixt rock and ice. It is so fine as to be scarcely depositable. No one who drives from Avignon to Vaucluse can fail to be struck with the contrast of the streams, artificially conveyed on one and on the other side of the road, in order to irrigate the parched plain of Provence. The one is the incomparably limpid water of Petrarch's fountain ; the other an offset from the river Durance, which has carried into the heart of this sun-burnt region the unequivocal mark of its birth amidst the perpetual snows of Monte Viso. This is the pulverizing action of ice.

Most erroneously have those argued who object to this theory that ice cannot scratch quartz—ice is only the *setting* of the harder fragments, which first round, then furrow, afterwards polish, and finally scratch the surface over which it moves. It is not the wheel of a lapidary which slits a pebble, but the emery with which it is primed. The gravel, sand, and impalpable mud are the emery of the glacier.

Although the contacts of ice and rock are very generally covered by moraines, an attentive examination of almost any glacier affords evidence to the wear of the lateral rocks. We shall shew in future chapters how unequivocally this appears on the Mer de Glace of Chamouni, and on the glacier of La Brenva, to which, in the meantime, we refer as evidence of the fact.

Having stated that the transporting and the abrading power of

---

* *Notice sur la cause probable du transport des Blocs Erratiques de la Suisse.* Par M. J. CHARPENTIER. (Extrait du tome viii. des Annales des Mines.) 1835.

glaciers is undoubted, we will now describe some of the phenomena at a distance from glaciers, which are supposed to give sure evidence of these powers having formerly been exerted. This evidence is so very remarkable (we speak now of Switzerland) as to deserve a most careful study before any hypothesis admitted to be *mechanically adequate* is rejected on grounds of indirect improbability or opposition to experience ; for the facts to be explained, if they rested on other evidence than that of eye-witnesses, would themselves be rejected as incredible and absurd.

A glance at any map of Switzerland will shew that it consists of three distinct portions—the great chain of Alps; the plain of Switzerland containing numerous lakes ; and the secondary chain of the Jura, which runs parallel to the Alps, and attains a very inferior elevation. The plain, or great valley, runs of course parallel to the two ranges which bound it, that is, in a direction from south-west to north-east, and having a breadth which may be roughly stated at 30 English miles, but the distance from the highest part of the Alps to the highest part of the Jura is not less than 80 English miles. Nearly opposite to the great gap in the main chain formed by the valley of the Rhone where it opens upon the Lake of Geneva, we have the Lake of Neufchâtel, with mountains of secondary limestone, corresponding to some parts of our Oolite formation, rising to a height of nearly 3000 feet above the valley.

Upon the slope of this range,—not at the level of the lake but considerably higher, and just facing the Rhone valley, lie extensive deposits of angular blocks of the kind of granite which especially characterizes the eastern part of the range of Mont Blanc, which is also the nearest point where the rock in question occurs *in situ*. It may be difficult to point out with certainty the locality whence these fragments are derived, as the kind of granite called *Protogine* (which contains talc instead of mica) of which they consist, is common in many parts of the Alps. But it is perfectly certain that no rock approaching to it in the remotest degree is to be found either in the Jura, or nearer than the part of the Alps which I have mentioned, and which may be from 60 to 70 miles distant as the crow flies. A great belt of these blocks occupies a line, extending for miles, at an average height of

800 feet above the level of the Lake of Neufchâtel, and above and below that line, they diminish in number, although not entirely wanting. They have been most extensively broken up and removed, for building purposes, or merely to disencumber the land, and many of them are concealed amidst the woods which clothe the mountain slope. But wherever seen they fill the mind with astonishment, when it is recollected that, as a matter of certainty, these vast rocks, larger than no mean cottages, have been removed from the distant peaks of the Alps, visible in dim perspective amidst the eternal snows, at the very instant that we stand on their debris. The most notable of these masses, called the *Pierre à Bot*, (or toad-stone,) lies in a belt of wood, not far from a farm-house, about two miles west of Neufchâtel, and near the road to Vallengin and La Chaux de Fonds. The first height above the lake being gained, (vine-clad on its lower slopes,) we come rather abruptly upon a well cultivated flat or terrace, where the farm-house just mentioned is situated. This hollow in the hill permits some accumulation of soil, which elsewhere is very thin and bare, and probably the configuration of the ground has had something to do with the deposition of the blocks, which have no doubt been carefully cleared away from this more level spot. Immediately behind, however, the hill again rises, covered with thick wood, in every part of which, not a few, but hundreds and thousands of travelled blocks may be found. Some small and rounded, but a vast number exceeding a cubic yard in contents, and perfectly angular, or at least with only the corners and edges slightly worn, but without any appearance whatever of considerable attrition, or of any violence having been used in their transport. Indeed, such violence would be quite inconsistent with their appearance and present position.

The *Pierre à Bot* is figured at the head of this chapter, from a sketch made on the spot. Its dimensions, according to Von Buch, are 50 feet long, 20 wide, and 40 high, containing 40,000 cubic feet (French.) It forms a stupendous monument of power. It is impossible to look at it without emotion, after surveying the distance which separates it from its birth-place. No wonder that geologists have vied with one another in attempting to account for so extensive and surprising a phe-

D

nomenon. If transported by water, why do these masses form a band so
high above the plain ?—why, rather, were they not buried in the depths
of the lake beneath ; and why do they show such slight marks of the
friction they must inevitably have experienced ?   If they slid down an
inclined plane, touching the Alps and Jura, of what was their plane
made, and what has since become of its material ?   Besides, how is it
possible that rough blocks could slide down any natural slope of 1° 8',
which is all that the relative positions of the blocks, and their origin,
permit ?*

Lastly, if these blocks were transported, like the erratics of the
arctic regions, upon floating rafts of ice, what was the extent, and
what the boundaries and barriers of the natural lake on which they
were transported ?   Such boundaries or barriers cannot be pointed out,
consistently with what has been said as to the unchanged condition of
the superficial deposit in Switzerland generally, since the period of the
transport of erratics.   Their orderly distribution with respect to the
nature of the rocks, those from the same origin being generally grouped
together, is inconsistent with the idea of icebergs floating hither
and thither, and wrecked or sunk by chance on any part of the lake.
Nor is this all : the supposition of a lake washing the base of the Jura
range, and cold enough to maintain a heavy fleet of ice-islands, is a
supposition as gratuitous, and very nearly, if not quite, as violent with
respect to change of climate, as that of Venetz and De Charpentier,
who attribute this transportation of rocky masses to a mere extension
of glaciers now existing, which are at this hour depositing terminal
moraines of blocks similar to those upon the Jura, but which are confined to the heads of the valleys which they formerly entirely occupied,
as well as the plains beyond.   Of course, this recession was not instantaneous, but went on gradually throughout a long series of years, so
that the moraines which commence on the Jura have covered by degrees
the whole intervening space between the former and the actual termination of the glaciers.

---

* CHARPENTIER, p. 174.

If this theory have any foundation, we ought to find confirmations of it in the valleys through which the supposed glaciers must have passed, and this we do in a most remarkable manner. Not to dwell too long on a general point, which would admit of much detail, I will confine myself to a few observations which I have had an opportunity of making, chiefly in company with M. de Charpentier himself, in the part of the Rhone valley between Martigny and the Lake of Geneva.

The narrow gorge through which the Rhone passes at St. Maurice is familiar to all Swiss travellers. If the glacier which then filled all the upper and tributary valleys whose waters now form the current of the Rhone, passed through this place, it must have been violently accumulated in this ravine, and pressed with excessive force upon the bottom and sides of the valley. The marks of glacier wear and polish are here extremely visible, especially on the rocks which occupy the bottom between St. Maurice and Bex ; and they extend to a very great height on the eastern side of the valley, exactly opposite to the village of Bex, where M. de Charpentier pointed out to me the most exquisitely polished surfaces of rock, quite as smooth as a schoolboy's slate, and displaying an artificial section of all the interior veins. After passing the defile of St. Maurice, the glacier spread itself over the enlarged basin immediately beyond, partly formed by the tributary Val d'Il-liers. The north-western face of that valley fronts the tide of ice which then flowed through the rocky defile (on the theory we are discussing), and which bore upon it with its lateral moraine. The result is not less surprising than what we have described upon the Jura. The rock here, too, is limestone, and not perhaps a fourth part of the distance of the Chaumont (above Neufchâtel,) from native granite, but the magnitude of the moraine is proportionally greater. The "blocks of Monthey," as they are called, from the village immediately below them, must be seen to be appreciated. I wandered amongst them for a whole forenoon, and though I had previously heard much of their magnitude, I had formed no idea of what I then saw. We have here, again, a belt or band of blocks—poised, as it were, on a mountain side, it may be five hundred feet above the alluvial flat through which the Rhone winds below. This belt has no great vertical height, but

extends for miles—yes, for miles along the mountain side, composed
of blocks of granite of thirty, forty, fifty, and sixty feet in the side—
not a few, but by hundreds, fantastically balanced on the angles of one
another, their grey weather-beaten tops standing out in prominent relief from the verdant slopes of secondary formation on which they rest.
They are thickest in the midst of a wood, and the traveller has
his admiration divided between the singularity of the phenomenon
and the exquisite *picturesque* of the spot.   For three or four miles
there is a path preserving nearly the same level, leading amidst the
gnarled stems of ancient chestnut trees which struggle round and
among the pile of blocks, which leaves them barely room to grow : so
that numberless combinations of wood and rock are formed, where a
landscape painter might spend days in study and enjoyment.   The
trees opening here and there display the valley of the Rhone beneath,
the exquisite meadows and orchards which surround the town of Bex,
surmounted by the lofty and imposing summit of the Dent de Morcle ;
whilst the snow-clad Alps, whose fragments are beneath our feet, close
the farther distance.   The blocks are piled one on another, the greater
on the smaller, leaving deep recesses between, in which the flocks or
their shepherds seek shelter from the snow-storm,* and seem not
hurled by a natural catastrophe, but as if balanced in sport by giant
hands.   For how came they thus to alight upon the steep, and there
remain ?   What force transported them, and when transported, thus
lodged them high and dry five hundred feet at least above the plain ?
We reply, a glacier *might* do this.   What other inanimate agent could
do it, we know not.

I have adverted to the marks of friction and polish visible upon the
fixed rocks near St. Maurice : I must add a word about another appearance higher up and which gave me a strong conviction of the impossibility of currents of water producing these effects which I examined

---

* One of these afforded shelter to a monomaniac, disappointed in love, whose sad story
is known to many of the inhabitants of the valley, who recollect him. The block, which
is figured in de Charpentier's work, is named from the poor man who lived I think for
forty years under it, *Pierre à Milan.*

carefully in August 1841. The cascade of the Pissevache between Martigny and St. Maurice, upon the left bank of the Rhone, is perfectly known to travellers, but few probably have taken the trouble to ascend to the level of the higher valley through which the stream (the Sallenche) descends before being finally precipitated. When by a toilsome climb the higher level has been gained—above 1500 feet above the Rhone valley—bare rocks are seen to rise almost precipitously on either side of the channel through which, at a great depth below, the stream leaps from crag to crag, and even the din of its greater fall is lost in the depth. Now these vertical precipices, which form the mural angle or buttress between the valley of the Sallenche and that of the Rhone (which are at right angles to one another,) are scored by *horizontal* stripes, or grooves, or fluting, evidently the result of superficial wear. But what could have worn it in this position? Could a current of water, of 1500 feet deep, have borne boulders on its surface which should leave these plain horizontal markings? What could have been moved with a steady pressure as a carpenter presses his cornice plane on the wood, or as a potter moulds with a stick his clay, pressed laterally too, with a perpendicular face of 1500 feet beneath? Nothing that I am acquainted with save a glacier, which at this day presses and moulds and scores the rocky flanks of its bed, extending to a depth often certainly of hundreds of feet beneath. A torrent however impetuous, a river however gigantic, a flood however terrific, could never do this.

The result of the attrition of fixed rocks attributed to glaciers is threefold. In the *first* place, the surface of rock, instead of being jagged, rugged, or worn into ragged defiles, is even and rounded, often dome shaped or spheroidal, showing the structure of the rock in section, and occasionally so smooth as to be difficultly accessible, as at the Höllenplatte near the Handeck. Such surfaces were called *Roches Moutonnées* by De Saussure.

Secondly, Subordinate to these general forms are the long, smooth, parallel grooves or flutings which have been already mentioned.

Thirdly, These polished grooves are often traversed by fine lines or striæ, cut as it were by a hard point, which often cross one another.

These various phenomena are observed both close to modern glaciers, and in the districts of the Alps and Jura which abound with erratics.

The striæ of the Pissevache are accompanied by the presence of erratic blocks. They are all, I think, from the neighbouring mountains to the westward: From this fact, and from the direction of the marks on the rock, I concluded, in 1841, that the Val de Trient was formerly occupied by a glacier which passed by the village of Salvent and joined the great Rhone glacier, by sweeping round the angle of the Pissevache. This conjecture will be found to be confirmed by more recent observations in the valley of Valorsine, which will be found in one of the later chapters of this work.

It remains to close this very brief sketch by referring to the two chief objections already mentioned, by which the glacier theory has been most ably opposed. And, (1.) that the cold supposed is contrary to received geological opinions, or to probability. To this I will briefly answer, first, that the opinion of geologists appears to have been far too exclusively grounded in this, as in some other parts of their science, on zoological evidence ; and in the present case that evidence appears to be both inconclusive and contradictory ; inconclusive, because new recent species (I allude to inhabitants of the ocean) are being continually found in climates to which they were not formerly supposed to belong, and contradictory, because, instead of a constantly warmer climate in former times appearing from the evidence, such as it is, of the fossil shells, it is affirmed, not without plausibility, by Mr. James Smith, Mr. Lyell, and M. Agassiz, that the shells of the particular epoch corresponding to the dispersion of erratic blocks have a decidedly arctic character. I answer, secondly, That the advocates of the theory of ice rafts require a much greater degree of cold than at present, and that all geologists, from De Saussure to M. Elie de Beaumont, admit that there are traces in certain glaciers of the Alps of their having formerly extended a certain way beyond their present limits. I observe, thirdly, that the depression of temperature need not probably be so very great, as might at first sight appear, in order to cover Switzerland with ice. It will be seen in the course of the present work that many glaciers have undergone surprising variations of extent,

and covered whole acres with their debris, within the memory of persons now living, and this due to causes which, though doubtless energetic, are not sufficiently developed to enable us clearly to define them.*
It would not be difficult to show, did space permit, that a great increase of glacier surface must result from a small depression of atmospheric temperature.

(2.) A more formidable objection has been drawn from the *small inclination* under which these primitive glaciers must have moved, and carried down their debris. The *mean* inclination of the entire glacier of the Rhone valley has been estimated by De Charpentier at 1° 8′; but the slope of a great part of its course must have been much less, and, comparing the height of the erratics near Martigny with those upon the Jura, it is estimated by M. de Beaumont at nearly 15′. The question then comes to be, can a glacier move at all under such slopes? Speaking from experience, we find the mean slope of glaciers to be much above what has been stated, but whether this is essential to their motion or not, is quite a different question; it may result merely from the actual inclination of the valleys to which the glaciers are confined by the present laws of climate. It seems impossible to give any just answer to the question,—" Under what slope would a glacier 100 miles long move?" Without first answering another,— " What is the immediate cause and mode of glacier motion?" It is hoped that something like an answer to the latter question may be found in this volume. We may then attempt to reply to the former.

Some farther illustrations of the subject of this chapter will be found in the *Edinburgh Review* already quoted, in which I have stated my opinions on several points more at large; as well as in the clear and able work of De Charpentier, where the rival theories are ingeniously handled.

In conclusion, I will call attention to two simple wood-cuts, of parallel and similar but very distant phenomena,—the one, of travelled

---

* See references to the glaciers *des Bois*, of La Brenva, of Val de Bagnes, of Leys, and Schwartzberg in the index.

blocks resting on an ice-worn surface, within a few fathoms of a modern glacier, by which they have been deposited,—the other represents a fragment of similar rock, upon a limestone surface, 90 miles in a right line from the preceding, and 60 miles from the nearest granite. The first figure is from blocks stranded by the Mer de Glace, near the Montanvert. The scene of the second is on the face of the Jura range, above Bienne, close to the great road.

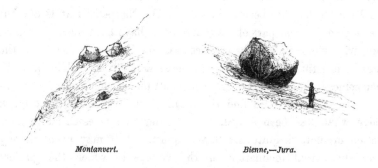

Montanvert.   *Bienne,—Jura.*

Granite Block on the Mer de Glace.

CHAPTER IV.

DESCRIPTION OF THE MER DE GLACE
OF CHAMOUNI.

PHYSICAL GEOGRAPHY OF THE ICE-WORLD——GLACIER DE LECHAUD——GLACIER
DU GEANT——SOURCE OF THE ARVEIRON——HAMEAU DES BOIS——COTE DU PIGET
——ANCIENT MORAINE OF LAVANCHI——LE CHAPEAU——MAUVAIS PAS——CATTLE
TRAVERSING THE GLACIER——BLUE COLOUR OF ICE——MONTANVERT——ACCOM-
MODATIONS——THE VIEW——LES PONTS——L'ANGLE——PASSAGE OF THE GLACIER
——LES CHARMOZ——TRELAPORTE.

Nec vidisse sat est :—durum calcavimus æquor.
OVID. *Trist.* III., x., 39.

THE glacier which occupies the vast gorge or system of valleys to the
east of Mont Blanc, is usually, and, I believe, correctly termed the
Mer de Glace;—the name of *Glacier des Bois* being confined to its lower
extremity, where, escaping from the rocky defile between the promontory of the Montanvert, and the base of the Aiguille du Dru, it
pours in a cascade of icy fragments, assuming the most fantastic forms,
into the valley beneath, between the fir woods of Lavanchi on the

one hand, and those through which the usually frequented path to the Montanvert passes, on the other. If I do not always use the *Glacier des Bois* to signify the lower, and the *Mer de Glace* the middle and upper part of this vast ice stream, I shall not probably incur any risk of being misunderstood.

It is proposed, in this chapter, to describe such peculiarities of structure, either in the valley in which the glacier lies, or in the ice itself, as may tend to illustrate the physical geography of the district; and especially the theory of existing glaciers, and of their former extension; and if the details into which I shall enter appear somewhat minute, it may be well to recollect, that the absence of such local knowledge has been the cause of much of the uncertainty under which we at present labour as to the past history of these wonderful masses. A permanent record of their present limits, condition, and phenomena, will be an important document for future times; and the conviction of this led me to incur the very great labour of constructing a detailed map of nearly the whole glacier. A more particular account of the survey will, as a matter apart, and less interesting to the general reader, be found in a separate chapter. The time required for such an undertaking, and for the minute inspection of every portion, was of the highest importance in forcing upon my attention facts which it is almost impossible not to overlook on a superficial glance; and the topographical detail I am now to give may aid the reader, in a similar manner, to transport himself in imagination to the scene of the experiments on glacier motion which I shall afterwards detail.

There is nothing more practically striking, or more captivating to the imagination, than the extreme slowness with which we learn to judge of distances, and to recognise localities on the glacier surface. Long after icy scenes have become perfectly familiar, we find that the eye is still uneducated in these respects, and that phenomena the most remarkable when pointed out, have utterly escaped attention amidst the magnificence of the surrounding scenery, the invigoration which the bracing air produces, and the astonishing effect of interminable vastness with which icy plains outspread for miles, terminated by a perspective of almost shadowless snowy slopes, impress the mind. I cannot now recall, without some degree of shame, the almost blindfold way in which, until lately, I was in the habit of visiting the glaciers. During three different previous summers, I had visited the Mer de Glace, and during two of them, 1832 and 1839, I had traversed many miles of its surface; yet I failed to remark a thousand peculiarities of the most obvious kind, or to speculate upon their cause, or else the clearer apprehension which I now have of these things, has wholly driven from my mind the previous faint impression. Of the existence of the moraines, generally, and their cause, as well as of the fact of the descent of the glaciers, I was aware, but I can scarcely recall another of the many singularities which they present, as affecting my imagination then in a lively manner—the wear and polish of the rocks—the vast masses of travelled stone thrown up high and dry far above the present level of the ice, like fragments of wreck, indicating, by their elevation on the beach, the fury of the past storm—the pillars of ice, with their rocky capitals, studded over the plain like fantastic monuments of the Druid age—or the beautiful veined structure of the interior of the ice, apparent in almost every crevasse,—these things, so far as I now recollect, were passed by unobserved.

Even in the summer of 1842, during which the present survey was made, I had abundant proof of how much remained unseen only for want of the faculty of concentrating the attention at once upon all the parts of so wide and glorious a field. We are not aware, in our ordinary researches in Physical Geography, or the natural sciences in general, how much we fall back upon our *general* knowledge and *habitual* observation in pursuing any special line of enquiry, or what would be our difficulty in entering *as men* upon the study of a world which we had not familiarly known as children. The terms of science are generally but translations into precise language of the vague observations of the uncultivated senses. Now the ice-world is like a new planet, full of conditions, appearances, and associations alien to our common experience; and it is not wonderful that it should be only after a long training, after much fatigue, and dazzling of eyes, and weary steps, and many a hard bed, that the Alpine traveller acquires some of that nice perception of cause and effect—the instinct of the children of nature—which guides the Indian on his trail, and teaches him, with unerring philosophy, to read the signs of change in earth or air.

But to return to the Mer de Glace. A glance at the map will show that this great ice river has near its origin two divided streams, derived from different sources. The westward branch, denominated the Glacier du Géant, or Glacier du Tacul, has its rise in a vast basin immediately to the eastward of Mont Blanc, confined between the proper ridge of the Alps extending to the Col du Géant, on the south, and the chain of Aiguilles of Chamouni* on the north, commencing nearest Mont Blanc with the Aiguille du Midi, and terminating with that of the Charmoz, round whose eastern foot the Mer de Glace sweeps. The other branch, called the Glacier de Léchaud, has its origin at the foot of La Grande Jorasse, one of the highest mountains of the chain which separates the Val Ferret from that of Chamouni. This glacier is smaller than its neighbour, although it is swelled before its junction by the tributary ice of the Glacier du Talèfre, which falls in upon its right bank from a detached basin, encircled by inaccessible pinnacles of rock, in whose centre is the spot called the Jardin, now so frequently visited. The length of the whole Mer de Glace is estimated by the guides of Chamouni at eighteen leagues, an enormous exaggeration, if leagues of the usual horizontal measure be reckoned. A league, however, is generally understood to mean an hour's walk amongst the mountains, and in that view the estimate will appear less absurd, although it conveys no correct idea of superficial extent. The distance from the foot of the Glacier du Bois to the top of the Glacier de Léchaud might probably be traversed in six or seven hours, and by the other branch to the Col du Géant, supposing that the state of the glacier permitted the traveller constantly to advance (which is not the case) in about nine. The shortest linear

---

* I retain De Saussure's spelling of this familiar name, although I am aware that the most correct orthography is Chamonix. But I have, in general, preferred De Saussure's authority, on the spelling of proper names, to all others, and that of Chamouni has been usual amongst English, as well as many Swiss authors. The second syllable is pronounced rather short.

distance from the foot of the glacier to the highest ridge of the Alps is by my survey about seven miles, and the breadth of the glacier seldom if ever exceeds two-thirds of a mile, but is generally much less. This does not give any idea of its apparent extent. The toil of traversing it, the endless *détours*, and the recurring monotony of its crevasses, exaggerate inconceivably the distance, even to those most experienced.

We commence our survey at the foot or lower end of the glacier, proceeding upwards.

The view of the lower end of the Mer de Glace, from the road leading from Chamouni to Argentière, is exceedingly striking. The valley of Chamouni is here broad and flat. Three hamlets of small size are planted in sight of one another, Les Praz, Les Tines, and the Hameau des Bois. The latter is almost in contact with the glacier; and, indeed, in 1820, it attained a distance of only sixty yards from the house of Jean Marie Tournier, the nearest in the village, when its farther progress was providentially stayed. The valley down which the ice pours, meets that of Chamouni at a considerable elevation : the western side of the glacier (in contact with the Montanvert,) presses right upon the verge of a precipice, down which fragments of ice are precipitated at all seasons, whilst the eastern stream, following a gentler slope of ground, sweeps more gently round the foot of the Aiguille du Bochard, and beneath the station called the Chapeau, when it is again diverted to the west, partly by the accumulation of its own moraine in front, and partly by a projecting rock of a remarkable kind, of which we shall immediately speak. From the village of Les Praz, then, this cascade of ice is seen directly in front, but the Source of the Arveiron, at its lower extremity, is hid by the mass of the moraines. The Source offers, however, nothing extremely remarkable, and the views which have been given of it are in general greatly exaggerated : It is an arched cavity, almost annihilated in winter, and gradually increasing as the season of waste and avalanches advances, until it forms an archway of considerable height and width, from which the turbid stream of the Arveiron flows. The quantity of water varies excessively at different seasons, and even, I have been assured, on different days. It is fullest, I think, in July ; and, in winter, though

small, I am assured by natives that it is very far indeed from altogether ceasing, retaining, I was informed, at least half as much water
as when I saw it in September, when I estimated the discharge very
rudely (it does not admit of exactness) at three hundred cubic feet per
second. The source of this water in winter, when the glacier is frozen,
may be partly from the heat of the ground in contact with the ice, as
supposed by De Saussure, but it must also be recollected that the ice
valley of the Montanvert may be supposed to have a due proportion of
springs taking their origin in the interior of the earth at a depth to
which even the cold glacier-contact does not communicate a sensible
influence, and the source of the Arveiron is the natural drainage of the
springs of that valley.

The final slope of the Glacier du Bois has a vertical height of at least
1800 feet, (the height of the summit called Le Chapeau, above the
valley at Les Tines,) down which, as has been said, the ice descends
half shattered, half continuous, twisted into wild shapes, and traversed
by countless fissures, whilst on the right the precipice above the source
of the Arveiron raises its bare forehead without even a stunted tree or
a blade of grass, for its surface is continually furrowed by avalanches,
and its hollows washed clean by foaming cascades, which both originate
in the diadem of jagged pinnacles of ice by which it is surmounted.
To the right and left the prospect is enclosed by the warm green fir
woods which touch either moraine of the glacier, and behind and aloft
the view is terminated by the stupendous granitic obelisk of Dru, which
has scarcely its equal in the Alps for apparent insulation and steepness—a monolith, by whose side those of Egypt might stand literally
lost through insignificance.

When we approach the foot of the glacier at the Hameau des Bois,
we are at no loss to perceive that the ice has retreated. The blocks of
the moraine of 1820, in which year the glacier made its greatest incur-
sion (in modern times) into the valley, lie scattered almost at the doors
of the houses, and have raised a formidable bulwark at less than a pistol-
shot of distance, where cultivation and all verdure suddenly cease, and
a wilderness of stones of all shapes and sizes commences, reaching as
far as the present ice. The limit of the moraine of 1820 is marked in

the map, whence it appears that the form of the extremity of the glacier was not very different from the present one, only that it swelled out more, and that it very nearly had divided itself into two streams, separated by the promontory marked Côte du Piget. This promontory commands an excellent view of the extremity of the glacier. Upon its southern face the glacier has spent its strength, heaping ridge upon ridge of its moraines against it. The northern slope is perfectly protected, and trees grow to the foot of it. One cannot help being reminded of the position of the Hermitage of St. Salvador, on Mount Vesuvius, round which the lava streams pass innocuous.

But this hillock has an especial interest. Its resistance to the pressure of the ice led me to suspect that it is composed of firm materials, and is not merely a heap of rubbish. And so it proved : but whilst the cliffs above the source of the Arveiron are of gneiss, whose beds dip inwards towards the axis of the chain at an angle somewhere about 30°, this hillock is of stratified limestone dipping similarly *under* the gneiss, and at about the same angle. We find it continued, in exactly the same circumstances, a little to the eastward, at the foot of the Aiguille du Bochard, on the path leading from the village of Lavanchi to the Chapeau. There is there a lime-kiln, and it is burned for use. The Côte du Piget is mentioned by De Saussure,* and he refers to its calcareous nature in his chapter on the secondary rocks of the valley of Chamouni. But he does not notice the section below the Chapeau.

The moraine of 1820 rises some way upon the slopes which border the east side of the terminal part of the glacier. But when we come to examine these slopes themselves, we find in them indubitable evidence of their being real moraines of a former age, left by the glacier when it had a greater extension than at present. This is a fact of which it seems scarcely possible to doubt. We find it admitted by Saussure,† and most, if not all, of his followers. There are circumstances connected with this moraine which renders it worthy of most

---

* *Voyages*, § 709.   † Ib., § 623.

particular attention, for it is a common ground on which the advocates of the former vast extension of the glaciers, and the opponents of that doctrine, are ready to meet, both admitting that this mass of debris, extending quite up the present glacier, has unequivocal marks of having been a former moraine.

Its form is not a little peculiar. It is the convex escarpment seen in the map to traverse the valley of Chamouni above the village of Les Tines, presenting its convexity towards Chamouni. Its length, reckoning from the existing glacier, was estimated by De Saussure at 1300 or 1400 feet; but by the map it would appear to be 6000 feet, or above a mile, reckoning from the rock of the Aiguille du Bochard to the opposite side of the Arve from Lavanchi. It has already been said that the valley opposite to the Glacier du Bois is flat and level; the road from Les Praz to Les Tines, a distance of above half a mile, is almost perfectly so. There we reach the foot of the convex escarpment of blocks, which are covered with soil and trees on the side next Chamouni, but its composition is abundantly testified by the appearance of its summit, and especially by the section in the ravine through which the river Arve, descending from the Col de Balme, and swelled by the glacier streams of Le Tour and Argentière, forces its way. The cut is a deep one, and we find the mound to be almost entirely composed of detached fragments of transported granite, similar to that of the chain of Mont Blanc, rough and angular, or only rounded at the edges by partial friction, and accumulated in the utmost disorder, mingled with sand, without any appearance of stratification. The embankment has been *cut through* by the river, so that a portion remains attached to the northern side of the valley, (the slopes below the Flégère and the Aiguilles Rouges,) upon which vast insulated granitic fragments may be found lying at a considerable height. There can be no reasonable doubt that this mound was once continuous, and obstructed the course of the river. Of this we have a farther evidence in the deposit of the alluvial flats which succeed it in ascending the valley towards Argentière, evidently formed by the waters of a lake; and just at the margin of these, close to the eastern side of the mound, the village of Lavanchi now stands.

The entire mound, I have already said, is composed of materials similar to those of the moraines of glaciers generally, and of the Glacier du Bois in particular. The arrangement of these materials is also the same. The escarpment to the west does not appear to be the result of erosion subsequent to the deposit, but to be the original form into which the materials have been wrought. The summit is a long narrow ridge, sloping rather steeply both ways, and garnished with huge blocks on its very top. The largest of these is marked on the map under the name of *la pierre de Lisboli*, and in some places these ridges are multiplied and parallel, exactly as in a modern moraine. It will be observed that the ground-plan of this mound is very singular, being convex towards the glacier, instead of concave as is usually the case. This is an important fact, and requires a special explanation, on the hypothesis (generally admitted) of its being due to the former extension of the Mer de Glace. The ice must have descended in such a mass as to have blocked up completely the whole valley, and abutted against the opposite slopes of the Flegère. So great was its mass, and so nearly level the valley of Chamouni into which it descended, that when resisted in front it spread laterally in both directions, and pushed its moraine up the valley as well as down. The presence of the glacier, obstructing the course of the Arve, produced a lake, as in other well known cases—such as the lake of Combal in the Allée blanche, formed by the Glacier du Miage, and that of Matmark in the valley of Saas, formed by the Glacier of Allalein. The almost entire disappearance of the moraine on the *western* or lower side of the glacier is no argument against its existence; on the contrary, we have direct evidence in favour of it, derived from vast blocks of granite which are met with as far down the valley as the village of Chamouni, and which were formerly very numerous indeed, but are every day disappearing with the progress of cultivation. In external and mineralogical characters they are identical with those already noticed. A farther confirmation will be found in the enormous transported blocks which lie some hundred feet above the level of the glacier on its west side near the Montanvert, and which are not, I think, alluded to by any writer. Possibly the glacier once filled the valley of Chamouni to a great extent, and thus formed its own barrier, and perhaps we are to look for the proper terminal moraine much farther down. This is indeed the more probable hypothesis, both owing to the appearance of the rocks below Les Ouches, which we shall hereafter notice, and because it would be difficult to account for the removal of a vast lateral barrier on the west side sufficient to produce such an accumulation on the east. De Saussure's remark on the smallness of the terminal moraines is one of the most acute in his work. He says, " Les blocs des pierres dont est chargé le bas de ce glacier, invitent une réflexion assez importante. Lorsqu'on considère leur nombre et que l'on pense qu'ils se déposent et accumulent à cette extrémité du glacier à mesure que ses glaces se fondent, on est étonné qu' il n'y ait pas des amas beaucoup plus considérables. Et cette observation, d'accord en cela avec beaucoup d' autres que je développerai successivement, donne lieu de croire, comme le fait M. de Luc, que l' état actuel de notre globe n' est point aussi ancien que quelques philosophes l' ont imaginé," *Voyages*, § 625. The reason which we would assign for this remarkable fact, is that the extremity of the glacier having a moveable position, the blocks have been gradually deposited as the glacier retreated from the lower end of the valley of Chamouni to its present position.

If we continue our survey of the glacier, ascending the ancient moraine of Lavanchi, we come in contact with the rock a little higher than the Pierre de Lisboli, and the rock here is limestone, as already mentioned. It is just in contact with the gneiss, whose beds lie sloping southwards exactly at the same angle with the limestone, namely, about 30°. This limestone is, no doubt, of the same formation with that which has been noticed in other parts of the valley of Chamouni, and especially by de Saussure, as underlying the gneiss of the Aiguilles opposite Chamouni, towards the hamlet of Blaitiere. Its position is very remarkable, thus interposed between two granitic masses, for the Aiguilles Rouges are also of gneiss or granite ; and the almost exact symmetry, in point of arrangement and stratification, which we shall find to exist on the southern side of the chain at Courmayeur, gives to it a peculiar interest.

When we begin to command the view of the glacier in approaching the Chapeau, we are struck by the size of the blocks which seem poised on the projections of the cliff, at a great height above the ice, and which are rounded and scored in such a way as to show that the detached masses were deposited here in the usual progress of the glacier when it attained this height. The view here of the Aiguille du Dru, and of the pinnacles of ice of the Mer de Glace itself is very striking. A portion of the moraine of 1820 is next crossed, and at length, after passing a torrent, we find ourselves at the foot of the hillock called Le Chapeau, on the precipitous side of which is a cavern affording some shelter, and an excellent view not only of the glacier, but of the valley of Chamouni which it commands, and the effect is extremely beautiful, especially in the evening. This spot, although extremely easy of access, is rarely visited by tourists, unless at seasons when the Montanvert is too much enveloped in snow to be conveniently reached; but the two views have very little resemblance, since the portion of the glacier seen from the Chapeau is the lower part, or Glacier du Bois, whilst the upper part, or Mer de Glace, is commanded from the Montanvert, and the other is nearly concealed.

Beyond the Chapeau, the precipices of the Aiguille du Bochard, actually meet the glacier, where it tumbles headlong from the rocks, and both seem to forbid farther passage. Nevertheless it is practicable, keeping the face of the rock, to continue the ascent along the east bank of the glacier; and indeed there is scarcely any part of this bank of the Mer de Glace, as high as the foot of the Aiguille du Moine, which I have not traversed. The rocky precipice alluded to would be very difficult to pass, were it not marked by rude steps cut here and there in the soft steatitic rocks, which mingle with the gneiss, and which, being continually wetted by trickling rills, are very slippery. The goat-herds are in the habit of continually passing, and there is nothing to daunt any tolerable mountaineer, although the spot has acquired the name of the *Mauvais Pas*, which it bears more frequently than its proper one of La Roche de Moré. This rock (which is exactly opposite to the extreme promontory of the hill of Montanvert on the west side) forms one of the barriers of the *Mer de Glace* above, past which it pours down the precipice in the manner already mentioned. Conse-

quently, when the height of the Roche de Moré has been gained, we have
a new reach of the glacier in view, and the ice begins to assume a connected and consistent appearance, although still so excessively full of crevasses as to be generally impassable but for a very short distance. But
the ice is here the real icy mass of the Mer de Glace, whilst below, it
has been tossed and twisted so as to be entirely remoulded, and to bear
none of its original impress. At the point at which we have now
arrived, the glacier may be compared to the inclined, dark, unruffled
swell of swift water, rushing to precipitate itself in a mass of foam over
a precipice, it has all the forms of a compact moving mass of ice,
although rent asunder across its breadth by the rapid depression of the
bed along which it is urged.

The promontory of the Roche de Moré gained and passed, the
slight bay behind has, as usual, been partly filled up by accumulated
moraines, upon which we now walk instead of on the solid rock. Somewhat farther on a noisy, foaming torrent, called Le Nant Blanc,
descends from a seemingly small glacier, called *Le Glacier du Nant
Blanc*, lodged in a ravine interposed between the Aiguilles of Bochard
and Dru ; this torrent is well seen from the Montanvert—it is most copious in July, and its appearance is a good index to the state of temperature in the higher regions, instantly diminishing with the first cold
nights of autumn. A second torrent descends farther on from the
glacier at the foot of the Aiguille du Dru, and beyond this are some
fine pasturages, which extend along the foot of the jagged and rocky
chain which extends from the Dru to the point of Les Echelets marked
in the map. Here, on the higher part of these grassy slopes, near the
promontory of Les Echelets, are the highest stunted pines and larches,
which occur on either side of the Mer de Glace. From amongst
them, now and then, some grand peeps may be obtained of the Aiguille
du Dru, which shoots almost vertically above the eye like some tall
steeple—pointing to the deep blue sky.

These pastures are worthy of notice from one circumstance, namely,
that they are grazed by *cows* for a good many weeks in summer. How a
cow can find footing among such rocks, or ascend and descend pathways
which might be pronounced disagreeably precipitous by even a not fasti-

dious traveller, and whose zig-zags are often not half the length of the animal's body, may appear sufficiently surprising ;—but it is nothing compared to the seeming impossibility of ever bringing them there at all or removing them. To traverse the Mer de Glace opposite the Montanvert, is at all times a feat of some difficulty for an unloaded man ; it is commonly said that there exists but a single practicable pathway amongst the crevasses. That this is not correct, and that it varies much at different seasons, I know from experience—but at all times it requires an expert iceman (a correlative word to seaman or rocksman, may perhaps be admitted) to effect this passage with certainty and alone. I remember to have found some stray goats, which had wandered from the shore, quite lost amidst the wilderness of crevasses, and bleating for help.* The only other access to this pasturage is by the Roche de Moré, and there, most certainly, no animal heavier than a goat or a man could make its way unaided. The most usual way of transporting the cows is by the glacier at the foot of the Mauvais Pas, where I have already said the ice is in the very act of tumbling headlong down. There, by the aid of hatchets and planks, a sort of rude pathway is constructed the day before the ascent or descent of the cattle is to be performed, and then about thirty peasants assemble to pass as many cows, and by the aid of ropes succeed, usually without any loss, in compelling the poor animals to traverse the rude gangways which they have prepared. The cows were taken to the valley in the end of September last, and I regretted extremely that I missed the opportunity of witnessing so singular a cavalcade.

I have traversed the Mauvais Pas frequently. On one of these occasions, I proposed to Auguste Balmat to attempt to cross the glacier diagonally from just above the promontory of the Roche de Moré to

---

* Cattle are sometimes taken across the glacier at this place, and one of the hotel-keepers at Chamouni recounted to me a curious history of the risk which he and a companion had run in transporting a mule. They were assisting him with ropes, and the animal slipping, pulled them both into a crevasse : they escaped with difficulty, abandoning the mule to his fate.

the Montanvert. The thing had never been done, he said, but there was no reason why it should be impossible, and we could but come back at the worst. We got upon the ice ; and after a long and circuitous progress, succeeded in reaching the other side as we proposed ; and I often crossed the glacier afterwards in nearly every direction (excepting just above the final *chûte*) where the guides declared that no one ever had passed, or could pass without ropes or a hatchet. The former we never used, and the latter rarely. Auguste, though he had lived three years at the Montanvert, had never been compelled to traverse the ice but in a few directions, and it was as new to him as to me ; but his intelligence and zeal were superior to the lazy dogmas of impossibility, which are no where more frequently heard than from the guides of Chamouni.

Speaking generally, the fissures of the glacier in this part (between the Montanvert and the Dru) are mostly transversal, though so interlaced, and forming so many compound fractures, that the solid part continually thins out into an edge, which at length becomes evanescent between two crevasses. It is evident that in this way, a glacier maintaining its continuity beneath, may become absolutely impassable, except by descending one vertical face and ascending another, which, owing to the depth and width of the crevasses here would always be a perilous attempt. The crevasses of the western and middle part of the Mer de Glace below the Montanvert are very continuous and straight, and some of them extend for at least half the entire breadth of the glacier. They are often 15 or 20 feet wide, with walls perfectly vertical, and to move at all parallel to the length of the glacier in this place requires immense *détours*.

It is the east side which is so excessively crevassed, and that during the whole length of the *united* stream of the Mer de Glace. Whenever we touch the medial moraine, (the mark of the junction,) there the multiplied and complicated crevasses begin. The reason I believe to be this : the glacier which forms the greater or western portion, which is derived from the Glacier du Géant, moves fastest, and has by far the greater mass. The other, from the Glacier de Léchaud uniting with it, is compelled to follow, or rather accompany it. It is therefore drawn out, and at the same time squeezed into very much narrower limits, as the united stream is forced through a space not greater than the larger alone had before occupied,—just as when two rivers unite, the smaller and weaker is thrown into turbulent eddies by the union with the swifter and more powerful.

Turning now to the western side of the Mer de Glace in its inferior part, but a few remarks occur. The usual path from Chamouni to the Montanvert, and the steep ascent of La Filia,* from the source of the Arveiron, require no particular mention, but the examination of the promontory north of the chalet of Montanvert is not without interest. It is possible there to get a little way upon the glacier, amongst the immense fissures which precede its abrupt descent; and from this icy platform a fine view of the valley is attained. The ice here is remarkably pure, and the fine blue caverns and crevasses may be as well studied as in almost any glacier in Switzerland. Of the cause of this colour I may observe once for all, that I consider it to be the colour of pure water, whether liquid or solid ; though there are no doubt conditions of aggregation which give it more or less intensity, or change its hue. But this has a parallel in very many cases not considered as paradoxical. Most bodies when powdered have a different hue than when crystallized and compact, the topaz amongst solid bodies, and the iodide of starch amongst fluids, change their colour with temperature, and many bodies change their tint with their consistency, or lose it altogether when mixed with grosser matter. During an expedition which I made upon the ice in the month of September, during a snow-storm, I observed that the snow lying eighteen inches deep exhibited a *fine blue* at a small depth (about six inches) wherever pierced by my stick. Nor could this possibly be due

---

* I do not know the origin of the name. Thinking that it might refer to some legendary story of a young woman lost at the source of the Arveiron, I once asked a native of Chamouni its meaning, to which he replied, simply enough,—" Je ne sais pas si ce n'est parce-qu'on y *file* tout droit," which all who have *descended* it will readily admit to be the case.

to any atmospheric reflection, for the sky was of a uniform leaden hue, and snow was falling at the time.*

The west bank of the Mer de Glace is here extremely steep, though not absolutely precipitous. It is clothed with grass and rhododendron, and in many places with spruce firs of considerable size. Amongst these lie fragments of transported granite, wherever a ledge exists sufficient to maintain them, and they are accumulated especially at the promontory at the foot of which the glacier still sweeps, though at a great depth below. On the steep side of the hill facing the valley of Chamouni, and therefore sheltered from the glacier, these masses are comparatively rare. They extend quite up to the dwelling of the Montanvert, a height of 240 feet above the glacier, and even somewhat higher ; but the limit is perfectly well marked ; for although the rocky ridge which descends from the Aiguille des Charmoz to the Montanvert (and which is here called simply Les Charmoz), is covered with vast debris,—these debris are all *in situ*, and in contact with the native rock, a slaty talcose gneiss. These blocks constitute, therefore, an undoubted moraine, corresponding to that of Lavanchi and Tines on the east side, and indicating the maximum level of the glacier in very remote times. I may add, too, for the sake of connection, that the fixed rocks in the immediate neighbourhood of the house of the Montanvert, exhibit clear traces of being rounded and furrowed, though too much weathered to exhibit any thing like polish. Such rocks occur on the descending path to the glacier.

The earliest habitation on the Montanvert is thus described by De Saussure :—" Mais oú couche-t-on sur le Montanvert ? On y couche dans un château ; car c'est ainsi que les Chamouniards, nation gaie et railleuse, nomment par dérision la chétive retraite du zerger qui garde les troupeaux de cette montagne. Un grand bloc de granit, porté là anciennement par le glacier, ou par quelque révolution plus ancienne,

---

* On the Colour of Pure Water see NEWTON, *Optics*, Book I. Part ii. Prop. 10 ; HUMBOLDT, *Voyages*, 8vo. ii., 133 ; DAVY, *Salmonia*, 3d edit., p. 317 ; ARAGO, *Comptes Rendus*, 23d July 1838 ; COUNT MAISTRE, *Edin. New Phil. Journal*, Vol. xv.

est assis sur une de ses faces, tandis qu'une autre face se relève en faisant un angle aigu avec le terrein, et laisse ainsi un espace vuide au-dessous d'elle. Le berger industrieux a pris la face saillante de ce granit pour le toit et le plafond de son château, la terre pour son parquet ; il s'est préservé des vents coulis, en entourant cet abri d'un mur de pierres séches, et il a laissé dans la partie la plus élevée un vuide ou il a placé une porte haute de quarante pouces et large de seize. Quant aux fenêtres, il n'en a pas eu besoin, non plus que de cheminée; le jour entre et la fumée sort par les vuides que laissent entre elles les pierres de la muraille. Voilà donc l'intérieur de sa demeure : cet espace angulaire, renfermé entre le bloc de granit, la terre et la muraille, forme la cuisine, la chambre à coucher, le cellier, la laiterie, en un mot, tout le domicile du verger de Montanvert."—*Voyages*, § 627.

This was in 1778. But it appears that things were soon improved ; for, in one of Link's excellent coloured views, (published at Geneva, and very superior to all the more recent ones,) entitled " Vue de la Mer de Glace et de l'Hopital de Blair, du Sommet du Montanvert dans le mois d'Aoust 1781," a regularly built cabin, with a wooden roof, is represented, with this inscription above the door :—

" BLAIR'S HOSPITAL.

UTILE DULCE."

from whence I conclude that this hut was built by an Englishman named Blair, between the years 1778 and 1781.

At a later period, a small solid stone house of a single apartment, was built at the expense of M. Desportes, the French Resident at Geneva,* having a black marble slab above the door, with the inscription, *A la Nature.* On my first visit to Chamouni this was the only building, but soon after a much more substantial and effectual shelter was erected at the expense of the *Commune* of Chamouni, and is let to

---

* Ebel gives the following account of it :—" M. *Bourrit* de Genève, l'aubergiste *Terraz* et les guides *Jacques des Dames* et *Cachat le Géant* ont exécuté le plan de M. *Desportes.* Le bâtiment offroit une grande salle pourvue d'une cheminée, de deux fenêtres, de quatre lits de sangle, avec des chaises, des tables, des glaces, &c. Les frais de l' établissement monterent à 95 louis." *Guide du Voyageur,* 1810. Tom. ii., p. 364.

the present tenant, David Couttet, (together with the grazing round,) for the considerable sum of 1400 francs. The principal floor consists of an ample public room, a small kitchen, a guide's room, and three bedrooms for strangers, besides accommodation below for the servants of the establishment, of whom two or three remain here for four months of the year. This establishment, though simple and unobtrusive, is sufficiently comfortable and cleanly, and I should be very ungrateful not to acknowledge the kindness and attention which I uniformly experienced during many weeks' residence in this house; cold and desolate it certainly was occasionally—in September the thermometer fell to 39° F in my bedroom, and there was little choice of provisions beyond the excellent mutton of the Montanvert; yet, on the whole, I preferred the tranquillity of the arrangements to the bustle of the hotels of Chamouni, whither I seldom resorted but under stress of weather.

We are almost tempted to forget that a view so universally seen, and so often described, as that from the windows of the Montanvert, loses none of its real majesty in consequence of the ease and familiarity with which it is visited by thousands of travellers. For myself, repeated visits, and a long residence, have only heightened my admiration of this, certainly one of the grandest of Alpine views. The Aiguille du Dru has in its way scarcely a rival, and there are very few glaciers indeed with a course so undulating and picturesque as the Mer de Glace, and with banks so wildly grand, of which the general effect can be so well seized from any one point.* Besides former visits, I have this year seen it under every circumstance which could enhance its sublimity;—under the piercing glow of the almost insupportable midsummer's sun, and again in the snowy shroud of premature winter, —in the repose of the stillest and serenest moonlight, and lit up at midnight by the brilliancy of almost tropical lightning.

The glacier immediately below the Montanvert is easily accessible, whilst it presents at the same time all the grander and more remark-

---

* It may be seen to most advantage from a station some hundred feet higher on the Charmoz, whence the view in Plate II., is taken.

able features of glacier ice. The moraine is abundant, and the crevasses moderately large. A few hundred feet farther down, there was this year (1842) a mass of travelling rock of enormous dimensions upon the ice. A sketch of it is given at the head of this chapter. Its position, which is accurately fixed on the map (where this block is marked D 7,) will define the motion of the glacier in future years. There is a footpath here along the moraine, which is a steep stony ridge, about thirty feet high on the landward side, and much more towards the glacier in its present state. The masses of which it is composed, and indeed of all the older moraines of this neighbourhood, are not larger than those which are at present to be seen on the surface of the glacier.

Proceeding upwards in our survey of the Mer de Glace, we find a footpath which conducts us from the house of the Montanvert, first nearly down to its level, and then parallel to its length. By and bye we come to pretty smooth faces of rock, which go down sheer under the ice, and whose exposed promontory is now visible, ground away by the friction of the ice, or rather of the mass of abraded rocks mixed with sharp stones and sand which it drags along with it. To cross this rocky face, some rude steps are cut in the slaty gneiss, and the two passes of this description are called the *premier et second Ponts*. De Saussure mentions (§ 628,) having employed two men to blast the rocks to facilitate this passage, and the marks may still easily be seen. Opposite to this promontory the glacier is greatly heaved and contorted, owing probably to the inequalities of its bed. It is not easy to estimate the magnitude of these icy hillocks or waves, as they have been termed. This arises chiefly from the enormous magnitude and great angular elevation of the peaks and wild rocks beyond. I had a proof of this one day on the rather rare occasion of a fog settling down to near the level of the glacier, which enveloped entirely the scenery of the farther bank. Then the ice inequalities seemed to rise to mountains, and it was difficult to persuade oneself that the glacier, like the ocean, did not now and then raise its billows in a storm, to twice or thrice the height which continual observation had made so familiar. It might be easily, and indeed is generally, supposed, that the glacier is here impassable; but on the 18th September last, I crossed it with Balmat, and found it less difficult than the oblique traverse we subsequently made to return to the Montanvert.

Having passed the second " Pont" the path descends to the moraine, which partly fills a sinuosity in the outline of the hill; and having followed this for some hundred yards we are met by a perpendicular cliff, the foot of which is abraded by the ice. This is the point marked *L'Angle* on the map, nearly opposite to the promontory of Les Echelets, formerly mentioned. Here there is no alternative but to descend upon the ice, and its contact with the rock offers some peculiarities worth observation. It is represented in Plate III. When the ice of the glacier, in the course of its progress downwards, has been forced against an opposing promontory of rock, and has passed it, it will easily be understood that a cavity will be left behind the promontory, which the ice does not immediately fill up. Here it is easy (occasionally at least) to descend into such a cavity, with a wall of ice on the left hand and of rock on the right. Between the two are wedged masses of granite, which have slipt from the moraine between the ice and rock, and which, pressed by the incumbent weight of the glacier, and carried along in its progress, evidently must, and really do, wear furrows in the retaining wall, which is all freshly streaked, near the level of the ice, with distinct parallel lines, resulting from this abrasion. The juxta-position of the power, the tool, and the matter operated on, is such as to leave not a moment's doubt that such striæ must result, even if their presence could not be directly proved.

The *Angle* is the point noticed by De Saussure as the junction of the true granite with the rocks of gneiss. It is a full half hour's walk from the Montanvert.

To advance higher up the glacier two courses may be taken; either to resume the moraine as soon as the promontory has been passed, and thus advance as far as possible along the foot of the Aiguilles des Charmoz, or to follow the glacier near its western border, by an intricate passage amongst the numerous crevasses by which it is traversed. The former is very fatiguing, and not without danger from the frequent fall of stones from the small glacier at the foot of the Charmoz. On one occasion I saw an immense discharge of stones and mud take place,

Pl. III.

Drawn from Nature by Professor Forbes _ J Pickers.

Day & Haghe Lith.rs to the Queen.

CONTACT OF THE ICE AND ROCK AT THE ANGLE, MER DE GLACE.

arising from some sudden change in the glacier, with loud noise, which continued for several minutes. The passage of the Mer de Glace almost requires an experienced guide. I know of no better instance of the confusing monotony of the glacier surface, and the kind of skill required to retrace one's steps on the ice, than the passage of the Angle. The crevasses are so multiplied, yet so similar, that each seems to rise endlessly "another yet the same." We continually fancy that we recognise a particular feature, which is perhaps a hundred times repeated, with the slightest possible variation of form. Once strayed from the right path it is difficult to find it again, because a false turn may separate us from the region we are endeavouring to reach by impassable crevasses. Consequently, the guides, who very frequently pass during the season in conducting travellers to and from the *Jardin,* resort to piling stones here and there upon the ice, or upon blocks, as landmarks, such as are used occasionally on moors or hills subject to fogs. Even one who has great facility in retracing a path once pursued on solid ground, or in discovering a track for the first time, finds himself here quite at fault; and I have frequently known experienced guides of Chamouni go astray, and lead travellers into difficult and embarrassing situations, or place landmarks in altogether wrong positions, so as to mislead future passers by. I suppose that I passed the Angle at least forty or fifty times last summer, and although I at last became pretty well acquainted with its intricacies yet it was impossible to extricate oneself mechanically, or without vigilant attention. M. Bourrit has given a just and not exaggerated description of similar difficulties. " Rién ne peut donner une idée du nombre prodigieux des crevasses de cette vallée, que la difficulté d'en sortir. Il n'est jamais arrivé de retrouver au sortir le même banc de glace par où l'on est entré; souvent, au contraire, l'on erre pendant trois quarts d'heure, et les guides étonnés, recourrent aux enchantemens pour expliquer cet effet de la multiplicité d'objets semblables et qu'une longue fréquentation n'apprend point à distinguer."* It deserves, however,

---

* Description des Glaciers, l. 107.

to be mentioned, as a point not only curious in itself, but highly important in considering the constitution of glaciers, that they present year
after year a surface so very similar, that an experienced guide will
make his way over the ice in the same direction, and seem to avoid the
same crevasses, whilst he is, in fact, walking upon ice wholly changed
—that is, which has replaced in position the ice of the previous year,
which has been pushed onwards by the progressive movement of the
glacier.

This is a fact which, though generally enough admitted, has not yet
excited sufficient attention. The surface of the glacier has, for the
most part, the same appearance as to the variations of level, the occurrence of moraines, the systems of complex crevasses, and the formation
of superficial water-courses, in any one season as in another. These
phenomena, then, are determined by the form of the bottom and sides
of the rocky trough in which the glacier lies, and by its slope at the
spot. Just as in a river, where the same molecules of water form in
succession the deep still pool, the foaming cascade, and the swift eddy,
all of which maintain their position with reference to the fixed objects
round which the water itself is ever hurrying onwards. The passage
of the Angle is more difficult in some seasons than others, but it probably varies much more in its character between spring and autumn of
any one year than between one year and another. This I have, on the
unanimous testimony of the guides, and my observations of three different years confirm it.

The Angle past, the most conspicuous object is the imposing Aiguille des Charmoz, which rises on the right. The rocky pinnacles of
which it is composed exceed in sharpness those which I have seen in
any other parts of the Alps. There is one which is conspicuous from
the Montanvert, and which has an unnatural and exaggerated appearance in most of the engravings, which is really as attenuated as it is
possible to represent it. The mass is of granite, in which sapphires
are found, though rarely, in the *Couloir* immediately beyond the Angle;
I have found a singular porphyritic rock amongst the fragments,
containing felspar and epidote, which it is difficult to refer to any class
of primitive rocks.

From the foot of the higher summits of the Aiguille des Charmoz, a small glacier, which has been already alluded to, takes its origin. It is one of those short limited glaciers termed by De Saussure, *glaciers of the second order.** They may be studied to advantage in these valleys, though the ice of which they are composed rarely descends so as to touch the principal glacier, which occupies the bottom of the valley. Their extent would hardly be conceived from the foreshortened view which we have in looking up at them. The map shows that they cover a large surface. They do not essentially differ in structure from other glaciers, but are shorter, owing in all probability to the little surface which they present for receiving snow and thus increasing their dimensions, as well as to the great angle of inclination of the beds on which they commonly rest. This is indeed such as to render their adhesion to the ground an astonishing circumstance. M. de Charpentier has very justly quoted several examples as proving, that if glaciers really slid over the soil as De Saussure supposed, these could not for a moment sustain their position at an angle of 30° or more. In the higher part of the Mer de Glace, or rather, on the great chain between the Grande Jorasse and Mont Mallet, there are some of the icy masses which seem to hold on to the face of the rocks by mere adhesion, presenting precipices certainly of several hundred feet in height. I have watched these masses day after day, when the sun shone so as to throw the deep shadow of the ice-cliff northwards, giving it a magnificent relief, when the stability of these glaciers appeared little short of miraculous. It would be of importance to ascertain the rate of motion of such glaciers. I had intended doing so, but the bad weather of the month of September 1842 put an end to this as to several other plans.

It is evident that the little glacier at the foot of the Charmoz has been more extensive and thicker, within no very long time. The former level of the ice remains perfectly well marked on the rock behind, showing its subsequent diminution; and occasionally these glaciers altogether disappear, and probably re-appear again after a series of cold

---

* Voyage, § 521, 529. See examples of glaciers of the second order in Plate IX.

seasons. I noticed on the Glacier d'Argentière, (near the Mer de Glace,) at the foot of the Aiguille of the same name, the vacant bed of a glacier which had melted away. De Saussure asserts the appearance of new glaciers, (§ 540,) though he does not give any instance of them within his own knowledge : but there is no reason to doubt the fact.

A rocky ridge, descending eastwards from the Charmoz, composes the massive promontory of *Trelaporte*, round the foot of which the Mer de Glace struggles more violently in its passage than at any other part. The result is a series of fissures, which immediately at the turn of the rock are quite impassable, and which extend radially outwards, like the joints of a fan, in the same way as M. Agassiz has figured in the great glacier of Gorner, at the north foot of Monte Rosa. To pursue the course up the glacier, these crevasses must be crossed nearly at right angles, until the centre of the glacier has been gained, or the great moraine descending from the promontory of the Tacul, which divides the glacier into two portions. We may, however, ascend the promontory of Trelaporte itself, which commands a very interesting view.

CHAPTER V.

DESCRIPTION OF THE MER DE GLACE—CONTINUED

TRELAPORTE—A TRAVELLER CRAG-FAST AMONGST PRECIPICES—THE MORAINES
OF THE MER DE GLACE—" MOULINS"—DISCOVERY OF DE SAUSSURE'S LAD-
DER—TACUL—LAKE—BIVOUAC UNDER A ROCK—THUNDERSTORM—THE
CHAMOIS HUNTER—SUPERB GLACIER TABLE—GLACIERS OF LECHAUD AND
TALEFRE—JARDIN—PIERRE DE BERANGER.

No part of the valley of the Mer de Glace shows better than the
Trélaporte the abrading action of the ice upon the rocks, or the height
to which the glacier has evidently once risen. The forms are every
where smoothed and rounded. Vast sheets of bare granite, nearly
vertical, and without a fissure, occur up to a great height, and a few
hundred feet above the glacier level is a sort of shelf, covered with
large detached masses of granite, which have formed an ancient moraine. On the top of one of these my surveying station G was actually
planted. There is something singularly desolate about the appearance
of these rocks, broken here and there by a tuft of grass, which adheres
in the midst of an inaccessible precipice ; and as a few sheep pasture
here every year, without any resident shepherd, these poor animals,
straying in search of food, perish in considerable numbers, from fa-
mine, or by falling down the cliffs. A singular incident occurred here
in the past autumn, which shows the danger of venturing into such
places without a guide, or at least an attendant.

On the 17th September, 1842, I walked up to this lonely promon-

F

tory, which, as it leads no where, is unfrequented, except by the occasional visit of the shepherd, to carry salt to his sheep.* Having stopped to sketch the bold outlines of the *Dru* and *Moine*, which form the opposite boundary of the glacier, I sent Auguste to seek some water, which, owing to the form of the rocks I have mentioned, it is difficult to find. I was not surprised that he did not immediately return, but when, having waited half an hour, and finished my sketch, I saw nothing of him, I began to fear that he had got entangled amidst these wild rocks, and proceeded in search of him. After some time I saw him coming up with two lads of Chamouni, whom we had seen start from the Montanvert in the morning, for the Jardin, and leading between them a man evidently exhausted, confused, and his clothes torn to rags. On approaching, I found Auguste scarcely less excited than the man he led, and to rescue whom from a ledge of rock, on which *he had passed the whole night*, he had placed himself in imminent danger. This person proved to be an American traveller, who had wandered all alone the morning of the day before over the hill of Charmoz, above the Montanvert, and scrambled as far as the solitary precipices of Trelaporte, unvisited, as we have said, except casually by a shepherd, and still more rarely by some Chamois hunter. Towards afternoon (by his own account) he had slipped over a rock, and being caught by the clothes on some bushes, had his fall checked, so as to gain a little ledge surrounded by precipices on every side, where he found himself lodged in a perfectly hopeless prison. Here he passed the whole night, which, fortunately, was not cold, and in the morning he succeeded in attracting, by his cries, the young men of Chamouni, who were on their way across the glacier, at a great distance below. The two boldest, with difficulty, climbed, by a circuitous path, so as to gain a position above him ; but their united efforts would have been unequal to rescue him had I not providentially gone, with my guide, the same morning, to this remote spot. Whilst he was on a search for the water which I

---

* Accordingly, here and elsewhere, a traveller may be incommoded by the importunate earnestness with which the sheep surround and follow him, supposing that he has brought salt with him. They are as tame as domestic animals.

required he came within sight of the boys, vainly attempting to extricate the traveller. Balmat instantly joined them, and by great personal courage, as well as strength, succeeded in dragging the man up by the arm, from a spot whence a Chamois could not have escaped alive. Balmat told me, that whilst he bore the entire weight of the man on the slippery ledge to which he himself clung he felt his foot give way, and for a moment he thought himself lost, which was the cause of the very visible emotion of which he bore traces when he joined me. I gave wine and food to the traveller, and the others, and especially applauded the humanity and courage of the lads, one of whom conducted the traveller back to Chamouni, for his nervous system was greatly affected, and for a time I doubted whether he was not deranged.* I returned with Balmat to view the exact spot of the adventure, and a more dreadful prison it is impossible to conceive. It was, as I have said, a ledge about a foot broad in most places, and but a few feet long, with grass and juniper growing on it. It thinned off upon the cliff entirely in one direction, and on the other (where widest) it terminated abruptly against a portion of the solid rock, not only vertical, but overhanging, and at least ten feet high, so that no man, unassisted, could have climbed it. The direction of his fall was attested by the shreds of his *blouse*, which were hanging from some juniper bushes, which he had grazed in his descent, but for which evidences it would have appeared to me inconceivable that any falling object could so have attained the shelf on which he was almost miraculously lodged. Immediately below the spot he fell from, the shelf had thinned off so completely that it was plain he must have fallen obliquely across the precipice, so as to attain it. The ledge was about twenty feet below the top of the smooth granitic precipice, to which a cat could not have clung, and below, the same polished surface went sheer down, without a break, for a depth of at least 200 feet, where it sinks under the glacier, whose yawning crevasses would have received the mangled body, and never would have

---

* I regretted to learn afterwards that he had not shown himself generously sensible of the great effort used in his preservation.

betrayed the traveller's fate. A more astonishing escape, in all its
parts, it is impossible to conceive. It is probable, that had the young
men not crossed the glacier at the fortunate moment, my guide and I
would have passed the rock fifty yards above him, (it was in the direction in which we were going) without either party having the remotest
idea of the other's presence.

The same day I climbed, with some difficulty, towards the ridge of
the Charmoz from this spot, intending to gain a remarkable cleft in
the rock, conspicuous both from the upper and lower part of the
glacier, and denoted on the map by the mark G.* There was fresh
snow on the rocks, which made the ascent very disagreeable, and the
secondary glacier, which extends for a long way on the south-eastern
foot of the Charmoz, facing the Tacul, sent down an intermitting fire
of stones by the passage which we chose to attempt, and rendered it
prudent to abandon the ascent until more favourable weather. This
never came, and I was obliged to quit the Montanvert without accomplishing it. This I regretted, for the station G* would command the
whole glacier, and would have enabled me to make observations of use
for the perfecting of my map.

To return to the Mer de Glace. The foot of the Trélaporte offers
several excellent contacts of the ice and rock, which is there, as at
the Angle, much worn by the abrasion of the stones or gravel. It is
quite practicable to traverse the glacier from hence to the Tacul, or
promontory at the bifurcation of the glaciers *du Géant* and *Léchaud*.
The usual course of proceeding is, as we have observed above, to cross
the glacier before reaching Trélaporte, until the principal medial
moraine is attained. The whole of the eastern part of the glacier is
here much lower than the western, which is heaped up against the
promontory, and the effect is to squeeze the moraines together into the
smaller or eastern portion of the glacier. The regular curvature and
general parallelism of these moraines, amidst all this confusion and dislocation, is exceedingly remarkable. From the point we have now
reached, upwards, four of them may be most distinctly traced. [See
Plate II.] Two descending the glacier de Léchaud, one from the
promontory of the Tacul, and one the principal medial moraine of the

glacier du Géant, which, descending from the promontory called La Noire, (See the map,) we shall designate by that name. Of the first two, one descends all the way from the foot of the Courtes, on the Glacier du Taléfre, and the other is the medial moraine of the Glacier de Léchaud. The two last have a remarkable *dislocation* or lateral displacement, opposite to Trélaporte, which arises from some cause which I am unable to determine. Nor do I know whether this apparent dislocation advances with the progress of the glacier.

Near the same spot are the " Moulins," which the guides always take care to point out to travellers going to the Jardin. They are deep and nearly cylindrical holes in the ice, into which the water accumulated in the rills which form the superficial drainage of this part of the glacier is precipitated in a more or less copious cascade according to the season. Sometimes these cascades are double in the same hole, or one stream separates into two cascades ; but always, *whatever be the state or progress of the glacier, these cascades or " Moulins " are found in almost exactly the same position*, that is, opposite to the same fixed objects on the side of the glacier. This is an evident proof of the continued renewal of the glacier as to its state of aggregation, the external forms remaining fixed, whilst the integrant parts are advancing.

I was greatly struck by the change which I perceived in this part of the glacier, between the month of June, when I first visited it last season, and the close of September, when I quitted it. At the former time the crevasses were comparatively trifling, and they continued to open more and more the whole summer, so that at the end many places were nearly impassable, which earlier I had traversed without difficulty. This is a most important fact, for it shows that during winter the glacier consolidates, and that every summer its crevasses open afresh, whilst its continued adaptation to the external constraint which its walls or bed impose, show that the glacier mass is far more passive and plastic than it has usually been supposed. I might have stated that in the lower part of the glacier this is perhaps even more striking, for there, the thaw beginning earlier, and being more complete, the crevasses which have opened in spring attain their widest extension in July and the beginning of August, and afterwards by the collapsing of their sides, and the general softening of the mass, they subside into rounder forms, and the cavities being partially filled are more easily crossed.

It was nearly opposite the " Moulins,"—that is, between the stations marked G and H on the map, that in 1832, on my way to the Jardin, my guide, Joseph Marie Couttet, pointed out to me some fragments of wood, evidently much wasted and rubbed, which he assured me were part of the identical ladder which De Saussure had used on his memorable journey to the Col du Géant, forty-four years before. I kept a portion of the wood as a relic, without, however, attaching very great faith to its history ; but the inquiries which I made this year (1842) dispose me to believe it probably correct. Couttet and his brother repeated to me exactly the same story as before, and mentioned the year 1832 as that in which the ladder reappeared, and pointed out the very spot where I had myself found it, without having the least idea that I had heard of the thing before. They farther mentioned that there was no question that it was a ladder, for that Captain Sherwill had seen and taken some of the steps still adhering to the lateral props. It was certain that the morsels in question had descended from the Aiguille Noire, or at least in that direction, for this, the most westerly of the medial moraines has its origin there ; * and it is quite certain that De Saussure descended the glacier on that side, and that he left a ladder there ;—for he tells us that he was unable to pass by the western side of the Glacier du Tacul, on account of the crevasses, and Couttet's father was himself on the expedition, and descended from the Col with the enormous load of 160 pounds after the termination of

---

* The fact that the *origin* of the medial moraines is familiarly known to the guides of Chamouni seems equivalent to a true theory of the origin of these moraines so strangely misapprehended by De Saussure and most of his followers. Since a medial moraine may always be traced up to a promontory, and there be seen to originate, or at least to be combined out of the two lateral moraines which there unite, it would seem impossible to ascribe to them any other than the true origin. And that the Chamoniards perfectly understand this, is plain from the fact that they seek in each moraine the minerals proper to the source whence it is derived ; for instance, the red fluor-spar in the most easterly moraine of the Glacier de Léchaud, which has descended the Taléfre, and has its origin at the foot of the rocks called Les Courtes, where this rare mineral is sought *in situ.*

the expedition, and he assured his sons that the ladder had been left there. Besides, among the few ascents to the Col du Géant since the time of De Saussure, perhaps every one has been performed by the western side of the glacier, which, as I have said, is the safer and more usual course; and had a ladder been left there it could never have reached the medial moraine of La Noire. There is indeed one other alternative,—that the ladder had been used by the crystal hunters who used to frequent the rocks of the Aiguille Noire, for the black quartz crystals which perhaps occasioned the name of the spot. But in this case it is more than probable that the Couttets themselves, the most experienced crystal hunters of the valley, would have been aware of the circumstance. On the whole, then, in the absence of any direct information of any other ladder having been left in this particular quarter besides that of De Saussure, it seems reasonable to admit that the ladder in question descended from the Aiguille de la Noire to the point in question, near the Moulins, between the year 1788 and the year 1832. The observation is interesting, as determining so far the mean motion of the glacier in the interval. By the map, the distance, allowing for the sinuosities of the glacier, appears to be about 16,500 feet, which being travelled in forty-four years gives 375 feet *per annum* for the mean motion of this part of the glacier. We shall afterwards consider the theoretical bearing of this fact.

A little higher up we stand in the centre of three valleys, and in the most extensive part of the Mer de Glace. The guides believe, and probably with reason, that it is here deepest. They assure me that they have sounded a "moulin" of above 350 feet deep. What is perhaps as good a proof as any of the mass and solidity of the ice is, that I have seen enormous crevasses and basins holding still water, and therefore completely closed below. The water was of an exquisite blue colour independent of the colour of the ice.* The view from the centre of the glacier in fine weather is one of the finest which can be conceived.

---

* I have described it in my Journal as "nearly or quite as blue as the Rhone at Geneva."

In order to reach the promontory of the Tacul, where the glaciers divide, it is usual to cross the fourth and third moraines (I will in future designate them by numbers counting from the east) and in the centre the glacier is here easily traversed. The Tacul is reached commonly in three hours from the Montanvert, but a practised walker will do it in two, and I have descended in much less. The union of the two glaciers is attended with some circumstances worthy of notice. That descending from the Géant is by far the most powerful one, and the other is forced to yield somewhat to its pressure. The mass of rock forming the lateral moraine of the Glacier de Léchaud is, however, the most considerable, and this is wildly tossed up into a lofty medial moraine at the meeting of the ice streams. The Glacier de Léchaud clings as it were to the rocky wall of the promontory,—the Glacier du Géant has thrown up a vast mound of debris, which prevents it from approaching the rock within some hundred feet, and leaves a hollow between, part of which is faced by a huge icy barrier, of considerable elevation and difficult to scale. In this hollow—between the edge of the Glacier du Géant and the promontory of Tacul—there exists at certain seasons of the year a small lake. I first visited it in 1842, on the 25th June, when it contained no water, but a few days of continued hot weather, by melting the ice, filled it, and it remained more or less full during the remainder of the season. I have seen it, however, vary exceedingly in level from one day to another, so that there can be no doubt that it has an outlet through the moraine under the glacier. Balmat affirms that the source of the Arveiron is seen suddenly to burst forth with great vigour, and that this is attributed to the emptying of the *Lac du Tacul:*—which is by no means impossible. It appears from the testimony of M. Bourrit, (*Voyages*, i., p. 90,) that De Saussure was the first stranger who reached the Tacul.

The point marked B on the promontory of the Tacul was one of my principal stations, commanding an extensive view of nearly the whole glacier. It was at a height of 277 feet above the lake, so that the view embraced not merely the three branches of the glacier, but that of Taléfre, the Jardin, and the mountains beyond, and a portion of the valley of Chamouni opposite the Montanvert, the range of the Aiguilles Rouges, and the snowy summit of the Buet peeping over beyond.

Near the side of the lake, at the foot of the promontory, lies an enormous block of granite belonging to the moraine of Léchaud. The cavity beneath its south-west side is a well-known refuge for Chamois hunters, and for the few travellers who pass the Col du Géant, who usually save from two to three hours of fatiguing walking by sleeping here instead of at the Montanvert. It is, in fine weather, a pretty, tranquil spot. The glacier is in a great measure concealed by its lofty embankments, which shelter it from the chillest winds. The slopes round are grassy, and diversified with juniper bushes, and the little piece of water, when unfrozen, has a cheerful effect. Here I spent two nights with Balmat, with a view to advance my survey and the experiments on the ice; for whilst pursuing my inquiries on the higher glaciers, it was found to make a most laborious day to ascend so far from the Montanvert (carrying instruments and food) before the day's work could be begun, and to return again in the evening. Day after day I have been out thus from ten to thirteen hours upon the glacier. A bivouac was, in favourable weather, a preferable alternative. The juniper bushes afforded a cheerful and serviceable fire, and with the aid of a Chamois skin to protect me from the damp ground, and a strong blanket hastily sewed into the form of a bag, in which I slept, the nights passed not uncomfortably. But, on both occasions, when I meant to have passed some days here, I was forced to descend from the bad weather, against which we had no sufficient protection, the cavity under the stone being quite open in front. The last time that we were driven from this poor shelter was on the 6th August, when a day of unnatural mildness was succeeded in the evening by the most terrific thunderstorm I have ever witnessed. We were overtaken by it, and thoroughly drenched, before we could reach the Montanvert; but after sunset it raged with the greatest fury. From the windows of the little inn, I watched with admiration the whole scenery of the Mer de Glace, lit up by the explosive lightnings which followed for some hours with

little intermission, whilst the frail building seemed to rock under the
fury of the gale, and vibrate to every peal of thunder Each tiny torent
now gave tongue increasingly, until the fitful roar became a steady din,
with now and then a crash arising from the discharge of stones hurried
along by the flood, or an avalanche prematurely torn from the glacier
of the Nant Blanc. It was Saturday night, and Balmat had gone
down to Chamouni to attend mass next morning. He told me afterwards that the dazzling effect of the lightning was such, that it was
with the utmost difficulty he could keep the familiar path from the
Montanvert, and that he wandered, drenched to the skin, as if blindfold, through the wood. Next day brought tidings of disasters from
the valley. The road at Les Ouches had been broken up by the torrents, so as to be impassable ; many cottages were filled with stones
and gravel, and deserted by the inhabitants ; and I believe some small
barns were carried away—but no lives were lost.

One night I had a guest in my rude shelter. It was a poor man of
Chamouni, who, impelled by an irresistible passion for the chase,
came to pass the night on the glacier, in hopes of finding his game in
the morning; a hopeless task,—for the Mer de Glace is now so completely bereft of chamois, that, during the whole summer, I do not
recollect to have seen more than two upon it, though on other less frequented glaciers I have seen whole herds. The *chasseur* was very poor,
and by no means young ; he gladly partook of the provisions which I
could spare ; and learning that he was a respectable man, though
unsettled in his habits, I could not but feel an interest in the singular
ardour with which he pursued his thankless toil : Truly might he say
with the hunter in Manfred—

>   her nimble feet
> Have baffled me ; my gain to-day will scarce
> Repay my break-neck travail.

The poor fellow owned the infatuation of what he called his " mal-
heureuse passion ;" but he seemed willing to die for it. Late on the

afternoon of next day I met him; his sport consisted in having seen a chamois' track, and killed a marmot. By his want of dexterity, however, he had very nearly made a victim of one whom I could ill have spared. Balmat, whilst employed for me on the ice, heard a ball whiz close past him, and, looking up, saw our guest of the previous evening behind a rock, whence he had taken aim at a marmot! These animals are very abundant in every part of the higher Alps. They emit a shrill cry like a whistle: they lie torpid in holes a great part of the year, and are valued for their fat. When young, they are eaten.

The chamois hunter seeks the limits of the glacier region in the evening; lies under a rock, as we did; and starts before dawn to watch the known avenues by which the chamois descend to feed. If alarmed, they take to the hill tops—to crags rather than glaciers; there he must follow them, heedless of danger, impelled alone by the excitement of the sport. The day is soon spent in fruitless ambuscades—night arrives—and his previous shelter is luxury compared to what he has now the option of;—a face of rock, or leafless bed of debris must be his couch, and his supper is bread and cheese. After a few hours rest, he repeats his meal, drinks some brandy, and starts again. If the chase be prolonged, physical endurance is pushed to the utmost. A most respectable man of the Canton of Berne, who had himself killed seventy-two chamois, assured me that he had wandered thus for three days together, tasting nothing but water; which would seem incredible, if we did not recollect that hunger is often repelled for a time by fatigue. De Saussure mentions three hunters, father, son, and grandson, who successively lost their lives in the chase;* but such accidents are, I conceive, now more rare. The value of a chamois is only from twelve to fifteen francs, including the skin, so that it offers little pecuniary temptation to the exposure of life. No doubt, as the historian of the Alps adds, the excitement is the real reward,

---

* *Voyages,* § 736.

as in the soldier, sailor, and gamester; and perhaps the naturalist has little reason to express surprise at the risks and privations of the hunter's life, when his own would appear to so many persons much less intelligible.

But to return to the glacier.—Following the eastern branch above the separation at the Tacul, we find ourselves on the glacier de Léchaud. Two conspicuous moraines belong to it, which I have called Nos. 1 and 2. The first is the medial moraine of the tributary glacier of the Taléfre; the other comes from the eastern side of Léchaud, above the union with the Taléfre. It is in connection with the former of these moraines, and nearly opposite the promontory of the Couvercle that there lies upon the ice a very remarkable flat block of granite, which particularly attracted my attention on my first visit in 1842 to this part of the glacier. It is a magnificent slab, (marked C on the map, being the position which it occupied in the month of June,) of the dimensions of 23 feet by 17, and about 3½ feet in thickness. It was then easily accessible, and by climbing upon it, and erecting my theodolite, I made observations on the movement of the ice. But as the season advanced, it changed its appearance remarkably. In conformity with the known fact of the waste of the ice at its surface, the glacier sunk all round the stone, while the ice immediately beneath it was protected from the sun and rain. The stone thus appeared to rise above the level of the glacier, supported on an elegant pedestal of beautifully veined ice. Each time I visited it, it was more difficult of ascent, and at last, on the 6th August the pillar of ice was *thirteen feet high*, and the broad stone so delicately poised on the summit of it, (which measured but a few feet in any direction,) that it was almost impossible to guess on what it would ultimately fall, although, by the progress of the thaw, its fall in the course of the summer was certain. On a still later day, I made the sketch in the frontispiece, when probably it was the most beautiful object of the kind to be seen anywhere in Switzerland. The ice of the pedestal presented the beautiful lamellar structure parallel to the length of the glacier. During my absence in the end of August, it slipped from its support, and in the month of September it was beginning to rise upon a new one, whilst the unmelted base of the first was still very visible upon the glacier.

The Glacier de Léchaud is on the whole pretty even on its surface— I mean that part which lies to the south-west of the medial moraines. On account of its great elevation, it is covered in its higher part with snow almost the whole year, and until the month of August it offers very disagreeable walking, on account of the half-melted snow on the surface, which likewise conceals the crevasses, and renders it somewhat dangerous. It is joined by some small tributary glaciers from the Mont Tacul. Opposite the Glacier du Taléfre occur two " Moulins," one of which was remarkable last summer for its great depth and perfect verticality. I had intended to ascertain the depth precisely, but was impeded by a fall of fresh snow, and broke the cord which I had lowered with my geological hammer attached as a weight for the purpose. About an hour's walk above the Tacul, is station E on the east side of the glacier, whence I watched its motion. It is here just passing into the state of *névé*, which defines the limit of perpetual snow on the surface of the glacier, whilst there is true ice beneath. The view here is very grand. The level is 7926 feet above the sea, and the glacier, almost free from crevasses, is spread out like a magnificent level floor from which rises the tremendous and inaccessible wall terminating the view to the southward, of which the Grande and Petite Jorasses form a part. The Grande Jorasse is the highest mountain of the range next to Mont Blanc, and its northern side is quite precipitous. From the point E it is seen under an angle of 30 degrees, the horizontal distance of its summit being less than two miles. The origin or feeders of the Glacier de Léchaud are derived from the right and left. From the immediate top or head it has no proper feeder except the fallen snows, which cannot adhere to the rocky precipice before mentioned. To the south-east there is a tributary glacier, which runs up to between the Petite Jorasse and the range of the Aiguilles de Léchaud, which separate the glaciers of Taléfre and Léchaud. It has its origin at a lofty and remote summit, considerably to the eastward of the Petite Jorasse, about which I made frequent enquiries of the guides, and I found that it was called (in translation from the Patois) the Montagne des Eboulements, or Des Ruines, being, as they assured me, and I daresay with reason, the summit at the head of the Glacier du Triolet, which descends into the Val Ferret in Piedmont, and of which the fall (partly of the glacier and partly of rocks) was attended with disastrous consequences about a century ago. The western feeder of the Glacier de Léchaud descends from behind the Mont Tacul from the serrated ridge which connects it with the great Alpine chain. This ridge is called *Les Périades*, and its culminating point, Mont Mallet, which, however, is to be distinguished from the Aiguille du Géant in its neighbourhood, which bears also the name of Mont Mallet on the Italian side of the Alps, from which it alone is visible. From the eastern foot of the pinnacle of Mont Mallet the tributary glacier descends. It is pretty extensive, and not wholly inaccessible, for the brothers Couttet assure me, that they have thus gained the summit of Mont Tacul from behind, which, at the best, must be a very long and difficult journey.

The higher part of the Glacier de Léchaud is scarcely ever visited, except by crystal and chamois hunters. Tourists who venture across the Mer de Glace always make their way to the Jardin, and with good reason, as it offers some of the grandest points of view any where to be found on this glacier; nor is there perhaps in the Alps any expedition so practicable in fine weather, which repays so completely the traveller who appreciates the wildest and grandest natural scenery. The traveller to the Jardin does not need to touch the Tacul at all. He crosses two of the medial moraines at the Moulins between Trélaporte and the Couvercle, and higher up he passes the other two, near the great stone, C. It is difficult to approach the lower part of the Couvercle much nearer. I have more than once ventured down the east side of the glacier, under the Aiguille du Moine, towards station F, but the passage is embarrassing, often impossible. When the two glaciers meet, (as I have already remarked), the eastern half is dislocated excessively, and is all but impassable. The promontory of the Couvercle itself, opposite C, may be easily reached, and offers some interest from the visible friction to which it is subjected by the descent of the glacier. Farther up, we stand in front of the descending ice of the Glacier du Taléfre, which presents a majestic and perfectly inaccessible accumulation of icy pyramids and fragments ejected through the narrow opening which gives vent to the basin of the glacier, which pours over the precipice in a solid cascade, presenting a perfect chaos of forms. Of the structure of the ice here I will afterwards speak, but I may observe that the preservation of the medial moraine in the midst of this mass of confusion is a very startling fact. It is indicated merely by a stripe of dirt, which discolours the centre of the icy cascade, but no sooner has it reached a comparative level, than the masses of rock dislodged from the side of the higher glacier are found on the surface, arranging themselves with admirable order along a line of no great breadth, which forms the medial moraine No. 1, which may be traced distinct from the others along by far the greater part of the Mer de Glace.

The ascent to the Glacier du Taléfre is usually accomplished by rocks of the Couvercle at the foot of the Aiguille du Moine. It offers no kind of difficulty. The ascent, where steepest, is called *Les Egralets.* Above these the view becomes wild, but very grand. On the left is the Aiguille du Moine,* one of the most elegant and uniformly conical summits of the chain ; at its foot are huge blocks of fallen rock, tenanted by marmots. Looking backwards, we command a large space of the Mer de Glace, and the grand view up the Glacier du Géant opens, and Mont Blanc begins to appear for the first time, fortified on this side by the impassable barriers of the Monts Maudits. The Aiguille du Midi, from its height, begins to overtop those of Grépont and Blaitière, and between it and Mont Blanc the rounded form of the Dome de Gouté is not to be mistaken. In front, the wide basin of the Glacier du Taléfre, in a great measure concealed from the Mer de Glace by its height, and the steepness of its outlet, begins to open. It has a singular and interesting appearance. It is shaped almost like a

---

* Called Aiguille du Taléfre by De Saussure.— *Voyages,* § 630.

volcanic crater with one side blown out, and it is surrounded by rocky pinnacles of the wildest forms, which appear, and for the most part are, totally inaccessible. It is certain that no one has succeeded in passing this serrated barrier at any point.* The topography of the Glacier du Taléfre is very ill laid down on all the maps and models. It is bounded on the west by the chain of peaks visible from the Mer de Glace, which connect the *Moine* and *Dru*. The north-east barrier, which is the longest, and tolerably straight, passes through the Aiguille verte, (which separates this glacier from that of Argentière), and a single range of very sharp Aiguilles called in succession Les Rouges, Les Droites, and Les Tours des Courtes, and it terminates in the south-east by a remarkable glacier summit marked [A] in the map, whence a tributary glacier descends. Though I have determined the position of this point, as well as that of another [B] to the south-west of it, I am unable to give them their proper names, as no one could inform me of them, but I apprehend that they belong to the axis of the great chain, and command the glaciers of the Val Ferret. It is also possible that the point [A] is the lofty white peak which I saw when I visited the extremity of the Glacier d'Argentière, of which it occupies the higher termination. It is perhaps the Mondelant. The south barrier of the Glacier du Taléfre is the range of the Aiguilles de Léchaud, which separates it from the glacier of that name.

The glacier of Taléfre is pretty even on its surface, and is covered for a great part of the year with snow; its level, according to De Saussure, is 1334 toises, or about 8500 English feet above the sea. In the centre of the snowy basin is a very large exposed surface of rock, of a triangular form, covered with soil on its lower part, sufficient to maintain a good turf, enamelled with the usual Alpine flowers, during the few weeks of the year that it is entirely uncovered with snow. This spot is called the Jardin (or *Courtil* in patois,) and is now a very frequent excursion from Chamouni. There is a spring of water near the lower part, and lying exposed, at a high angle

---

* The Couttets wished to pass to the glacier of Argentière behind the Aiguille Verte, but having gained the ridge, they were unable to descend.

towards the south, it is anything but cold in fine weather. Indeed I scarcely ever remember to have found the sun more piercing than at the Jardin. On three different occasions I have visited it, and on all under the most favourable circumstances, in 1832, in 1839, and in 1842. The reflection of the heat from the snowy basin by which it is surrounded, and its comparative shelter from the wind, probably cause this intensity. On each visit I have found the scenery if possible more admirable than before. On the last occasion I climbed to the very summit of the triangular rock forming the Jardin, a task of more labour than it would appear to be, as it is both long and steep. The top is at a level of 9893 English feet above the sea (trigonometrically determined), and commands an admirable range of view. From thence I took a number of magnetic bearings for the plan of the glacier. The Glacier du Taléfre presents two medial moraines, marked on the map ; one takes its origin from the Jardin itself, the other is derived from Les Droites already mentioned. These become commingled during the precipitous descent of the ice, and reappear as one on the Glacier de Léchaud.

From the Jardin it is not difficult to descend to the Glacier de Léchaud by the south margin of the Glacier du Taléfre. The passage of the last named glacier is, however, almost always wet, and the foot perpetually bursts through the frail superficial coating of ice formed in the night, and plunges ancle-deep into the snow-cold sludge beneath. The lateral moraine gained, it presents a steep and uneasy descent towards the Glacier de Léchaud. At about two-thirds of the descent is a grassy shelf upon which some of the debris of the moraine have accumulated. One mass is of enormous size, and from its peculiar form is well seen from the Glacier de Léchaud in all directions. It is a useful landmark, and is called La Pierre de Béranger, no doubt from a M. de Béranger who is frequently mentioned in M. Bourrit's narrative, though I am not acquainted with any particulars respecting him. The Pierre de Béranger is marked on the map ; it is sometimes used as a shelter for the night by hunters ; thence the glacier may be more easily gained by the rock than by the moraine, a bypath not generally known to the guides.

This concludes what I have to say respecting the topography of the Mer de Glace generally, and of its tributaries of Léchaud and Taléfre. The other great branch, the *Glacier du Géant* or *du Tacul*, remains to be described, but that may properly form a part of the narrative of the passage of the Col du Géant, to which I shall devote a separate chapter.

CHAPTER VI.

ACCOUNT OF A SURVEY OF THE MER DE GLACE AND ITS ENVIRONS.*

OBJECT OF THE SURVEY——THE INSTRUMENTS——THE BASE LINE——THE TRIAN-
GULATION——HEIGHTS OF THE STATIONS REFERRED TO MONTANVERT——SLOPE
OF THE GLACIER——HEIGHTS OF THE NEIGHBOURING MOUNTAINS ABOVE THE
SEA——CONSTRUCTION OF THE MAP——GEOGRAPHICAL POSITIONS.

IT was the especial object of my journey in 1842 to observe accurately the rate of motion of some extensive glacier at different points of
its length and breadth.

In order to draw precise or valuable conclusions from these experiments it was necessary to be able to assign accurately the relative
positions and distances of the points observed. It was also highly
desirable to ascertain the *slope* of the glacier in its various parts, which
can only be exactly done by a combination of horizontal measures with
vertical heights. These, and some other proposed experiments required
of necessity a geodetical apparatus, with which I provided myself, and
I soon found that the only satisfactory mode of proceeding would be to
add to my direct observations on the movement of the ice, a general
survey of the glacier in its various portions. This being resolved
on, it followed as a matter of course to lay down the neighbouring
mountains, and to determine their height, and hence the construction

---

* It may be right to observe, that this Chapter, which is little likely to interest the general reader, may be omitted by those who would avoid such details. The two following
Chapters on the Motion and Structure of the Ice may also be slightly passed over by those
who are willing to take for granted the results which will be found in the concluding
Chapter of this volume.

of the map which accompanies this work, and which is based exclusively* upon my own observations.

The instrument on which I chiefly depended, as well for the determination of the movement of the glacier as for its triangulation, was a Kater's astronomical circle, made for me by the late Mr. Robinson. It is of the larger size of such instruments, having both the horizontal and vertical circles of four and a half inches diameter, the former with three microscopes and verniers, the latter with two. The construction of the horizontal part is, as usual, most carefully executed, and the readings incomparably better than the vertical ones. The telescope is provided with five fine vertical wires, like a transit instrument, and one horizontal, and a diagonal eye-piece; there is a spare telescope, for *keeping watch*, the working of which was not, however, satisfactory, and it was seldom used, the steadiness of the instrument being scrupulously verified by a return to the starting or zero point. It is farther provided with a good level, which was frequently verified, and which fortunately escaped any accident. The whole, mounted on Robinson's excellent portable tripod, was as steady as could reasonably be desired, where the ground for planting it was favourable. The use of it upon the ice was always attended with some embarrassment owing to the unequal sinking of the supports. The working of the tangent screws was perhaps the least satisfactory part of the apparatus; but, on the whole, it would be difficult to imagine an instrument more perfectly adapted for the various uses to which it was put, or to find one which could have better answered to the character of its maker throughout a season of most trying work. The instrument, as mounted for use, is seen in the Frontispiece.

In selecting a station for pursuing my inquiries, I had no reason to regret my choice of the Mer de Glace of Chamouni. Its vast surface presents every variety of glacier structure and arrangement; its great length, and its remarkably uniform breadth, for some distance, adapted

---

* It is scarcely necessary to mention as an exception a single point (the position of the Col du Géant) which has been determined from De Saussure's data, and the sparing and cautious use of some published engravings in laying down some natural features.

it peculiarly to my investigations, which had reference to the velocity of movement as affected by the former circumstance and without reference to the latter. Its easy accessibility, and the ready communication with places whence any requisite implements might be procured, aided in determining my choice;—but beyond all other circumstances, the certainty of finding a permanent shelter during all weathers, and in the immediate vicinity of the ice, such as the cottage of the Montanvert affords.

The form of a glacier like the. Mer de Glace, presents almost every difficulty which can be experienced in conducting a survey. The only spots nearly level (I mean on the surface of the ice itself) are unfit for measuring a base, or performing any important part in the triangulation. The walls of the glacier are excessively rugged, often inaccessible. The stations are difficult to choose so as to be visible from one another, owing to the intricate windings of the ice stream, and the enormous height of the rocks. The fundamental triangulation must be carried up a *valley*, whose extremities, independent of *mountains*, differ in level by 4400 feet. Finally, the observer—solitary and without the assistance of any one understanding the use of instruments—is exposed, even in the finest weather, to alternations of the most intense, almost stupifying, solar heat reflected from the snow and threatening him with blindness, with the chill winds and sudden storms of these great heights:—he must often have the alternative of walking for seven or eight hours of the day over ice, merely to go to and return from a single station to gain perhaps a single observation, or else of camping out under some rock, exposed to the chances of weather. These considerations, it is hoped, will be taken into account in estimating the kind of accuracy to be expected in a survey like this, and the amount of criticism of which it may be thought deserving in respect of minuteness of detail. It will appear, however, (I hope) from what is to follow, that the basis or skeleton of the map is sufficiently accurate as to the relative positions of all the great points, which are in general determined with more exactness than the scale permits in laying them down.

Besides the instrument just mentioned, I had taken with me from

England, a small sextant and horizon (of which I made no use), one of
Troughton and Simms' standard steel tape measures, an ordinary var-
nished tape measure, waterproofed string and silk lines ;—a telescope
by Tulley, $2\frac{1}{4}$ inches aperture, and a Kater's compass, which I found
of admirable service.*

I had originally intended measuring a base line upon the ice, near
the confluence of the Glaciers de Léchaud and Tacul, which I intended
at once to be the foundation of my triangulation, and to decide the
question of the supposed dilatation of the ice during its progress. I
subsequently devised a better scheme of ascertaining the latter point,
and soon became convinced of the inconveniences which would attend

---

* In addition to these I had, for general purposes, six thermometers, a maximum and
two minimum thermometers, one Bunten's barometer, two improved sympiezometers, two
photometers, two actinometers, a Russian furnace, with alcohol for boiling water and a
thermometer for ascertaining its temperature, a measuring chain of 10 metres, borrowed at
Chamouni, as well as a *jumper*, 30 inches long, for making holes in the ice ; two hammers,
three screw drivers, a hatchet, numerous staves for station pointers with red and white
flags, chisels, red and white paint and brushes for marking the stations and other points
permanently on the rocks, two clinometers and compasses, a portable telescope (*Feldstecke*)
by Plössl, gimblets, pliers, a hand vice, a protractor, scale, drawing materials and colours,
colours for injecting the ice, viz. logwood, carmine, litmus, and lithographic ink. So ample
had been my provision of apparatus that I had no occasion for any thing in addition which
was not readily procurable at Chamouni.

It may not be useless to add that in point of clothing, though my wardrobe was very far
from bulky, I seldom suffered from cold. Without any voluminous cloaks or furs, I found
flannel next the skin (doubled if necessary,) surmounted by a complete suit of soft Chamois
leather to be warmer than it was generally agreeable to wear even in sleeping in the open
air near the glacier in summer. To avoid the continual risk of exposure to cold during
protracted observations after a laborious walk, I generally found a woollen waistcoat
with sleeves, over the one I usually wear, a sufficient protection ; and together with a
light Scotch woollen plaid, or a common single-breasted greatcoat, sufficient for the worst
weather during exercise. Long worsted stockings over short ones, and fur-lined gloves
without separation for the fingers, which, however, I very seldom wore. A straw hat, on
all occasions during exercise, with a velvet cap under, or exchanged for a soft fur cap
during observations. For shoes nothing is equal to London ones with double soles, continually supplied with fresh nails as the old ones are knocked out. Square headed nails
or anchor headed nails, such as are used in Switzerland, to defend the edges of the leather, are the best. Of late years I have never habitually used spikes or *crampons* of any
kind for crossing the ice.

such an operation. I ultimately acquiesced in the advice which Mr.
Airy had already given me to measure my base in the valley of Chamouni, and to continue the triangulation up the glacier. The main
obstacle to this was the extreme rapidity of the Montanvert, and the
difficulty of finding any point from which the glacier and the valley
could be at once commanded to any extent. This was however at
length overcome, and stations, though not affording by any means the
most eligible forms of triangles, were at last decided upon. The
measurement of the base was one of the later operations, and the connection of it with the triangles on the higher level of the glacier the
very last. I shall commence, however, with an account of

The Base Line.

It being decided that the trigonometrical base should be in the valley, there was no difficulty in selecting the most proper place for it. A
perfectly straight road, leading from the village of Les Praz to that of
Les Tines—a distance of about an English mile—parallel to the length
of the valley of Chamouni, and nearly opposite to the foot of the Glacier
du Bois, at once suggested itself. A nearer inspection confirmed my
opinion as to its fitness. From a little wooden bridge 500 yards east of
the village of Les Praz to the woods near the hamlet of Les Tines, which
begin to conceal the glacier and parts adjacent, is a clear space of 1000
yards, which seemed a sufficient length for my purpose. A mathema-
tically straight line could be drawn throughout this space upon the road,
which was sensibly level to the eye, (its mean inclination being 44' slanting upwards towards Les Tines,) and which was formed of dry well compacted gravel. In short, a more eligible spot could hardly be desired.

The *termini* of the base marked N and O on the map were fixed by
long pins of hard wood driven into the ground, and the tops marked
by a nail driven into each as a starting point. These pins may probably remain for some years. The first is exactly at the eastern
end of the beam which forms the south side of the little wooden bridge
near Les Praz already mentioned. The station O is marked by a very
conspicuous solitary larch tree, with its lower branches lopped off, on
the south side of the road, standing quite apart from the wood near
Les Tines. The terminus at O may be referred to the *centre* of this

larch tree by co-ordinates perpendicular and parallel to the line O N. It is *on the road* 5 feet 7 inches nearer to N, and 4 feet 1 inch north of the said larch.

The interval was twice measured—on the 14th and 23d September 1842—by the aid of a ten metre chain kindly lent to me by M. Blanc of Chamouni ; and the chain was compared with Troughton's steel tape divided into English feet and inches, in the interval of time between the two measurements. The chain was a good iron one, united by very solid brass rings. The first measurement was intended only as an approximation, but the second one confirmed it so well that I thought it unnecessary for the purpose in view to undergo the labour of a third. The distance was 91½ chains, or, more correctly, 914 metres. With the assistance of my intelligent and useful guide Auguste Balmat, and another person, I fixed the distances of 30 and 60 chains, also by hard wood pins, and on the second measurement the difference was under one inch at both places, and a temporary mark at 80 chains appeared to give as small an error. Therefore, though on the whole measurement there appeared a difference of 5 inches, I considered myself as justified in attributing it to some mistake in the mode of counting the links, which, in my first or approximate measurement, was the only method I had used of estimating the fractional part of a chain. Having measured the space between the 80th chain and the end of the base twice over without sensible difference, and having measured the fractional residue with Troughton's tape, I think myself entitled to conclude that the probable error of the whole does not exceed materially the observed difference at 30, 60, and 80 chains—that is, one inch, or little more, or about 1-36000 of the length of the base,—a quantity much under the other inevitable errors of observation in the subsequent triangulation. The length of the base was,

$$91 \text{ chains} + 26 \text{ English feet} + 2.50 \text{ inches.}$$

The comparison of the 10 metre chain and the steel tape was very carefully performed under the wooden gallery of the Hotel de l'Union at Chamouni. Both the chain and the tape were stretched with a weight of about 14 lbs., which is that indicated as the stretching weight of the tape, and the chain was found not to yield sensibly to any moderate force, after being once fairly stretched. Its length was found to be

32 feet 10.675 inches,

the temperature of both being 41° F. The probable error of the measurement might amount to one-fiftieth inch. Hence the length of the base will be found to be

2992.952 English feet.

This ought to be reduced in the proportion of the cosine of 44′ to radius, that being the mean inclination of the base to the horizon. This amounts to a shortening of only *one twelve thousandth* part; and the calculations having been commenced without adverting to it, I have thought it superfluous to alter the numbers as to the distance and elevation of the various points of the survey by this small quantity, which, however, may easily be applied, if desired, to the results which are to be given.

The Triangulation.

The plan proposed for the survey was simple enough. A series of points were taken on the rocks on either side of the glacier, and at some height above it, so as to command a view of two or more of the others. The circumstances which, as already observed, compelled the selection, and the labour of access and of observation at many points prevented a great multiplication of the triangles, or the avoidance of some very oblique ones. Still, as the object was only to lay down correctly a very limited range of country, the accuracy of the instrumental methods was sufficient to allow me to dispense with the selection of stations which would have been requisite in an extensive survey. But every pains was taken to ascertain that the triangles were accurately determined, of which the three angles were in most cases measured; and considering the circumstances under which they were observed, it will probably be considered as a fair proof of their exactness, that the sum of the angles seldom differs 1′ from 180°, notwithstanding the great deviations from a horizontal plane. The sides of the system of triangles thus determined along the valley served as bases from which to measure the altitude of the mountains, and the bearings of the various points which were to be laid down on the glacier; whilst various subordinate stations were marked, from whence, as secondary stations, fresh angles might be taken with the compass for the better comprehension of details.

The stations were taken with particular reference to the observations which were to be made upon the ice, and they were all marked in such a way as may enable them to be recognized at a distance of some years. They were in every case (excepting at the extremities of the base, which have been already described) on rock, or, at least, the instrument was placed on some fixed stone, little likely to be removed. A cross was cut with a hammer, or chisel, immediately below the centre of the instrument, determined by a station pointer, and that cross was painted red, with a large capital letter also in red, painted beside it. I will briefly describe the position of the stations which served for the main triangulation, and which are marked on the large map.

Commencing in the valley, the extremities of the base line marked N and O on the map have been already described. From these two stations the rock above the Chapeau could be distinctly seen, and that formed the third station marked I. It is at a height of no less than 1800 feet above the other two. It is on the summit of a green saddle-shaped slope, above the rock and cave already mentioned, quite above the few trees which fringe the precipice. The cross is on a flat stone, sunk in the grass, very near the first fixed rock, at the head of the grassy ridge. This station commands the glacier to a considerable extent, even as far as the Tacul; and I was able to observe the signal which I had planted there (station B.) It was, therefore, a most eligible connecting point between the valley and the upper level of the Mer de Glace.

The observed angles of this triangle were,

$$
\begin{array}{lr}
\text{N I O} & 18° \ 55' \ 50'' \\
\text{I O N} & 127 \ \ \ 54 \ \ \ \ 0 \\
\text{O N I} & 33 \ \ \ \ 8 \ \ \ 40 \\
\hline
& 179° \ 58' \ 30''
\end{array}
$$

The next station L is on the rocky ridge, extending above the Montanvert, towards the Aiguille des Charmoz, which ridge is itself familiarly called Les Charmoz. It is 867 feet above the Montanvert, and therefore a full half hour's ascent along the southernmost part of the rugged ridge in question. It was found impossible to obtain a station here which should command at once the two extremities of the measured base and the middle portion of the Mer de Glace. I was obliged to select a projecting mass of rock, which commanded the eastern extremity of the base, and the station I. It is at a height 3600 feet above the former and 1800 above the latter. It is marked with a cross and pyramid of stones and the letter L. The angles are

O L I	36° 18′ 20″
L I O	84 56 21
L O I	58 45 40
	180° 00′ 21″

The Montanvert itself was not found to be a commanding enough position to be selected as a principal station, although it has been accurately connected with four or five others.

Station F is on the eastern side of the glacier, on the promontory of Les Echelets, and is at about 150 feet above the level of the glacier, near the ridge of the hill, but rather facing the north. It commands a view at once of the Chapeau and Tacul. It is marked like the others, and connects with stations I and L, as follows,—

L F I	48° 30′ 15″
I L F	102 40 - 10
F I L	28 48 50
	179° 59′ 15″

Station G is on the ridge of Trélaporte, or Entre-la-Porte, on the west side of the glacier, at the foot of the Charmoz, and its elevation may be 300 feet above the level of the ice. It is on the little level space formerly alluded to, occupied by a series of blocks which appear to mark the former boundary of the glacier. It is on one of these large blocks (but to the south of the *largest*) that station G is marked by a red cross and pyramid of stones. It is one of the most conspicuous stations, and commands the glacier both ways. Only two angles of the triangle connecting G with L and F were observed, as follow,—

L F G	122° 37′ 45″
F L G	28 37 50
	151° 15′ 35″
L G F	28 44 25
	180° 00′ 00″

Station B, on the promontory of the Tacul, commands a more general view of the glacier than any other station. I could see up the ice streams of Léchaud and the Géant, and down to the station of Le Chapeau, whence the station B was distinctly observed, and served as a verification.* The promontory is composed of rather loose rock. A pretty solid pedestal was built up of the larger masses on the ridge, at a height of 277 feet (barometrically measured) above the little lake below. The cross was cut in the centre stone under the pyramid which supports the flag, and the pyramid was removed as in the other cases, when the instrument was planted. From this station, by observations right and left, the movement of either glacier was determined.

The position of B is determined by a very oblique triangle, the precipitous nature of the rocks opposite affording no convenient station whence F and B could at once be seen. Its angles are

G B F	11° 42′ 0″
F G B	151 52 25
G F B	16 25 5
	179° 59′ 30″

---

* The fluttering of a *white* flag seemed superior to any other mark of the stations. Where too distant to allow the staff bearing the flag to be seen distinctly, its motion served to distinguish the flag from any mark on the rock, or patches of snow which, in the latter part of the season, often rendered the observations difficult and uncertain. It was, for instance, with the greatest trouble that I succeeded in observing station L from I, partly owing to the badness of the light, partly to the mark not being seen against the sky or distant hill, and partly from the numberless patches of snow. I was obliged to return repeatedly, and on this and many other occasions was struck with the far more perfect eyesight of Balmat for distant objects than my own.

Station H was at the foot of the Couvercle, opposite station B, and beneath the Aiguille du Moine. It is on a grassy shelf, above rocky precipices, which rise perhaps 200 feet from the level of the glacier. It is accessible only by a small ravine or watercourse to the south, and even there not easily. The vegetation here is more luxuriant than on most parts of the glacier banks at so high a level. I observed particularly a luxuriant specimen of the alpine rose without thorns. It is 1177 feet above the Montanvert, or 7487 above the sea. The station is marked on a great stone imbedded in the turf. From it neither the Montanvert nor Les Echelets could be seen. The position of H is determined from G and B as follows,—

G B H	66° 40′ 10″
B H G	87 48 5
H G B	25 32 30
	180° 00′ 45″

Finally, another station was selected as high up the Glacier de Léchaud as a fixed rock conveniently accessible could be found for establishing it. It is on the eastern side of the glacier, exactly opposite a remarkable rock, called *Capucin du Tacul*, and about half an hour's walk above the junction with the Glacier du Taléfre. It is marked station E, and will be more readily found by the outlines on the map, which were carefully sketched, than by minute description; it is on a rocky promontory, perhaps about sixty feet above the level of the ice, which is here extremely crevassed, so as to render this spot somewhat difficult of access in the latter part of the season. From it several of the higher summits were measured, and the motion of the ice near its origin determined. The following triangle connects it with B and H.

B E H	21° 24′ 1″
H B E	67 50 40
E H B	90 45 40
	180° 00′ 21″

The sides of these various triangles have been calculated by two computers, and compared by a third. The result gives the following numbers (subject to the reduction of one 12000th, on the ground stated above, page 105.)

Feet.		Feet.
N O = 2992.95	G L =	8208.28
N I = 7275.78	F B =	10853.1
I O = 5043.04	G B =	6508.75
L O = 8484.49	G H =	5980.79
L I = 7282.61	H B =	2808.01
I F = 9485.56	H E =	7127.97
L F = 4686.50	B E =	7695.66
G F = 4670.13		

The station E is distant from the base line about 28,600 feet.

As these stations were to be used for the determination of the actual height of various mountain tops observed from them, as well as of various points of the glacier surface, it became necessary to ascertain their elevation above some known point, such as the Montanvert. I accordingly observed carefully the elevations, and usually also the corresponding depressions of the different stations, so far as they were visible from one another. The distances being known, the relative heights of the stations, taken by pairs, becomes easily known from the following data, on which it must be remarked, that the observations with an asterisk prefixed are double ones, or the mean of an observed elevation and depression, which eliminates at once the effects of the curvature of the earth, atmospheric refraction, and the error of collimation, while the single observations (those of elevation are marked +, those of depression —) are corrected by the following simple but abundantly accurate formula for such small distances, which includes the effects both of curvature and refraction.

Correction in English feet, (always +) $= .5714 \, D^2$.
D being the distance between the stations in English *miles.*

The error of collimation in a vertical direction was ascertained by observation, and found to be

$$\pm \; 1' \; 15''$$

the readings being too low, when the microscope A was uppermost, and too high when the microscope B was uppermost.

Stations.	Distance. Feet.	Angular Altitude.			Observed Height in Feet.	Calculated Height in Feet.
* O above N	——	0°	44′	0″	38.3	40.0
* I —— N	——	14	15	45	1849.5	1847.8
* I —— O	——	19	42	15	1806.0	1807.8
* L —— I	——	13	49	45	1792.6	1789.8
* F —— I	——	8	40	0	1445.6	1445.5
☾ L —— O	——	22	58	45	3597.7	3597.6
* L —— F	——	4	11	0	342.8	343.3
* G —— F	——	4	48	15	392.5	396.2
G —— L		+ 0	22	15	54.5	52.9
G —— I		− 7	27	30	1843.1	1842.7
* B —— F	——	3	56	45	748.6	750.6
* B —— G	——	3	5	30	351.5	354.4
* B —— H	——	1	53	0	92.3	96.6
* H —— G	——	2	26	22	254.7	257.6
E —— G	——	+ 3	3	15	707.9	703.0
* E —— B	——	2	26	30	350.5	348.6
* E —— H	——	3	32	30	441.2	445.4
M† —— I	5590	+ 9	20	45	920.6	922.9
F —— M	4754	− 6	17	45	523.5	523.6
G —— M	9006	− 5	51	45	922.0	919.8
B —— M	15450	− 4	43	45	1273.2	1274.2

† M stands for the *Montanvert*, of which the position and elevation was determined by reference to these four stations ; but as no observations were taken from the Montanvert itself, these are all single altitudes.

The numbers marked "calculated" were obtained as will be immediately stated.

These results plainly form so many equations of condition for the determination of the height of the stations amongst one another. There are nine stations independent of M, which we take as a starting point, or absolute level, and there are twenty-one equations of relative height. These have been combined by the method of least squares, so as to give the most probable height of each point referred to M. Double weight has been given to the observations of elevation and

depression combined. The mode of proceeding is to combine in a
single equation all those which contain any one unknown quantity.
From the nine equations thus formed, the unknown quantities are
successively removed, by a somewhat tedious elimination.   Great precautions were taken to avoid numerical errors, and the results are as
follows :—

			Feet.
N below M	.	. . .	2770.7
O —— M	.	. . .	2730.7
I —— M	.	. . .	922.9
L above M	.	. . .	866.9
F —— M	.	. . .	523.6
G —— M	.	. . .	919.8
B —— M	.	. . .	1274.2
H —— M	.	. . .	1177.4
E —— M	.	. . .	1622.8

With these numbers, those in the last column of the preceding table
have been computed, and the comparison appears satisfactory.

Besides the main points of the triangulation, several other stations
were assumed, either for determining the motion of the ice, or for extending the subordinate triangulation, which was performed partly
with the theodolite, and partly by means of Kater's compass.   The
determinations with the latter are manifestly inferior to the former.
Some errors, perhaps, were owing to a local magnetic disturbance of
no great amount, and others to a want of perfect mobility in the
compass, which, bearing a heavy card, is not, after some use, so delicate
as might be desired.   It is hardly necessary to observe that Kater's
compass is provided with a vertical wire, and slit for taking azimuths
which are read off at the instant of observation by the aid of a re-
flecting prism, which conveys the image of the divisions of the scale to
the eye, at the same time and in the same direction as that in which
the distant object is viewed.

Of the subsidiary stations, the chief were—

1. Station A, on the glacier opposite L'Angle, marked on the map
near the foot of the Charmoz; it was the earliest station employed for

measuring the motion of the glacier. It is determined in position by reference to stations L, F, and M.

2. Station C, on a block of granite on the ice, between the promontory of the Couvercle and the Tacul. It has already been described in Chapter V. Its position was ascertained by angles from B and G.

3. Station D, near the Montanvert, on a very large granite block, forming part of the ancient moraine, nearly on a level with the Inn, and distant 60 yards N. 40° E., magnetic, from the south-east corner of the building. From this point the motion of the glacier was frequently determined; and, by means of a small base line, its breadth in this part, and the distances of several marks upon it.

4. A station near the source of the Arveiron, upon the moraine of 1820, near the hillock of Côte du Piget. From it a number of magnetic bearings were taken, for laying down the details of the lower part of the glacier.

5. La Croix de Flégère; a small building, exactly opposite to the Montanvert, and at the same level, on the north side of the valley of Chamouni. Its position is fixed by reference to stations I and M. A number of magnetic bearings were taken from hence especially for the position of the Aiguilles, and of Mont Blanc, which is laid down on the map from its bearing from hence, and from the summit of the Jardin.

6. The summit of the Jardin, fixed by reference to station B and the Aiguille du Moine.

7. A station on the border of the Glacier du Taléfre, above the Pierre de Béranger, marked [γ]. It was fixed by reference to the Aiguille du Moine and the top of the Jardin. This, and the last station, afforded angles for determining in some measure the form of the basin of the Glacier du Taléfre.

8. Station K, marked on the south-east side of the Glacier du Géant, was fixed and accurately marked for the purpose of ascertaining the motion of this part of the glacier. The cross is cut in a triangular imbedded stone at a considerable height above the glacier, and which is attained by ascending a steep couloir. Unfortunately, no second observation was ever made to determine the motion of the ice, the premature winter in September rendering it impossible.

9. For the same reason an intended station, marked G*, between

H

Trélaporte and the Charmoz was never attained.　By angles taken from thence the topography of the upper part of the Glacier du Géant would have been considerably improved.

The topography of the valley of Chamouni has been partly filled-in for the purpose of connection, but without very adequate materials.

Most of the Swiss models represent the glacier of Argentière as extending altogether behind the basin of the Taléfre, and touching the mountains at the head of the Glacier de Léchaud!　In fact, the upper part of the Glacier d' Argentière is scarcely ever visited.　I apprehend that the extreme boundary of these three glaciers, as laid down in the map accompanying this work, forms very nearly the real axis of the chain, and that the glacier of Triolet, in the Val Ferret takes its origin from the mountain to the east of the Petite Jorasse, which is marked Montagne des Eboulements.　The mountains at the head of the Glacier de Léchaud,—the range called *les Périades*, which connects the Tacul with the main chain,—the group of the Géant and Mont Mallet, and the rather complicated and slender ramifications in the neighbourhood of the Col du Géant, are all pretty satisfactorily understood and de-termined.　The summit of Mont Blanc is entirely cut off from the upper ice basins of the Mer de Glace by the inaccessible ridge extend-ing from the two Flambeaux to Mont Blanc du Tacul.　Of the range of the Monts Maudits, and the origin of the glacier of La Bren-va, I have only vague topographical information ; it is therefore not detailed.　Between the elevated and wild rocky summits of the Aiguille du Midi and Mont Blanc du Tacul, there is a sort of depression (probably quite inaccessible on its western side) which separates the glacier of Bossons from the tributaries of the Mer de Glace.　The range of Aiguilles of Chamouni then follow.　I have considered them as four in number ;—1. The Aiguille du Midi, with its great subordi-nate glacier on the north slope.　2. The Aiguilles de Blaitière or du Plan—a great group, which some authors subdivide into two, and which has a third appendage on its southern side, to which I shall else-where advert.　3. The Aiguille de Grépon—which is more definite : between it and the last there is a small glacier, as well as to the east-ward between it and, 4, the Aiguilles des Charmoz, a many-headed group of sharp points difficult to define. I have marked on the map the position of two of the chief heads. The Aiguille des Charmoz branches into two, one arm descending steeply to Trélaporte, the other embracing the Mer de Glace all the way to the Montanvert.

Elevations above the Sea.

It has already been observed that the heights of the various stations have been referred to the level of the Montanvert, which thus becomes our starting point. This level has not been very accurately ascertained. De Saussure (*Voyages*, § 607,) calls it 428 toises above the valley of Chamouni, or 954 above the sea, without stating upon what this measure is founded,—probably on a single barometrical observation. It gives

6101 English feet.

Berger, quoted by Alphonse Decandolle, * gives

994 toises = 6357 English feet,

whilst Shuckburgh † obtained 5001 English feet above the Lake of Geneva, or above the sea

6231 English feet.

If we add 32 feet to De Saussure's measure for the admitted error of his earlier estimation of the height of the Lake of Geneva, ‡ it becomes 6133, and taking the mean of these three not very concordant observations, we obtain,

Height of Montanvert, mean of Berger, Shuckburgh, and De Saussure,
6242 English feet.

But the height usually adopted is the smallest, or De Saussure's.

During my stay at the Montanvert, in June, July, and August, I frequently observed my barometer near the hours of regular observation at the Geneva Observatory, as well as the attached and detached thermometers. The barometer having been broken on the 6th August, an end was put to these comparisons. It had been carefully compared

---

* Hypsometrie des Environs de Genève, p. 63. A most useful work.

† *Ibid.* There appears to be some error in Decandolle's reduction of this observation to the level of the sea. *Compare Phil. Trans.* LXVII. p. 592.

‡ *Ibid.* p. 8.

by M. Plantamour with the Geneva barometer, and found (after fourteen comparisons) to stand only 0.08 millimètre higher, a quantity so small, considering the uncertainties of a Syphon barometer, (it was on Bunten's construction,) that I have left it out of account.

Now the mean of twenty-seven comparisons gives

	m.m.	Att. Ther.	Det. Ther.
Montanvert,	610.35	12°.82 Cent.	9°89 Cent.
Geneva Obser.	728.50 (reduced to)	0 Cent.	19°04

whence we obtain for the height,

By the tables of the " Annuaire,"   4960.39 English feet.

By Baily's Table,   .   .   .   4960.64   —

The height of the barometer in the observatory at Geneva is 407 metres above the sea,* or .   .   .   1343 English feet.

Montanvert above the sea,   6303

A result nearly 200 feet greater than De Saussure's, and 60 feet greater than the mean of the three old observers. Nevertheless, as the previous observations were, so far as I know, single and isolated, I feel bound to adopt the new value, considering the number and advantageous circumstances of the observations. It is also confirmed in a striking, though no doubt partly accidental manner, by the measurement of the Col du Géant, which, adopting the height just given for the Montanvert, by comparison with the barometer at the Montanvert, is 11,144 feet above the sea, whilst, by direct comparison with Geneva, I find 11,146. The barometer at the Montanvert was hung at the usual height above the floor of the inn, or " pavillon," which may, therefore, be reckoned at exactly 6300 feet. But the point taken as station M is the eave of the roof, and about ten feet higher, wherefore,

Level of station M above the sea,   .   6310 English feet.

I observed that the door-step of the pavillon of the Croix de Flégère on the opposite side of the valley, viewed from the Montanvert, was depressed 32′ 45″. Its distance (by the map) is 13,200 feet. Hence

------

* Meteorological Tables in Bibliothèque Universelle.

the (corrected) level is 122·2 feet lower, or it is 6181 feet above the sea. The mean level of the base line appears also to be 3552 feet above the sea, which is, therefore, the height of the valley of Chamouni in that place.

Slope of the Glacier.

The source of the Arveiron, according to Decandolle, is 805-6 metres, or 2643 English feet below the Montanvert. The ledge of rock over which the western side of the glacier discharges itself, is near the level of the station of the Chapeau, which is 923 feet below the Montanvert. The height of that precipice is, therefore, about 1700 feet. The length of the final sweep of the glacier, from the top of the precipice to the source of the Arveiron, is about 4500 feet, following the curve, hence the mean slope of this part of the glacier, (which is completely dislocated and confused,) is   .   .   .   20° 41′ 44″.

The level of the ice near the western bank, a little below the position of the Montanvert, at the point marked D 2 on the map, is 285 feet (determined trigonometrically) below station D, which is on the level of the Montanvert nearly. Hence, between the west edge of the precipice and the Montanvert, the surface of the glacier rises by about 638 feet, on a distance of 3000 feet, being an inclination of 12° 0′ 22″.

Again, from Montanvert, or rather the point D 2, reckoned along the glacier to station A opposite the Angle, is a distance of 5200 feet. The level of station A is (barometrically) 131 feet above Montanvert, therefore 416 feet above D 2, giving a mean inclination of 4° 34′ 26″.

From station A to the foot of the Tacul, reckoning parallel to the axis of the glacier, is a distance of about 10,520 feet. Now, by three barometrical observations, the Cabane at the Tacul, near the level of the lake is 1003, 990, and 997 feet,* mean 997, which may be increased by 30 feet, to bring it to the level of the glacier in this place, making 1027 feet, or a rise of 896 feet from station A, which gives a mean inclination of   .   .   .   .   4° 52 5″.

If we reckon along the glacier, from station A to station C, the

---

* One day's observations have been neglected which had not corresponding ones, and of which the result was 70 feet above the mean of the others.

" Pierre Platte," we have a distance of 10,600 feet. The level of C, by two barometrical observations, on different days, is 1076 or 1097 feet above Montanvert; mean 1086, which gives 955 above station A. The mean inclination from A to C is, therefore, 5° 5′ 53″.

From station C to station E is a distance of 6800 feet along the ice. A single barometrical observation gives 1668 feet above Montanvert for the level of the centre of the glacier there. I prefer, however, to deduce it from the trigonometrical height of station E, or 1623; and we may suppose the highest part of the glacier there to have a height of 1600 feet, or 514 above station C, giving an inclination of
. . . .   4° 19 22″.

On the whole, from the precipice of ice opposite the Chapeau, where the ice becomes broken and discontinuous to this point E, not far from the proper origin of the glacier de Léchaud, is a distance of 25,600 feet, and a rise of 2520 feet, giving a mean inclination of   5° 37′ 19″.

I should have multiplied these observations much more had not my barometer been unfortunately broken. In particular, I regret not to have taken the height of the glacier du Géant at the foot of the Aiguille Noire. But if we reckon the entire course of that glacier up to the Col du Géant, we find the distance of the latter from the Tacul to be 24,700 feet, and its rise 3814 feet, giving an inclination of
. . . .   8° 46′ 40″.

Taking the entire length of the Glacier des Bois, Mer de Glace, and Glacier du Géant, from the source of the Arveiron to the Col du Géant, we have a distance of about 47,920 feet, with a rise of 7484 feet, being a mean inclination of .   .   .   .   8° 52′ 36″,
but the higher and lower parts are precipitous.

From the various stations several of the principal mountain tops which admitted of more accurate triangulation were observed, and their elevations taken, which were then reduced to the Montanvert, and thence to the level of the sea. The differences in the following table for the same summit arise almost entirely from the impossibility of determining the geometrical summit of the hill accurately; not one of these peaks being accessible. A very small error in the distance entails a very sensible one on the height, on account of the great angles of elevation.

Mountain.	Station whence observed.	Distance.	Corrected Elevation. Deg.	Min.	Sec.	Height above Station. Feet.	Height above Montanvert.	Mean height above the Sea.
Aiguille du Dru, No. 1......	L O		25	27	45	5002	5868	12178
Aiguille du Dru, No. 2......	G B	9541 12916	27 19	42 50	15 15	5012 4663	5932 } 5938	12245
Aiguille des Charmoz,........	B H	7750	25	16	15	3660	4634	10944
Aiguille du Moine,............	B G E	7553 7059 9250	24 28 19	45 45 11	15 15 45	3684 3875 3222	4758 4794 } 4845	11109
Aiguille Verte,........	G [γ]	12350	26	39	15	6202	7122	13432
Jardin, highest point,.........	B [γ]	11750	11	6	15	2309	3583	9893
Tours des Courtes,............	B [γ]	18000	14	7	15	4535	5809	12119
Aiguille de Léchaud,.........	H B	8400 9760	22 18	18 42	45 45	3449 3308	4626 } 4582	10914
Petite Jorasse,...............	G H B	20174 14100 14628	13 18 17	58 31 43	45 45 15	5030 4730 4678	5950 5907 } 5952	12246
Grande Jorasse,...............	G H E	19786 14556 9356	17 22 30	32 44 18	15 45 45	6261 6106 5472	7181 7283 } 7095	13496
Mont Tacul, east summit,...	G H	11400 6560	18 28	10 21	15 45	3744 3542	4664 } 4720	11002
Mont Mallet,.................	E G	11730	23	37	45	5135	6758	13068
Aiguille du Géant,............	G H B	18306 15111 12695	17 20 23	31 7 35	15 45 15	5920 5546 5530	6839 6820 } 6707	13099
Croix de Flégère,............	M I	13200	0	32	45	122	122	6188

The Aiguille Verte is called the Aiguille d'Argentière in some Maps, I think erroneously, as by Sir George Shuckburgh. Phil. Trans., vol. 67.

It may be doubted, whether the Mont Mallet is not as high as the Aiguille du Géant.

The Aiguille du Moine is the Aiguille du Taléfre of De Saussure.

The Promontory of Trélaporte is called Entre la Porte by the older writers.

The Aiguille du Dru, No. 1, is that visible from the Montanvert. No. 2 is a higher summit in the direction of the Aiguille Verte, which is invisible from the Montanvert.

These heights vary in several particulars from those previously given, especially by De Saussure. Those of the Grande Jorasse and Aiguille du Géant in the preceding table have been verified with especial care. The former is 300 feet greater, the latter no less than 800 feet less than the previous determinations. According to Pictet and De Saussure the order of elevations of mountains next Mont Blanc would be

Géant,   Verte,   Jorasse,

By the preceding table the heights of the Grande Jorasse and the Aiguille Verte are almost the same, and come next to that of Mont Blanc, whilst the Aiguille du Géant and Mont Mallet are 400 feet lower. The following table contains the heights above the sea of points mentioned in this chapter from my own observations, and the best of those formerly made ; the latter I have taken chiefly from Decandolle's *Hypsométrie des Environs de Genève*, and reduced them to English feet.

	J. D. F., 1842.	Other Authorities.	
	English feet.	Eng. feet.	
Mont Blanc, . . .		15744	French Engineers.
Grande Jorasse, . .	13496	13192.5	Pictet.
Aiguille Verte, . .	13432	13402	Shuckburgh.
Aiguille du Géant, .	13099	13875	Saussure.
Mont Mallet, . .	13068		
Aiguille du Midi, .		12822	Saussure.
Aiguille du Dru, No. 2.	12245		
Aiguille du Dru, No. 1.	12178	12520	Pictet.
Petite Jorasse, . .	12246		
Tours des Courtes, .	12119		
Col du Géant, . .	11146	11172	Saussure.
Aiguille du Moine, .	11109		
Mont Tacul, . . .	11002		
Aiguille du Charmoz,	10944	9131	Pictet.
Aiguille du Léchaud,	10914		
Grand Mulet, . .		9996	Barry.
Jardin, (highest point,)	9893		
Jardin, ...		9042	Pictet.
Glacier du Taléfre, .		8530	Pictet.
Station E ; Léchaud,	7933		

	J. D. F., 1842.	Other Authorities.	
	English feet.	English feet.	
Station B ; Tacul, .	7584		
Tacul, Lake, . . .	7300		
Station H ; Couvercle,	7487		
Station C; Pierre Platte,	7389		
Station G ; Trélaporte,	7230		
Station L ; Charmoz,	7177		
Station F ; Echelets,	6833		
Station A ; L'Angle,	6434		
Montanvert, level of floor,	6300	6242	Mean of 3.
Croix de Flégère,	6188		
Station I ; Chapeau,	5387		
Source of the Arveiron,		3667	Decandolle.
Station O ; Tines, .	3578		
Station N ; Praz, .	3539		
Chamouni, . . .		3425	Mean of several.

A very few words remain to be said respecting the construction of the map accompanying this work.

The sides of the fundamental triangles were, as we have seen, computed from the observations. They were then laid down on a scale of $\frac{1}{10000}$ of nature. It is much to be regretted that the practice of adopting a *natural* instead of an arbitrary scale has not yet found its way into maps laid down in this country. It has the very obvious advantage of at once conveying to the mind an idea of the real scale on which the map is drawn, which cannot be the case otherwise, unless the person is acquainted with the local measures used as a standard. The scale of six inches to a mile, offers no idea to a person accustomed to a scale of metres or of French leagues, but a fractional scale of nature is the same in Germany, France, or England, and maps drawn to any scale may very readily be reduced to another, provided a convenient series of sub-multiples of the natural magnitude be used. A scale of $\frac{1}{10000}$ of nature is 6.336 inches to a mile, or somewhat greater than the recent Ordnance Survey of Ireland. The details of the ice and outlines of the glacier as far up as the Tacul were also filled-in on the same scale. The more general map was drawn out roughly by myself on a scale of $\frac{1}{25000}$. It was then reduced by the pentagraph to $\frac{1}{25000}$, and drawn out, under my own eye, in a very artist-like manner by Mr. Knox, the principal draughtsman of Messrs. Johnston, engravers in Edinburgh. It has been also engraved by them. The aim has been to preserve distinct the three characters of glaciers, moraines, and solid rock; these were distinguished by colouring in detailed eye sketches taken on the spot, and an endeavour has been made, I hope not ineffectually, to preserve their character in the engraving.

I need not say that I am very far from considering my map as even approaching to perfection. The parts, however, most open to criticism are perhaps those of the least interest, I mean the details of the valley of Chamouni, and the low grounds adjoining, of which I took no regular survey as being beside my principal object. I hope, however, that the candid and well-informed traveller will find the details of the glacier and its ramifications as faithful as the circumstances under which they were made fairly admitted of. Information and corrections I shall be most anxious to receive.

The geographical position of the district of country contained in the map, may be most correctly inferred from the position of Mont Blanc, as determined by the French Engineers. According to them it has for

Latitude, ... 45° 49′ 58″.84 N.

Longitude, East of Paris, . . 4° 31′ 42″.52 „

Greenwich, 6° 52′ 6″.5 „

In 1832 I made observations for the latitude and longitude of Chamouni by altitudes of Polaris for the former, and by a chronometer for the latter. The details are given in a paper in the *Philosophical Magazine* for 1833,* and the results are,—

Chamouni, Lat. 45° 55′ 54″ N.

Long. 6° 51′ 15″ E. of Greenwich.

---

* *New Series*, ii. 61. To these we may add De Saussure's determination of the latitude of the Col du Géant, viz. 45° 49′ 54″.—*Voyages*, § 2036.

I regret very much that I neglected during my late visit to determine the position of the meridian with reference to the points of my map, which might very easily have been done. It is laid down on the paper magnetic north and south, but there is some uncertainty as to the magnetic variation. It appears, however, in the time of De Saussure's visit to the Col du Géant to have been 19° W. which was its value about the same time at Geneva, and as that is also the present variation at Geneva, it has been assumed to be the same at Chamouni.

CHAPTER VII.

ACCOUNT OF EXPERIMENTS ON THE MOTION OF THE ICE OF THE MER DE GLACE OF CHAMOUNI.

GLACIER MOTION A MECHANICAL PROBLEM—CONTRADICTORY OPINIONS RE-
SPECTING IT—EXPERIMENTS COMMENCED—DAILY MOTION DETECTED—MO-
TION BY DAY AND BY NIGHT—HOURLY MOTION—CENTRE MOVES FASTEST
—TABLE OF RESULTS—LAWS OF GLACIER MOTION FROM OBSERVATION—AS
RESPECTS THE LENGTH—THE BREADTH OF THE GLACIER—THE SEASON OF
THE YEAR, AND STATE OF THE THERMOMETER—CHANGES OF LEVEL OF THE
ICE AT DIFFERENT SEASONS.

> The glacier's cold and restless mass
> Moves onward day by day.
> BYRON.

From the time of my being introduced to the Theories of the
Formation and Maintenance of Glaciers, maintained by M. M.
De Charpentier and Agassiz, it had struck me as very singular, that
no *numerical tests* had been applied to ascertain their insufficiency,
or to prove their correctness. A careful perusal of the writings of
these and other ingenious authors had left on my mind no clear
demonstration of any fact connected with the *cause of progression*
of glaciers. Yet this surely lies at the very basis of any speculation
respecting the causes of their existence and perpetuation, as well as
their formerly greater extent and geological agency. Most of the
arguments in favour of the progression being due to water absorbed by
capillary fissures, and then frozen so as to produce dilatation in the
whole mass, were deduced from considerations either *à priori*, or at

least indirect. In 1841, I suggested to M. Agassiz the use of coloured liquors, to act as the injections of an anatomical preparation, in showing the fissures and capillary canals of the ice ; but no such satisfactory mode occurred to me of proving directly the fact of congelation in the mass ; to which, besides, there appeared to me objections arising from first principles so insurmountable as to render any thing short of the most unequivocal demonstration, unsatisfactory. On studying the subject at home and leisurely, I satisfied myself that experiments could be made upon the motion of the ice, which should, in a good degree, throw light upon the question. The question is reduced to one of pure mechanics, and should be treated as such by a rigorous analysis. The *motion* is the thing to be accounted for. Have the laws of the motion been determined? Have we the data of the problem of which we seek for the solution? Had not observatories existed for centuries, and empirical astronomy arrived at a very great degree of precision, the theory of Newton would have been a baseless speculation. In fact, it never could have existed at all, its very essence consisting in its conformity to certain facts respecting the motions of the planets far from obvious, and the result of elaborate observation and still more elaborate efforts of combination and reduction. No doubt, many problems are so simple as not to require so elaborate a mechanism as that by which the theory of planetary motion was successfully reduced to law; and under the supposition that the glacier problem was a very simple one, and only required a general knowledge of elementary mechanics to explain it in an obvious way, De Saussure and other writers first treated it—and others followed with unhesitating assent—until Venetz and De Charpentier had the courage to expose the difficulties which the sliding theory involved. But, from this moment, the glacier theory required a more elaborate analysis than had yet been given to it. The mere fact of motion seemed explicable in various ways, but to each, substantive, if not unanswerable, objections, might be urged. From this moment, it became necessary to submit the phenomena to analysis, and to ascertain the Law of Variation of the Quantity of Motion, in terms of some of those varying agents which were supposed to influence that motion. Such a comparison, such a reduction of *circumstances* to *mea-*

sures, has operated, in every science, in the most wonderful manner, in reducing guesses to certainties. In geology, it is unfortunately not possible to ascertain the measure of effects in terms of their supposed causes, because (in the opinion of most geologists) the effects and their causes having ceased or been greatly modified, and not reproducible by human agency, nor recurring at known periods, a looser kind of induction must be tolerated in that science. The glaciers have had their phenomena more closely linked to the sciences of observation than to those of experiment—to natural history rather than natural philosophy ; and hence a problem which is, in the first instance, one of pure mechanics —the motion of a mass, the nature, intensity, and direction of whose cause of motion is to be ascertained,—has been left to be discussed on mere grounds of probability, or the adequacy of supposed causes to produce a certain kind of effect, of which the degree and circumstances had remained almost unstudied.

It was, accordingly, matter of the greatest surprise to me, to find that those ingenious persons who had been engaged for years in the study of glaciers, and in maintaining their theories of their motion by many ingenious analogies, and observations of structure, and the like, should not have thought of determining the motion accurately, with reference to season, weather, inclination of surface, alternation of day and night, and at different points of the length and breadth of the glacier. I suggested one experiment of this kind, which seemed to me to be critical :—

" If De Saussure's theory be true, the glacier moves onward without sensibly incorporating new matter into its substance—continually fed by the supplies from behind, which form a new and endless glacier. The mechanism may not inaptly be compared to that of the modern paper machine, which, from the gradually consolidated material of pulp, (representing the névé), at length discharges, in a perpetual flow, the snowy web. The theory of De Charpentier, on the other hand, represents the fabrication of the glacier going on within the glacier itself, so that each part swells, and the dilatation of each is added to that which acted upon itself, in order to shove on the section of the ice immediately in advance. *In the former case, then, the distance between two* *determinate points of the glacier remains the same ; in the latter, it will continually increase.* Again, *on the former hypothesis, the annual progress of any point of the glacier is independent of its position ; on the latter, it increases with the distance from the origin, (the transverse section of the ice being the same.)* The solution of this important problem would be obtained by the correct measurement, at successive periods, of the spaces between points marked on insulated boulders on the glacier ; or between the heads of pegs of considerable length, stuck into the matter of the ice, and by the determination of their annual progress."*

The more that I revolved the subject in my own mind, the more clearly was I persuaded that the motion of glaciers admitted of accurate determination, and must lead to definite conclusions.

We have seen that the *motion* of glaciers has been for much more than half a century universally admitted as a physical fact. It is, therefore, most unaccountable that the *quantity* of this motion has in hardly any case been even approximately determined. I rather think that the whole of De Saussure's writings contain no one estimate of the annual progress of a glacier, and if we refer to other authors we obtain numbers which, from their variety and inaccuracy, throw little light on the question. Thus, Ebel gravely affirms† that the glaciers of Chamouni advance at the rate of 14 feet a-year, and those of Grindelwald 25 feet a-year ; whereas, as we shall see, such spaces are actually traversed by most glaciers in the course of a few days. This statement is quoted by Captain Hall,‡ and other recent writers, and even by M. Rendu, (now Bishop of Annecy,) the author of a most ingenious paper on Glaciers, too little known.‖ Hugi perceived the errors arising from a confusion between the rate of *apparent* advance of an increasing glacier into a warm valley, whilst it is continually being shortened by melting, and the rate of motion of the ice itself.§

---

* *Edinburgh Review*, April 1842, p. 77.
† Guide du Voyageur, Art. *Glacier*.   ‡ Patchwork, I. 109.
‖ Mem. de la Société académique de Savoie, X. 95.
§ *Alpenreise*, p. 371.

He points out the correct method of observation; and although his work contains no accurate measures, he was perhaps the first who, by observing the position of a remarkable block upon the glacier of the Aar, indicated how such observations might be usefully made, instead of trusting (as appears to have been the former practice) to the vague reports of the peasantry. Hugi's observations on the glacier of the Aar give a motion of 2200 feet in nine years, or about 240 feet per annum.* Now, in contradiction to this, it would appear from M. Agassiz' observations, that from 1836 to 1839, it moved, as far as in the preceding nine years—that is, three times as fast.† There is reason, however, to think, that M. Hugi's estimate is the more correct.

Bakewell ‡ assigns 180 yards per annum as the motion of the Mer de Glace, and De la Beche§ 200 yards, on Captain Sherwill's authority.‖ But both of these were hearsay estimates by the guides. M. Rendu seems to have been more aware of the importance of the determination of the rate of motion of glaciers than any other author; but the best information which he could collect in 1841, did not much tend to clear up his doubts. He gives the following rates of motion of the Mer de Glace, or Glacier des Bois, without being able to decide upon which is the most trustworthy: 242 feet per annum; 442 feet per annum; a foot a day; 400 feet per annum; and 40 feet per annum, or *one-tenth* of the last!—a difference which he attributes to the different rates of motion of the centre and sides.** De Charpentier, so far as I recollect, offers no opinion in his work on glaciers as to what is to be considered as their rate of motion. I was not therefore wrong in supposing that the actual progress of a glacier was yet a new problem when I commenced my observations on the Mer de Glace in 1842.

I had myself been witness to the position, in 1841, of the stone whose place had been noted by Hugi fourteen years before, and it was manifest that it had moved several thousand feet. In conformity with the prevalent view of the motion of the ice being perceptible chiefly in

---

* AGASSIZ, *Etudes*, p. 150.   † *Ibid.*
‡ Travels in the Tarentaise, i. 365.   § *Geological Manual*, p. 60.
‖ 100 yards, in *Philosophical Magazine*, Jan. 1831.   ** *Memoires, &c.*, X. 95.

summer, I made the hypothesis that the annual motion may be imagined to take place wholly during four months of the year with its *maximum* intensity, and to stand still for the remainder. With this rude guide, and supposing the annual motion of some glaciers to approach 400 feet *per annum* (as a moderate estimate from the previous data,) we might expect a motion of at least 3 feet *per diem* for a short time in the height of summer. There appeared no reason why a quantity ten times less should not be accurately measured, and I, therefore, felt confident that the laws of motion of the ice of any glacier in its various parts, and at different seasons, might be determined from a moderate number of *daily* observations.

I went to Switzerland, therefore, fully prepared, and not a little anxious to make an experiment which seemed so fruitful in results, and though so obvious, still unattempted.

The unusually warm spring of 1842, gave me hopes of commencing my operations earlier than the glaciers are usually frequented ; and it was evident, that, to detect the effect of the *seasons* on the motion of the ice, they could not be too soon begun. I left Paris on the 9th of June, by the *malle poste* for Besancon. After spending a day at Neufchâtel, I proceeded to Berne to visit M. Studer, and from thence I went to Bex, to make the acquaintance of M. de Charpentier, with whose geological and other writings I had so long been familiar. I only allowed myself a hasty visit to my friends at Geneva, and left that town with lowering weather, on the 23d June, for Chamouni, determined to await its clearing, and then proceed at once to the Mer de Glace. No patience was, however, required. The weather cleared that very day, and reaching Chamouni early on the following one, I made the requisite arrangements at the village, and leaving my baggage to follow, I proceeded straight to the Montanvert.

I resolved to commence my experiments with the very simple and obvious one of selecting some point on the surface of the ice, and *determining its position with respect to three fixed co-ordinates*, having reference to the fixed objects around ; and, by the variation of these, to judge of the feasibility of the plans which I had laid out for the summer campaign. One day, (the 25th,) was devoted to a general *recon-*

naissance of the glacier, throughout a good part of its length, with a view to fixing permanent stations ; and the next I proceeded, at an early hour, to the glacier opposite to the rocky promontory on the west side of the glacier, called *L'Angle*, thirty minutes' walk from the Montanvert, which presented a solid wall of rock in contact with the ice, so that upon the former, as upon a fixed wall or dial, might be marked the progress of the glacier as it slid by.

The instrument destined for all these observations was the small astronomical circle, or $4\frac{1}{2}$ inch theodolite already described, supported on the portable tripod. A point of the ice whose motion was to be observed, was fixed by a hole pierced by means of a common blasting iron or *jumper*, to the depth of about two feet. At first, I was much afraid of the loss of the hole by the melting of the ice, and the percolation of water from day to day ; but I soon found that very little precaution was necessary on this account, and that such a hole is really a far more permanent mark than a block of stone several tons in weight resting on the ice, which is very liable to change of position, by being raised on a pedestal, and finally slid into some crevasse.

An accurate vertical hole being made, the theodolite was nicely centred upon it by means of a plumb line, and levelled. A level run directly to the vertical face of rock, gave at once the co-ordinate for the *vertical* direction, or the height of the surface of the glacier. The next element was the position or co-ordinate parallel to the length or direction of motion of the glacier. This was obtained by directing the telescope upon a distant fixed object, nearly in the direction of the declivity of the glacier, and which object was nothing else than the south-east angle of the house at the Montanvert, distant 5000 feet. The telescope was then moved in azimuth exactly 100° to the left, and thus pointed against the rocky wall of the glacier, which was here very smooth and nearly perpendicular, owing to the friction of the ice and stones. My assistant (Balmat) was stationed there with a piece of white paper, with its edge vertical, which I directed him by signs to move along the surface of the rock until it coincided with the vertical wire of the telescope. Its position was then marked on the stone with a common pencil, and the positions of successive pencil marks were carefully measured by a tape or ruler from day to day. Marks were then indented in the rock with a chisel or pick-edged hammer, and the mark painted red with oil paint, and the date affixed. These marks, it is believed, will remain for several years. The station on the ice, (marked A on the map,) was distant 250 feet from the rock, and, by repeating the observation frequently, I found that it could be depended on to about one-fourth or one-third of an inch.

The third co-ordinate, or that which should measure the distance of A from the rock was not so accurately ascertained. No ready means offered itself for ascertaining with quickness and accuracy any variation of distance in respect to the breadth of the glacier. Whilst I admit that this would have been an advantage, I may observe that in most cases there is no reason to doubt that the motion of the ice is sensibly parallel to its length, and that any small error in the direction would scarcely affect the result. The direction of motion of the ice is unequivocally proved by the direction of the moraines, which are an external indication of that motion. In general, therefore, I have measured the movement of the ice parallel to the moraines where they were well marked. I am of opinion, however, that a check of some kind, such as the measurement of a third co-ordinate, would be advantageous where applicable.

It was with no small curiosity that I returned to the station of the "Angle" on the 27th, the day following the first observation. The instrument being pointed, and adjusted as already described, and stationed above the hole pierced in the ice the day before, when the telescope was turned upon the rock the red mark was left far above, the new position of the glacier was 16.5 inches lower (that is, more in advance) than it had been twenty-six hours previously. Though the result could not be called unexpected, it filled me with the most lively pleasure. The diurnal motion of a glacier was determined, (as I believe,) for the first time, from observation, and the methods employed left no doubt of its being most accurately determined. But a question of still greater interest remained behind. Was this motion a mean and continuous one, or the result of some sudden jerk of the whole glacier, or even the partial dislocation of the mass of ice on which I stood? This could only be tested by successive days' trial, and I awaited the result with doubt and curiosity. Of this I was persuaded, that if the motion should appear to be continuous, and *nearly* uniform, it could not be due to the mere sliding of the entire glacier on its bed, as De Saussure supposed ; for, admitting the possibility of gravity to overcome such intense friction as the bed of a glacier presents, it seemed to me quite inconsistent with all mechanical experience that such a motion, unless so rapid as to be an accelerated one, and that the glacier should slide before our eyes out of its hollow bed, (which would be an avalanche), could take place, except discontinuously, and by fits and starts.   To this most elementary question no answer founded on direct experience is to be found, so far as I know, in any work ; and although the whole theory might turn upon so simple a point, as whether the glacier *flows* down evenly, or moves by jerks, opinions seem hitherto to have been divided.*   On the 28th June I therefore hastened with not less interest to my post, and found that in $25\frac{1}{2}$ hours the advance had been 17.4 inches, nearly the same, though somewhat more rapid, than on the previous day. I no longer doubted that the motion was continuous, but I hastened to put it to a still more severe test, and likewise to make an experiment critical for the theory of congelation and dilatation.   I proposed to compare the *diurnal* and *nocturnal* march. I fixed its position at six P.M. on the 28th, and next morning by six o'clock I was again stationed on the glacier. It had moved eight inches, or exactly half the mean daily motion already observed. The night had been cold ; the ice was still frozen, though the temperature of the air had already risen to 40° ; a thermometer laid on the ice stood at 36°. If congelation had resulted during the night, so as to freeze the water in the capillary fissures, nearly the whole motion of the twenty-four hours ought to have taken place whilst the glacier froze : but not at all : from six A.M. to six P.M. of the 28th, the glacier advanced 9.5 inches, giving a total motion of 17.5 inches in twenty-

---

* I have found the true opinion, that of constant, insensible motion, to be held by almost all the intelligent mountaineers with whom I have spoken ; a majority, I think, of whom also declare that the glacier advances during winter as well as in summer.

four hours, somewhat greater than either of the preceding days, the motion appearing to increase as the warm weather continued and increased in intensity : At least so I interpreted it. The same afternoon I had no difficulty in detecting the advance of the glacier, during an interval of *an hour and a half.* The continuity of motion was thus placed beyond a doubt. The marks on the rock indicated a regular descent in which time was marked out as by a shadow on a dial ; and the unequivocal evidence which I had now for the first time obtained, that even whilst walking on a glacier, we are day by day and hour by hour imperceptibly carried on by the resistless flow of this icy stream, with a solemn slowness which eludes our unaided senses, filled me with an admiration amounting almost to awe, whilst I foresaw with lively interest the definite and satisfactory knowledge of *laws* which would result from these methods of observation.

The following morning (30th June) at six o'clock, the glacier was 8.5 inches in advance, and during the succeeding twelve hours of day, 8.9 inches, making together 17.4 inches for the twenty-four hours, a result not differing sensibly from that of the day before.

I observed distinctly the progress of the glacier on the 30th from five to six o'clock P.M., and on this occasion, as on the day before, it appeared to me that the motion at that time of day was *more rapid* than the mean motion. The motion in twenty-four hours for these four days had been :

$$15.2 :— 16.3 :— 17.5 :— 17.4 \text{ inches.}$$

A variation which I believed (and am persuaded) to be by no means accidental, but due to the increasing heat of the weather.

These results were the more interesting, (and with respect to their regularity the more unexpected) because the spot where they were made was a part of the ice deeply crevassed. It had been selected on account of the proximity of the naked rock ; but though the most solid accessible part of the ice was chosen for station A, it was surrounded by chasms in every direction, and the glacier in nearly all its breadth between the Angle and the Echellets is (in ordinary language) impassable on account of its dislocated and shattered condition. Yet amidst all this turmoil' and confusion there were no fits of advance, no halts, but an orderly continuous progression.

But during the last week of June, in which, stimulated by the extraordinary fineness of the weather, and the fresh interest of every day's experiments, I spent from twelve to fourteen hours daily on the glacier, —I was able to make other observations of interest to the theory, and not less consistent with one another.   I fixed two points in the ice by bored holes a little way below the Montanvert, one near the side, the other near the centre of the glacier.   Most authors, I believe, have asserted, that the *sides* of the glacier move faster than the *centre*.* But this seemed worthy of proof.   Stationing my theodolite, not upon the ice, but upon the lofty western bank at the station D, on a great boulder 60 yards in a direction north, 40° east (magnetic) from the south-east corner of the house of the Montanvert, I levelled it carefully, and then turning the telescope so as to point *across* the glacier to the rocks on the opposite side, by unclamping the telescope I caused it to describe a vertical great circle.   I caused a tall cross to be painted in red bordered with white on a face of rock oppo- site, making an angle of 118° with the corner of the Montanvert already mentioned, and distant from D 2898 feet.   The cross is marked D 1, and is a little to the north of a small cascade laid down in the map.

By pointing the telescope upon the cross, and then causing it to describe a vertical circle (like a transit instrument adjusted upon a meridian mark) the velocity of the different parts of the glacier could be determined as they flowed past.   Two stations, as has been said, were first fixed upon and marked by vertical holes in the ice renewed from time to time ; the first D 2 (see the map) was about 300 feet from the west bank of the glacier, therefore, nearly corresponding in position to station A, which was 5200 feet higher up ; the other, marked D 3, was 795 feet farther east, or rather beyond the centre of the glacier, being within 150 feet of the first moraine.   It is, however, very near the centre.

	Side (D 2.)	Centre (D 3.)
From 29th June to 1st July the motion in		
24 hours, was	17.5 inches.	27.1 inches.

---

* See, for example, Agassiz, *Etudes*, p. 167.

Here, then, was a difference not to be mistaken, and the near coincidence of the side station with the result at station A, I considered at the time confirmatory of its accuracy. Henceforth, I entertained no doubt that the generally received opinion is incorrect, and that the glacier stream, like a river, moves fastest towards its centre.

In the same line across the glacier with D 2 and D 3, several other stations were afterwards fixed with a view to test the modification of velocity depending on the distance from the bank or edge of the glacier. These measures proved that the velocity of the central parts is nearly alike, and that the greatest differences in velocity are close to the side, where friction may be expected to act exactly as in a current of water.

My next object was to ascertain the rate of motion of points of the Mer de Glace higher up and nearer its origin. For this purpose I fixed upon the remarkable large flat stone, or glacier table, formerly described, and marked C on the map. It lay on the Glacier de Léchaud, between the promontory of the Couvercle and the Tacul. Trusting to its apparent solidity, I did not apprehend that its position was likely to be speedily disturbed, and I fixed my instrument upon it over a red cross with the letter C. In this, however, I was deceived, for three weeks later, I could no longer mount upon the stone by any effort, or even see its upper surface, and in the month of August it slipped off its pedestal of ice. This did not, however, alter the character of exactness of the observations made by means of it. It was but for a few days that I used the stone as I had done station A at the angle, planting the theodolite upon it, fixing the azimuth by a distant object in the direction of the length of the glacier, and then turning the telescope through a certain number of degrees, and marking the progress of the ice upon the clean bare face of rock at the Couvercle. I soon, however, abandoned this plan, and stationing myself on a commanding spot of the promontory of the Tacul (station B), I directed the telescope upon a cross marked C 1, painted upon the opposite rock of the Couvercle, and causing the telescope to describe a vertical circle, I noted exactly as at station D, the progress of the great stone C, as it flowed on with the glacier. Thus, between the 27th and 30th June, this mark had advanced 30¾ inches, or about 10.2 inches per diem, instead of 17 inches as at the *Angle*, and near the side of the glacier below the Montanvert, or 27 inches as at its centre. Hence, it was quite certain that in this particular case the higher part of the glacier moved more slowly than the lower. Well aware, however, that this might be due to the variable section of the glacier, I made preparations for confirming it at different points, and still nearer the origin of the glacier, but at the season I have mentioned the higher parts were almost inaccessible from the quantities of half-melted snow which concealed the crevasses.

A week of singularly fine weather enabled me to obtain all the results which I have mentioned, and several others, previous to the lst July, on which day I left the Montanvert, in order to proceed to Courmayeur and Turin, for the purpose of witnessing the total eclipse of the sun. In these few days I had satisfied myself of the applicability and certainty of the methods I had employed; and the marks which I left in the ice, with directions to Auguste Balmat to watch them in the interval of my absence, and to renew the holes, if necessary, promised me fresh results of interest on my return. In the meantime, I communicated to my friend, Professor Jameson, the leading results which I had then obtained, in a letter, dated the 4th July, from Courmayeur, which was afterwards printed in his Journal.*

After my return, the measurements were renewed on the 28th July, the holes having been all deepened in my absence by the care of Balmat. The rate of loss at the surface will be mentioned by and bye. There was thus determined the motion, during one month, of the ice at station A, *L'Angle*, already mentioned; of two points in the breadth f the glacier at the Montanvert; of station C, and another point in the breadth of the Glacier de Léchaud, marked B 1, and of a point marked B 2 on the Glacier de Géant, opposite the Tacul, and at about the same distance as C and B 1 from the lower end of the glacier; B 1 and B 2 were observed in the way already described from station B.

---

* Re-printed in Appendix, No. II.

At later periods, there weré added the following points of observation, all of which are accurately laid down on the large map, but which will be more readily understood at a glance from the annexed wood-cut, which shows their relative position.

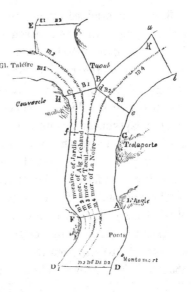

D 4, D 5, and D 6, opposite station D, near the Montanvert; intermediate between D 2 and D 3, as shown on the map. D 4 and D 5 were on the same part of the glacier, but the mark, D 4, was suffered to disappear during one of my absences. Thus the order of stations, and their distances were—

E. edge glacier. { .. about 350 yds. D 3. . 60 yds. . D 6. . 75 yds. . D 5. 130 yds. . D 2. . 100 yds. } W. edge glacier.

On account of the steepness and discontinuity of the glacier farther down, it was not thought advisable to attempt to observe its velocity at a lower point.

Station A at the *Angle* was observed during the whole season, and the corresponding marks in red paint were continued on the rock. Between the end of June and the end of September, the motion of the ice amounted to 103 feet.

After the great stone at station C had slipped from its icy pedestal, the velocity of the glacier was observed by means of a hole driven into the ice as at the other stations. In addition to mark B 2 on the small moraine of La Noire on the Glacier du Géant another point in the breadth of the same glacier was taken near its centre, and in the same right line from station B with the mark B 2. This line was determined by pointing the theodolite from station B upon the mark C 1, painted on the rock of the Couvercle, and then causing it to revolve through 115° to the left. The mark in the centre of the Glacier du Géant was distinguished as B 3. The position of the marks B 2 and

B 3, were determined by angles from H. The only observation of velocity made during the season 1842 which I have rejected, is one of B 3. The daily velocity from August 2 to August 4 was 32 inches, whilst that of the neighbouring mark B 2 was only 14 inches. Although the proximity of the latter to the side of the glacier was a good reason that it should move slower, this disproportion seemed unlikely, and the experiment was immediately repeated. The next two days gave a velocity of 18 to the first, and 14·25 to the second, a proportion which was nearly preserved during the remainder of the season. The mark B 3, was fixed on the 2d August.

The next station, E, formerly described as being at the higher part of the Glacier de Léchaud, was used to determine the velocity of the ice at the side and centre of the glacier in that part. A mark, E 1, was placed 210 feet from the east edge of the glacier, and another, E 2, about 645 feet farther. These marks were established on the 29th July. The observation was made by fixing the instrument at station E, and fixing the azimuthal wire of the telescope upon the sharply-defined apex of the singular rock opposite, called Le Capucin du Tacul ; the telescope being then moved in a vertical circle, passed over the marks E 1 and E 2. A check reference was made to the prominent edge of the Pierre de Béranger, which, on one occasion, proved most serviceable. Owing to their great distance from the Montanvert, these points were not often visited.

I also established two corresponding stations on the higher part of the Glacier du Géant, on the 6th August, which were marked K 1 and K 2. Most unfortunately, after my return from a journey to Monte Rosa, in the beginning of September, I was prevented, by incessant bad weather and snow, from reaching this remote station, and repeating the observations, which therefore led to no result.

During the absence just alluded to, the holes at *all* the stations were visited by Auguste Balmat, and deepened, so as to preserve them for my return. I must record my gratitude for his zeal in accomplishing this fatiguing and not very agreeable task during his recovery from a rather severe illness, from which he suffered during my absence, brought on partly, I fear, by the fatigues and exposure which he underwent in my service.

My habit was to enter my observations in a journal, and reduce them immediately to the mean daily velocity of each point since the last observation, allowing for variations in the hours at which they were noted, if such occurred. The first of the following tables contains the actual progress made by each part of the glacier, from the commencement of observations upon it, in inches and feet; the second contains the corresponding velocities, or motions in twenty-four hours, expressed in English inches.

TABLE I.

GLACIER MOTION.

RECKONED IN EACH CASE FROM THE COMMENCEMENT OF THE OBSERVATION.

Near Montanvert, (1.) D 2.			Near Montanvert, (2.) D 4.			Near Montanvert, (3.) D 6.			Near Montanvert, (4.) D 3.			L'Angle. A.		
1842.	English Inches	Feet	1842.	English In.	Ft.	1842.	English In.	Ft.	1842.	English In.	Ft.	1842.	English In.	Ft.
June 29.	0	0	July 28.	0	0	Sept. 17.	0	0	June 29.	0	0	June 26.	0	0
July 1.	35.0	2.9	Aug 1.	84.0	7.0	,, 20.	59.1	4.9	July 1.	54.2	4.5	,, 27.	15.2	1.2
,, 28.	518.5	43.2	,, 9.	278	23.2	,, 26.	179.7	14.9	,, 28.	774.2	64.5	,, 28.	31.5	2.6
Aug. 1.	585.5	48.8	D. 5.			,, 28.	227.1	18.9	Aug. 1.	861.2	71.7	,, 29.	49	4.1
,, 9.	715.5	59.6	Sept. 17.	0	0				Sept. 16.	1962.7	163.6	,, 30.	66.4	5.5
Sept. 16.	1399.5	116.6	,, 19.	37.2	3.1				,, 17.	1986.4	165.5	July 28.	460.9	38.4
,, 17.	1416.4	118.0	,, 20.	57.5	4.8				,, 20.	2047	170.5	Aug. 1.	515.4	42.9
,, 18.	1430.2	119.2	,, 26.	172.7	14.4				,, 26.	2169	180.7	,, 9.	633	52.7
,, 19.	1443.3	120.3	,, 28.	223.1	18.6				,, 28.	2215	184.6	Sept. 16.	1127	94.
,, 20.	1459.6	121.6										,, 26.	1238	103.2
,, 26.	1544.8	128.6												
,, 28.	1583.8	132.0												

Glac. de Léchaud. Pierre Platte. C.			Glacier de Léchaud. B 1.			Glacier du Geant. B. 2.			Glacier du Géant. B. 3.			Glacier de Léchaud. E 1.			Glacier de Léchaud. E 2.		
1842.	English In.	Ft.	1842.	English In.	Ft.	1842.	English In.	Ft.	1842.	English In. Ft.		1842.	English In.	Ft.	1842.	English In.	Ft.
June 27.	0	0	June 30.	0	0	June 30.	0	0	Aug. 4.	0	0	July 29.	0	0	July 29.	0	0
,, 30.	30.6	2.5	Aug. 2.	355	29.6	Aug. 2.	454	37.8	,, 6.	36	3	Aug. 2.	45	3.7	Aug. 2.	54	4.
Aug. 2.	359	29.9	,, 6.	395	32.9	,, 4.	482	40.2	Sept. 17.	564	47	,, 8.	131	10.9	,, 8.	152	12.
Sept. 17.	758	63.2	Sept. 17.	803	66.9	,, 6.	510.5	42.5				Sept. 25.	672	56			
						Sept. 17.	949	79.1									

TABLE II.

MEAN DAILY MOTION.

D 2.		D 3.		B 1.	
	Inches.		Inches.		Inches.
June 29 to 1 July,	17.5	June 29 — 1 July,	27.1	June 30 to 2 Aug.	10.8
July 1 — 28 ,,	17.3	July 1 — 28 July,	25.7	Aug. 2 — 6 ,,	10.0
,, 28 — 1 Aug.	16.2	,, 28 — 1 Aug.,	21.0	,, 6 — 17 Sept.	9.7
Aug. 1 — 9 ,,	16.6	Aug. 1 — 16 Sept.,	24.0		
,, 9 — 16 Sept.	18.0	Sept. 16 — 17 ,,	23.7	B 2.	
Sept. 16 — 17 ,,	16.9	,, 17 — 20 ,,	20.3		
,, 17 — 18 ,,	13.8	,, 20 — 26 ,,	20.4	June 30 — 2 Aug.	13.8
,, 18 — 19 ,,	13.1	,, 26 — 28 ,,	22.5	Aug. 2 — 4 . ,,	14.0
,, 19 — 20 ,,	16.3			,, 4 — 6 ,,	14.25
,, 20 — 26 ,,	14.2	A.		,, 6 — 17 Sept.	10.4
,, 26 — 28 ,,	19.5				
		June 26 to 27 June,	15.2	B 3.	
D 4.		,, 27 — 28 ,,	16.3		
		,, 28 — 29 ,,	17.5	Aug. 4 — 6 Aug.	18.0
July 28 — 1 Aug.	21.0	,, 29 — 30	17.4	,, 6 — 17 Sept.	12.6
Aug. 1 — 9 ,, ;	24.7	,, 30 — 28 July	14.		
		July 28 — 1 Aug.	13.6	E 1.	
D 5.		Aug. 1 — 9 ,,	15.4		
		,, 9 — 16 Sept.	13.0	July 29 — 2 Aug.	11.3
Sept. 17 — 19 Sept.	18.6	Sept. 16 — 26 ,,	11.15	Aug. 2 — 8 ,,	14.3
,, 19 — 20 ,,	20.3			,, 8 — 25 Sept.	11.3
,, 20 — 26 ,,	19.2				
,, 26 — 28 ,,	25.2	C.		E 2.	
D 6.		June 27 — 30 June,	10.2	July 29 — 2 Aug.	13.5
		,, 30 — 2 Aug.	9.9	Aug. 2 — 8 ,,	16.3
Sept. 17 to 20 Sept.	19.7	Aug. 2 — 17 Sept.	8.7		
,, 20 — 26 ,,	20.1				
,, 26 — 28 ,,	23.7				

From the preceding tables, especially the second, we may gain a great deal of practical information. The consistency which may be shown to subsist between their parts, inspires great confidence as to the results for this particular glacier, and shews that *a very few* experiments, as carefully made, would suffice to determine all that it is important to know respecting any other. A convenient way of representing the results to the eye, is to project the velocity of any point of the glacier by a vertical line, whilst the lapse of time is expressed by a horizontal line, whence the space moved over in the interval of any two times will be denoted by the area of the shaded spaces represented in the opposite figure. Had the velocities been measured daily, we

DIAGRAMS OF THE MEAN RATE OF DAILY MOTION OF DIFFERENT PARTS OF THE MER DE GLACE OF CHAMOUNI, AND THE CORRESPONDING MEAN TEMPERATURE.

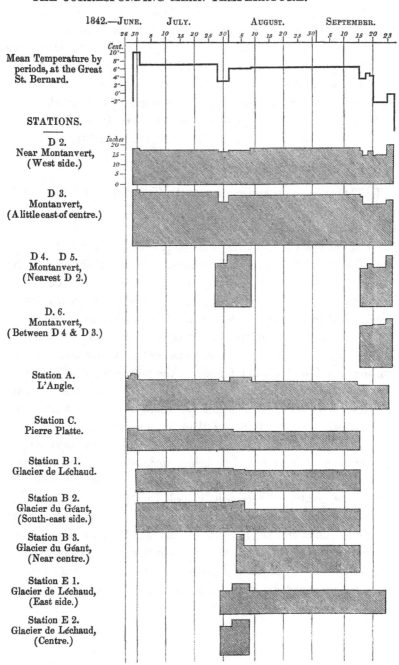

should have had a curve whose height would have been constantly
varying. As it is, we, of course, give to the velocity its mean value,
and suppose it constant during the intervals of observation. An increase in the rapidity of motion of any part of the glacier, will be
indicated by a rise in the serrated line; a decrease by a fall. A careful examination of the second table, and of the diagrams, will confirm
the following deductions, more full and explicit than those which my
first week's observations afforded, and which lay down, I believe for
the first time, the General Laws of the Motion of a Glacier deduced
from observation.

I. *The motion of the higher parts of the Mer de Glace is, on the whole,*
SLOWER *than that of its lower portion; but the motion of the middle
region is slower than either.*

I had not failed to point out, when I proposed the determination of
the velocity of different points of a glacier, as a test of the cause of its
motion, that this must depend materially upon the form of its section
at different parts. The velocity of a river is greatest where it narrows,
and is small in the large pools. Just so in the Mer de Glace. It is
truly a vast magazine of ice, with a comparatively narrow outlet, as
the map distinctly shows; the two glaciers of the Géant and Léchaud,
uniting just above the strait formed by the promontories of Trélaporte
and the Couvercle. Hence results, as we have seen, the great ice basin,
where we have reason to conclude (as before observed), that the glacier attains a greater thickness than at any other part, and thus,
though the breadth of the two confluent glaciers taken separately is
greater than after their union, being, undoubtedly, much shallower
there, their area of section is smaller, and therefore the velocity of the
ice will be greater. There will, indeed, be always a *condensation* of
the ice within the triangle BHG, owing to the resistance opposed to
its egress; and here, accordingly, the surface of the ice is most level.
It is not indeed strictly true, that the quantity of ice passing through
any section of the glacier in a given time, is exactly equal; because
there is fusion and evaporation, amounting to an actual loss of sub-

stance, between any two sections, and this becomes especially obvious near the lower extremity of the glacier. It is like the well known problem of the distribution of heat in a bar of iron come to a steady temperature, where the transfer of heat across any section of the bar is equal to the transfer across any other section nearer the source of heat, diminished by the amount radiated by the surface in the interval. There is, therefore, no ground for surprise at the fact, that the middle part of the glacier moves forward slower than the higher parts. Had the glacier *continued* to expand in breadth, as very many glaciers do, no check would have occurred, and the anomaly would have disappeared.

Since we have no accurate means of gauging the section of the glacier in any part, can we form any judgment of what would be the motion of the ice in a uniform canal, or draw any conclusion as to the *cause* of glacier motion? I think we can; but first let us place the observed law of velocity in a more concise shape.

The first station in order,—that at the mark D 2, a little below the Montanvert, and at 100 yards from the western edge of the glacier,—has been that, on the whole, most constantly observed throughout the whole season. Taking its motion as a standard of comparison, we may compare it with the motion of any other part of the glacier during the particular season when the latter was observed, and thus we shall obtain an *approximation* to the relative velocities of the different points of the glacier to D 2, taken as a standard. That this ratio depends in some degree on the season, will be shown farther on; still it affords the most ready way of obtaining a practical comparison. Thus, for example, it will be found, from Table I., that the point C of the glacier moved, between the 27th June and the 17th September, over 757.6 inches, whilst D 2 moved over 1579.8 inches, or more than twice the former. The exact ratio is 479 to 1000, which may be conveniently expressed by the decimal fraction .479 for station C; and so of the others. Thus, the velocity ratios of the different points are, as in the annexed Table,—

TABLE III.

Names of the Stations.	Relative Veloci-ties of the Ice.
D 2,	1.000
D 4, D 5,	1.375
D 6,	1.356
D 3,	1.398
A,	0.770
C,	0.479
B 1,	0.574
B 2,	0.678
B 3,	0.722
E 1,	0.674
E 2,	0.925

We may select from amongst these, the points most fitted for our purpose of comparison, those, for example, along the Mer de Glace and Glacier de Léchaud, not very distant from the edge, and therefore all retarded by the friction of the sides,—

D 2,	. . .	1.000
A,	. . .	0.770
C,	. . .	0.479
E 1,	. . .	0.674

Let us observe then,—the mere mechanical constraint to which the glacier is subjected, by the form of its valley, would necessarily, and irrespective of all theory, infer a quicker motion at D 2 and A, than at C, where the glacier is near its greatest width, or at E, near its origin. These facts are, then, so far in conformity with the mechanical necessity alluded to. But again, if the cause of glacier motion were the expansion of the superior portion of the ice forcing down the lower end, that velocity (supposing the section constant) would be proportional to the distance from the upper end or origin. Now, to take a most extreme supposition, let us imagine the Glacier of Léchaud (see the map) to take its origin at the very foot of the Grande Jorasse, which is 8000 feet beyond station E ; then, on the Dilatation theory, the motion at E would be due to the expansion of 8000 feet of ice, by the congelation of infiltrated water. This, we will suppose, produced the mean daily motion of 14.2 inches in the height of summer. Then, considering only the influence of length, irrespective of section, the station D 2, is 23,000 feet farther down, or nearly four times as far from the Grande Jorasse,—the velocity ought, therefore, to have been four times greater, or fifty-six inches per day. It was only 16.6 inches, or *one-seventh* part greater. And yet we have seen that the influence of *section* must have been to accelerate the motion in the lower part. I do not mean to say, that the reasoning just used is rigorous, but the results to which it leads, are so wholly opposed to the truth, as to be, it seems to me, quite conclusive against the theory of Dilatation. We have two powerful glaciers uniting, forming a great *ice-pool*, which issues by a channel not wider than the smaller of its feeders; making all allowance for evaporation, we conclude, without difficulty, that, in order that the ice-stream shall discharge itself, it must accumulate above the contraction, diminish in velocity there, and then rapidly increase in swiftness, as it issues through the opening, where it will certainly move faster than in either of the original tributaries, whose united breadth is far greater than the single channel of efflux. All this happens, as the simple mechanical theory of discharge, without indefinite accumulation, would indicate; but if we come to combine with this, a theory of glacier motion, which would require a velocity in the lower part of the glacier three times greater than we find it to be, we are entitled to reject the theory as inconsistent with facts, even although the mere statement, that the lower end of a glacier, on the whole, moves fastest, may appear to confirm it.

At present, we have to do with the conclusions of our own observations, and not with other or hypothetical cases. I may observe, however, that if a glacier widens uniformly, the *mere law of discharge without accumulation*, or change of volume, would give a diminishing velocity at the lower extremity. Such an occurrence would, evidently, be still more opposed to the theory of Dilatation.

II. The Glacier du Géant moves faster than the Glacier de Léchaud, in the proportion of about seven to six (compare B 1 with B 2 and B 3, in Table III.) The vast mass of the former glacier tends to overpower the other, in some measure, and it takes the lion's share of the exit through the strait between Trélaporte and the Couvercle, squeezing the

ice of Léchaud and Taléfre united, into little more than one-third of
the breadth of the whole. It is to this circumstance that I impute
the excessively crevassed state of the eastern side of all the Mer de
Glace, which renders it almost impossible to be traversed ; the ice is
tumultuously borne along, and, at the same time, squeezed laterally
by the greater velocity and mass of the western branch.

III. *The centre of the glacier moves faster* (as we have seen) *than the
sides.* When two glaciers unite, they act as a single one in this
respect, just as two united rivers would do. Now this variation is
most rapid near the sides, and a great part of the central portion of
the glacier moves with no great variation of velocity. Thus we find
that four stations taken in order, from the side to the centre of the
glacier (or a little beyond it), have (by Table III.) the following rates
of motion.

1.000   1.375   1.356   1.398.

Or if we compare observations made all at the same season of the
year (September), we shall find the increase of velocity in every
case,

1.000   1.332   1.356   1.367.

The first point was 100 yards from the edge of the glacier; the
next 130 yards farther. In this short space the velocity had increased
above a third part.

The explanation which we offer of this, as due to the friction of the
walls of the glacier, would lead us to expect such a law of motion.
The retardation of a river is chiefly confined to its sides ; the motion
in the centre is comparatively uniform.

Similar reasoning would lead us to expect that (supposing the
glacier to slide along its base) the portions of ice in contact with the
bed of the valley will be retarded, and the superficial parts ought to
advance more rapidly. The change of velocity in this case also, will
be greatest near the bottom.

IV. *The difference of motion of the centre and sides of the glacier varies
(1) with the season of the year, and (2) at different parts of the length of
the glacier.*

(1.) The following numbers show the velocity ratios of the centre and side of the glacier, near the Montanvert, at the marks D 3 and D 2, during different parts of the season 1842 :—

Relative Velocity, D 3 : D 2.

June 29—July 1, ...   1.548
July 1—July 28, ...   1.489
July 28—September 16,   . . .   1.349
September 16—September 28,   .   1.367

In general, therefore, *the variation of velocity diminished as the season advanced ;* we shall presently show that it was very nearly proportional to the *absolute* velocity of the glacier at the same time.

(2.) *The variation of velocity with the breadth of the glacier* is least considerable in the higher parts of the glacier or near its origin. Thus, if we compare the velocities of station C, and the mark B 1 on the Glacier de Léchaud near the Tacul, the former being near the side, the latter near the centre of the glacier, we find

Relative velocity, B 1 to C.

June 30—August 2, ...   1.09
August 2—September 17,   . . .   1.12

Again, higher up the same glacier, opposite E, we have the velocity ratios at the centre and side of the glacier—

E 2 : E 1.

July 29—August 2,   . . .   1.19
August 2—August 8,   . . .   1.14

This ratio is indeed a little greater than the preceding, which corresponds with the fact which we have already found, that the absolute velocity of the glacier is greater at E than at C. Hence, it is highly probable in every case that *the variation of velocity in the breadth of a glacier is proportional to the absolute velocity, at the time, of the ice under experiment.* This is farther confirmed by the velocities of the Glacier du Géant at the marks B 2 and B 3, of which the former is near the side and the latter near the centre—

Velocity ratio, B 3 : B 2.

August 4—August 6, ...   1.30
August 6—September 17,   . .   1.21

Now the absolute velocity of this glacier is greater than that of Léchaud, but less than that at the Montanvert.

V. *The motion of the glacier generally, varies with the season of the
year and the state of the thermometer.* Perhaps the most critical consideration of any for the various theories of glacier motion is the influence of external temperature upon the velocity. In this respect my
observations, though confined only to the summer and autumn, are
capable of giving pretty definite information. Indeed, one circumstance which on other accounts I had much reason to regret, I mean
the rigorous weather of the month of September, which hindered many
of my undertakings, gave me an opportunity of observing the effect of
the first frosts, and thus establishing some important facts as to the
influence of cold and wet upon the glacier. This I apprehend to be
clearly made out from my experiments, *that thawing weather and a wet
state of the ice conduces to its advancement, and that cold, whether sudden
or prolonged, checks its progress.* I may appeal generally to the curves
of page 141 as showing the variations of velocity with the season. It
is to be attended to in looking at these figures that they only represent
the *mean* motion during certain intervals which are not exactly the
same at the different points, and that, therefore, the rises and falls do
not appear always to coincide when they might actually do so, being
lost in the average of a distinct period. A careful examination of them
will, however, show that the variations of velocity have been remarkably general and simultaneous, and that we are entitled to look for a
common cause. This cause seems clearly to be found in the *temperature*
of the air combined with the degree of moisture which on a glacier
usually accompanies· a rise of temperature. The rapid movement in
the end of June which is perceptible at D 2, D 3, A and C, is due to
the very hot weather which then occurred, and the very marked reduction at the end of July, to a cold week which occurred at that period.
The striking variations in September, especially at the lower stations,
which were frequently observed, prove the connection of temperature
with velocity to demonstration.

During the continuance of the cold weather, accompanied by snow, from the 18th to the 27th September, it will be observed that the glacier motion was visibly retarded at all the lower stations which were then observed. During this period the thermometer fell at the Montanvert to 20° Fahr.; but when mild weather set in again, the glacier became clear of snow, (which took place in the lower part on the 27th,) and being thoroughly saturated with moisture, it resumed a march as rapid as that of the height of summer.

This fact is surely most important as showing that we cannot possibly ascribe the motion of the glacier to the effect of congelation; for, saturated as the ice was by the effects of the damp and changeable weather of the month of September,—when a week of frost set in every thing must have been exactly in the condition to acquire a rapid increase of velocity, exactly in proportion as the cold penetrated the mass of the glacier, supposing that it did penetrate to a considerable depth, which I will afterwards endeavour to prove clearly was not the case then, and à *fortiori* never can be the case in the height of summer, when the glacier motion is most rapid.

But I would farther request attention to a still more direct proof of the dependence of the velocity of the glacier upon the external temperature. I have taken from the register, kept at the Great St. Bernard, the mean daily temperature during the summer months of 1842. I have divided them into periods corresponding to those intervals at which the progress of the glacier at the point D 2 was ascertained; and I have taken the mean temperature of those periods. I find, that in almost every instance a change of increase or diminution of mean temperature is accompanied with an increase or diminution of the glacier's motion. And when we consider the difference of position of the stations, the coincidence seems quite as perfect as we can reasonably expect. The convent of St. Bernard is 21 English miles distant from the Montanvert, in a right line, and 1900 feet higher; but as many parts of the ice of the Mer de Glace have a still greater elevation, it may be supposed to represent pretty truly the conditions of climate to which the entire glacier was subjected.

A comparison of the first curve, or serrated line, in page 141, which represents the mean temperature of certain periods, with the curve

immediately below, which shews the glacier motion for the same intervals, will fully justify the assertions just made.

I do not say that the velocity is always the same at the same temperature. In autumn the velocity was as great with a temperature of
0° centigrade, as in summer with a temperature of 10° c. This was
the case, however, only at the *side* of the glacier. Near its centre,
as at D 3, it will be seen by the diagram that the motion is still more
nearly conformable to the change of temperature. All that I infer
from the comparison is, that a rise of temperature was generally
accompanied with an increased rate of motion of the glacier, and the
converse. If the state of *imbibition*, or wetness of the glacier, be the
main cause of the increased velocity, as I believe it is, we can readily
understand how mild rain, or thawing snow, produces the same effect
as intense sunshine.

Whilst it appears probable, or, indeed, certain, from these facts, that
the motion of the ice depends upon the temperature of the air in contact with it, and that it is greater in warm, and least in cold weather,
it does not at all follow, as has in general been too hastily assumed,
that the glacier *stands still* in winter. On the contrary, I have long
believed that it continually advances, although in a less degree. The
circumstance just mentioned, that though hot and cold weather produce
relatively the effect of accelerating and retarding the movement of the
ice, the velocity is in no direct proportion to the temperature, confirms
this. The opinion of many of the most intelligent peasants, whom I
have consulted on the point, are also in favour of this view. They
generally believe, that the glacier pushes itself forward under the snow
in winter; and when I have applied to them for the evidence, they
assure me, that they have seen the ice, at the lower extremity of a
glacier, pressing the snow onwards. I do not, indeed, lay great stress
upon this testimony, considering the facility with which such persons
often adopt wrong opinions; but its generality amongst the peasantry,
and its coming in direct corroboration of the same conclusion to which
I have been led from other sources, entitle it to some weight. These
grounds will be stated more particularly when we come to consider the
question in another place in a more general form; but I may add, that

the best conjectures which I can at present form, in the absence of direct experiments, as to the *annual* motion of the Mer de Glace, would give a result so very much exceeding that which can reasonably be attributed to the progress, during the *summer* months alone, that it is highly probable, that the motion is continuous, though unequal, throughout the year, and is far from being nothing at any season.

I will give one example of my meaning. The motion of the Glacier du Géant, at the mark B 2, has been shown above to be .678, or about *two-thirds* of the motion of the ice at mark D 2, near the side of the glacier below the Montanvert. Now, let us admit, for a moment, the story of De Saussure's ladder, which would assign, if true, a velocity of 375 feet per annum to this part of the glacier; consequently, the comparative advance of the lower ice would be a half more, or 563 feet. Now, of these 563 feet, only 132 (see Table I.) were performed during the three hottest months of the year, which barely amounts to the proportional rate of motion of a quarter of a year. Now, this estimate may be thought a very rude one, from the nature of the authority whence it is derived. And so no doubt it is. Yet it will be, on the whole, confirmed from other sources; and as we have seen, Mr. Bakewell has estimated, from the information of his guides, the movement of the ice of the Mer de Glace at 180 yards, or 540 feet per annum, a quantity, it will be observed, singularly agreeing with the previous one. But without supposing these facts to be more than presumptive evidence, they, at least, give strong reason for believing, that the velocity of motion is not excessively small, even in winter.[*]

From information which I have received since my return home, I find, that my guide, Auguste Balmat, has, at my request, watched the progress of the great block of stone below the Montanvert, (marked D 7 on the map,) and has found that it moved

From October 20 to December 12, 1842, 53 days, . 70 feet.

Its daily velocity was therefore, . . . 15.8 inches,

or very nearly its average summer velocity.

---

[*] The preceding pages were written before I possessed the direct proofs of the winter motion of the glacier contained in the succeeding paragraphs.

From December 12, 1842, to February 17, 1843, it moved 76 feet.

Or daily, ... 13.6 inches.

From February 17 to April 4, 1843, it moved . 66 feet.

Or daily, ... 17.2 inches.

I have perfect confidence in the fidelity of these observations ; as, however, in the first and last case, Balmat observed that the stone had rolled onwards, so as to fall upon a new side, and has attempted to estimate its rolling progress, there may be a slight error on this account. The measurements are in English feet, made with a line which I left at Chamouni, on purpose.

I presume that the immobility of glaciers in winter, so long received as an undoubted fact, as a basis of theory, will now be admitted to have been as gratuitously assumed, as the greater velocity of the sides of a glacier compared to its centre.

The continuity of glacier motion, even in winter, might have been inferred from the well known instances on record of the fall of great avalanches of ice during that season : Such, for instance, was the fall of the glacier of Randa, in the valley of St. Nicolas, on the 27th December 1819 ; * and such is the direct testimony of De Saussure in these words : " Les glaciers mettent aussi en mouvement et chassent devant eux les terres et pierres accumulées devant leur glaces, à leur extrêmité inférieure. Je vis ce phénomène en 1764, de la manière la plus évidente, et j'eus en même tems *la preuve que ce mouvement avait lieu même dans une saison qui est encore hiver pour ces montagnes.*† Comme le glacier et tous ces alentours étaient en entier couverts de neige ; lorsqu'il poussait en avant les terres accumulées devant ses glacons, ces terres, en s'éboulant se renversaient par dessus la neige et mettoient en evidence les plus petits mouvements du glacier qui se continuerent sous mes yeux pendant tout le temps que je passai à l'observer."‡

---

* Agassiz, page 158.

† This appears most probably to have been in the month of March 1764, from a parallel passage in § 520 of *De Saussure.*

‡ De Saussure, *Voyages,* § 538.

On the change of level of the Mer de Glace.

It has already been observed that one of my first cares on reaching the glacier in June, was to ascertain the level of the ice at station A. These levels were taken from time to time, and afford unequivocal proof of the depression of the surface of the glacier during summer, to an extent which has probably not been suspected.

1842.	From	To	The level had lowered Feet. Inches.		Daily depression in the interval. Inches.
June 26	to	June 30,	. 1	9.0	—— 4.1
——	——	July 28,	. 10	11.0	—— 3.6
——	——	Aug. 9,	. 14	10.0	—— 3.7
——	——	Sept. 16,	. 24	6.5	—— 2.5

Now this depression is not necessarily the result of superficial waste alone. I doubt whether it is even mainly due to that cause,—and not rather to a subsidence of the entire mass of the ice, which visibly collapses as the warm season advances. Such a collapse may be due to several circumstances : 1. The undermining of the glacier by the excavating action of the water streams which flow beneath it in summer : 2. The fusion of the ice in contact with the soil, due to the earth's heat : 3. The lower extremity of the glacier moving faster than its higher portions, and thus extenuating the mass, a cause which acts with energy at those seasons when the difference of motions of the two parts is a maximum. The superficial waste is not so easily measured as at first sight it might appear to be. M. Escher de la Linth measured it in 1841, and the glacier of Aletsch, by the exposure of stakes inserted to a certain depth in the ice,—as the ice melted, the stakes were exposed. M. Martins measured it by the geometrical depression of the surface. The last method we have seen measures several effects instead of one ; the former may lead to the most inaccurate results. When the stakes have been exposed to a certain depth, the apparent result is actually inverted— the hole is *deepened*. The irregularities resulting from this mode of observation will appear from the following facts :—*

---

* The holes were examined, and the sticks notched by Balmat on the 16th July, when they were also deepened, and the variations were afterwards measured by myself.

Inches.

At the mark D 2, the stick rose out of the hole in the ice from
    July 1 to July 16, ...   17
    From July 16 to July 28, only   5
At the mark D 3, during the first period the stick rose .   22½
    During the second period it actually *sunk*, showing, that
    from some cause the hole had deepened faster than the
    surface wasted.
At station A, from June 30 to July 16, the stick rose .   29
    From July 16 to July 28, it *sunk*   2
At the mark B 1, from June 30 to July 16, the stick rose   28
    From July 16 to July 29, it *sunk*   5

The cause of this anomalous action it is not difficult to explain.
It is, I apprehend, the same as we have pointed out on page 27, as
occasioning the formation and perpetuation of holes in the ice, owing
to the less density of freezing water than that some degrees warmer.
The holes by which my stations were marked always contained more
or less water. Whilst the stick fitted them accurately, it nearly or
completely obstructed the fluid currents; but, in proportion as the
holes widened, the water circulated more freely, and the cavities spontaneously deepened, which is one cause of their preservation.

It is evident that the apparent loss of surface of the ice in this experiment will be generally too small and never too great. Thus it
appears that at stations A and B, the superficial loss of ice was, *at
least*, 29 and 28 inches respectively, during the first sixteen days of
July, or about 1¾ inches per day. The actual *fall* of the surface at
this time was, as we have seen, twice as great; but this I attribute
mainly to the general subsidence. A method, which seems the only
sure one of determining the superficial loss (an important datum,)
would be to drive *horizontal* holes in the vertical walls of conspicuous fissures, and to measure their distance from the surface of the
glacier. At those stations which did not conveniently admit of running a level to the side, I employed a different and very simple
method of measuring the *absolute* depression of the surface. It had the

advantage of being applied at the same time that the motion of the glacier was measured, and with little additional trouble. Thus, at station D, whence the various marks across the glacier were observed in succession, the progress was noted by causing an assistant to descend to the ice with a deal rod, a chisel, hammer, and pegs of wood. If the motion since the last observation was small, he was directed to lay the deal rod parallel to the length of the glacier, and to push it up or down as directed by signals until the extremity was in the exact azimuth of the opposite mark beyond the ice. A pencil mark and number were then made upon the deal rod, in order to fix the distance from the previous station. If the distance moved over was greater, the extremity of the rod was moved parallel to the glacier as before, and when duly placed, a hole was made in the ice with a chisel, and a peg inserted, until I had time myself to descend and measure with a line, and in a carefully determined direction, the whole motion from the last fixed point. Now, in addition to this, in order to ascertain the change of *level* of the ice, I had only to observe from my elevated position, the angular depression of the marks in succession on any particular day. These were, for example, on the 20th September,

D 2.$=22°$ 0′ 0″ D 6.$=11°$ 31′ 45″ D 3.$=10°$ 6′ 0″

A vertical rod being placed at each of these points on any future day, the telescope being depressed to the same degree, pointed, of course, to a height upon the rod equivalent to the former level of the ice, which was determined by my assistant sliding up or down a slip of paper in obedience to my signals. It is to be recollected, that the ice was here much crevassed ; and though its onward movement was wonderfully regular, it was liable to local subsidences. Occasionally, I have found as great a depression as a foot per day during wet mild weather in the later part of the season. During frost, when the glacier had more consistence, the subsidence was evidently diminished.

This much is certain respecting the level of the ice, that the glacier undergoes a surprising waste during the summer, and that there is not the slightest reason for believing that any process, whether of congelation or other, assists in its renewal during that season. The comparison of a glacier to a mass of leavened bread expanding upwards, and thus supplying the superficial waste, appears to involve an assertion wholly unsupported by evidence, and contradicted by my experiments. And as I readily admit, that such a swelling or vertical dilatation of the mass would be a necessary result of the theory which ascribes the motion of the glacier to the expansion of water frozen in its fissures, I must consider the fact, that no such dilatation is apparent at the season when the motion is most rapid, to be in itself conclusive against the dilatation theory of glacier progression.

CHAPTER VIII.

ON THE STRUCTURE OF THE ICE OF GLACIERS, AND OF THE MER DE GLACE IN PARTICULAR.

GENERAL FACTS OF STRUCTURE—DISCOVERY OF WAVE-LIKE BANDS ON THE SURFACE OF THE GLACIER—FIGURES OF THE STRUCTURE, AND SECTIONS OF THE MER DE GLACE—DETAILS—GLACIER DU TALEFRE—CREVASSES OF GLACIERS—THEIR MONTHLY CHANGES—MINUTE FISSURES OF THE ICE—ITS PERMEABILITY TO WATER—VEINED STRUCTURE EXPLAINED.

SOME account has already been given, in page 28 of this work, of the structure of ice, which was noticed by M. Guyot of Neufchâtel, in 1838, in the Glacier of the Gries, and which I rediscovered in 1841, on the Glaciers of the Aar, the Rhone, and others, and described as being one, probably general, and certainly important in the consideration of the mechanism and functions of glaciers.

It has already been said, that I am disposed to regard the problem of the cause of glacier motion as a purely mechanical one, and that it should be treated, like other problems of motion, by a consideration of the manner and degree in which that motion varies with seasons and circumstances, rather than by endeavouring to deduce, *à priori*, the motion from the circumstances, and from a hypothetical structure of ice, or any peculiar functions of its molecular constitution. I am far from denying, however, that a knowledge of that internal constitution will be of the utmost consequence in modifying or confirming our mechanical theories. From an early period, I felt convinced, that the veined structure of ice, described by me in December 1841, was an important, though obscure, index of the mode of glacier progression ; and when I proceeded, in 1842, to obtain definite information to bear upon my speculations, I proposed to myself, as a chief problem, *to endeavour to combine the direct evidence which the observation of the*

velocity of the ice in different parts of its mass might furnish as to the
cause of motion, with the statical or permanent evidence which the forms
of the veins or ribboned structure pervading its mass, undoubtedly bear,
to some change operated or operating in its interior.

I am inclined to think, that I have arrived at a result which com-
bines these independent evidences; and I feel the more confidence in
it, because I am conscious of having commenced my researches with no
bias in favour of one theory of glacier motion rather than another, or
one cause of veined structure rather than another; indeed, I might
rather say, that I commenced them in 1842, with an equal distrust of
all theories proposed to account for the former, and in ignorance of any
theory worthy of the name which should account for the latter. Farther than this, I spent some weeks amongst the glaciers in June and
July 1842, without even approximating to a theory either of motion
or of structure, until at length I began to fear, that days and months
of incessant observation, or patient thought, would leave me no wiser
about this great problem, than when I commenced. But, as has been observed to be the process of discovery in all complicated questions,—when
the confusion seems greatest, and the mind is so imbued with the subject, that the very multitude of details confounds, and the antagonism
of conflicting speculations sets order at defiance, then from some un-
suspected corner springs up a light, unsought, and seemingly casual,
but which struggles into more perfect evidence by being dwelt upon,
and at last, throws a complete illumination over the scattered elements,
which appeared undecypherable and unmeaning, only because they were
dimly seen.

Such information respecting the theory of the glacier structure, I
acquired first on the 24th of July, and again a fortnight after, on the
7th August, 1842. One half hour on each of these days seemed to
teach me all that I learnt during my stay upon the ice. All before
was preparatory to knowing, all after was simply confirmatory, or
proving what I knew.

But before I can make the reader aware of the nature of the observations and reflections which then came home to me with so much
force, I must endeavour to describe what I had previously observed

with respect to the structure of the ice of the Mer de Glace in particular.

The external form of the ice, the crevasses by which it is fissured, and often divided into transverse slices, or pyramidical blocks, and the finer net work of fissures, which we shall also find to pervade its interior, all these may be described as, in some sense, the "structure" of the ice. But what we here mean by "structure," is something anterior to, and more fundamental than all these,—it is the intimate arrangement of the very particles of the frozen water, and which constitutes as properly its structure, as the pattern of a piece of curious damask does, or as the veins of a woody fibre do, in a piece of mahogany. The proximate cause of the *ribboned structure* of the ice, it has been seen, that I ascribed to the alternation of bands, or parallel veins of ice, of different textures. These bands or veins were conspicuously distinguished (on the glacier of the Aar and others) by two characters, 1. Difference of hardness ; 2. Difference of colour. The former distinction causes the harder (which are also the *bluer*) veins to stand up in ridges, as the ice melts by the action of the sun or rain, and allows the comminuted sand from the moraines to lodge in the intervening linear hollows, which led, as we have seen, some persons to suppose that the heat of the sun, acting upon the sand, *caused* the hollows in which it lay. This peculiarity is admirably seen on many parts of the Mer de Glace ; and no where better than upon the common route from the Montanvert to the Jardin, where it passes by the foot of the Aiguille des Charmoz, between the *Angle* and Trélaporte. Here the whole surface seems striated with fine lines ; and where groups of the harder bands occur, there are projecting ridges, with grooves between, continuous for very many fathoms along the ice, resembling the cart ruts of a much travelled road, when covered with stiff mud, which was the accurate comparison of an English traveller, whose attention was directed to them last summer for the first time. This appearance is most conspicuous after rain. The other characteristic, that of *colour*, requires an attentive examination to perceive its immediate cause ; but in any glacier, where the structure is well developed, there is no difficulty in deciding upon it. The phenomenon is not one of those which, like the colour of water, or of air, can only be seen in vast masses. I have often detached *hand specimens* of the ice, which, if they could be preserved in cabinets, would convey the most perfect idea of the structure ; there, to be sure, the depth of colour has nearly vanished, but the bands and the cause of colour remain. If we attempt to look through such a piece of ice *across* the direction of the ribboned structure, it looks opaque ; but if we look *parallel* to the veins, we perceive that semi-opaque bands alternate with others of glassy purity ; the former appear greenish white on a great scale,—the latter blue. If we examine them closely, either with the eye, or with a magnifying glass, we find, that the blue and glassy part is pure smooth ice, whilst the intermediate portion is, not granular or snowy, as I myself at one time supposed, but simply frothy or full of air-bubbles of various forms, disseminated through the pure ice, and always arranged in parallel planes, of more or less abundance, producing greater or less opacity. These cavities do not *appear* to communicate, though we shall see reason to believe that they generally do so. It is a general fact, that, as ice loses the perfection of its crystalline structure, it passes from blue, *through green*, to white, which is always its colour when granulated. It is for this reason, that the transition from ice to snow, in the higher glacier regions, is *usually* through shades of green ; but when even common snow has acquired a certain degree of imbibition by moisture, and is no longer dry and powdery, but allows a pretty free passage to the light, it becomes distinctly *blue*, by transmitted light, and of as great or greater intensity, than I have ever observed in pure ice or water at the same thickness. I attribute it to the free admission of light, in consequence of moisture filling the cavities between the snowy granules. I have elsewhere observed, that I consider that no farther explanation of the blue colour is required, or can be given, but that it is the colour proper to pure water, both in its solid and its liquid form.

It has been said, in a former chapter, as well as in the paper which will be found in the Appendix, No. I., that the *direction* of the bands depends materially upon the configuration of the glacier, and the nature of its boundaries. In a long, canal-shaped glacier, like that of the lower Aar, it was nearly parallel to its length, and nearly vertical, but inclining upwards and outwards where the ice was supported by the lateral rocks. On the glacier of the Rhone, on the other hand, which has a not very elongated form, and which enlarges itself suddenly, these bands described oval lines upon the surface of the ice, as we have already seen, and dipping inwards at angles more nearly perpendicular, as the centre of the glacier was approached, might be compared to sections of inverted cones, having a common apex pointed downwards, but whose angles continually diminished towards the centre. Not, indeed, that the ovals were complete all round the glacier, but they were complete, or nearly so, for two-thirds or three-fourths of the circumference, as shown in page 29. Guided by what I saw at the glacier of the Rhone, I ascribed the *apparent frontal stratification* of the lower extremity of the glacier of the Aar to the same cause, namely, the twisting round of the planes of structure which cropped out (to use a geological phrase) on the slope of the lower end of the glacier, with a continually diminishing dip, as the level of the ground was approached.

Evidently, then, the one of these structures was but the *limiting* case of the other; the canal-shaped glacier is but the oval glacier drawn out longitudinally, its lower or unsupported part invariably assuming the depressed conoidal structure.

In the course of my numerous crossings and recrossings of the Mer de Glace, I observed a general confirmation of the disposition of the ice to a parallel structure, sometimes vertical, sometimes leaning against the walls of the glacier, and often, where one side of the glacier was heaved up in its progress against some opposing promontory, the whole structure (preserving the general trough-shaped section) appeared to lean over in one direction, as shown in the figure No. V., page 166. At the same time, I found so many anomalies, as to make me cautious of hazarding the assertion, that the trough-shaped structure was rigorous and general, and I determined, by patient observation, and laying down on a sketch the bearing of the veins or bands, and their dip at a great number of points, to obtain an empirical representation of the

L

structure in question, over as large a portion of the surface as possible. The labour would have been great, without some better clue to guide so extensive an enquiry ; fortunately it had hardly commenced before I obtained one.

On the evening of the 24th of July, the day following my descent from the Col du Géant, I walked up the hill of Charmoz to a height of 600 or 700 feet above the Montanvert, or about 1000 feet above the level of the glacier. The tints of sunset were cast in a glorious manner over the distant mountains, whilst the glacier was thrown into comparative shadow. This condition of half illumination is far more proper for distinguishing feeble shades of colour on a very white surface like that of a glacier, than the broad day. Accordingly, whilst revolving in my mind, during this evening's stroll, the singular problems of the ice-world, my eye was caught by a very peculiar appearance of the surface of the ice, which I was certain that I now saw for the first time. It consisted of a series of nearly hyperbolic brownish bands on the glacier, the curves pointing downwards, and the two branches mingling indiscriminately with the moraines, presenting an appearance of a succession of waves some hundred feet apart, and having, opposite to the Montanvert, the peculiar form which I have attempted to show upon the map, where they are represented in the exact figure and number in which they occur. They were evidently distinguished from the general mass of the glacier by discolouration of some kind, and indeed they had the appearance of being supernumerary moraines of a curvilinear form, detached from the principal moraines, and uniting in the centre of the glacier. Although this was my first idea, I was satisfied, from the general knowledge which I then had of the direction of the " veined structure" of the ice, that these discoloured bands probably followed that direction ; and accordingly next day I carefully examined the surface of the ice, with the view of determining, if possible, their connection and cause, being well satisfied that this new appearance was one of great importance, although, from the two circumstances of being best seen at a distance or considerable height, and in a feeble or slanting light, it had very naturally been hitherto overlooked, both by myself and others.

I had often observed that some parts of the ice were dirty, and some parts clean, but it was not until I examined its surface minutely on 25th July, that I discovered that the " dirt bands," as I called them, had a definite position upon the glacier and a regular recurrence. I had no difficulty now, whilst examining the ice when on its surface, in deciding whether I was standing upon one of the " dirt bands," or on the clean ice, although, from the inequalities of the surface and local effects of light, it would have been almost impossible to have traced out, step by step, the forms of these discolourations. They are like what are called " blind paths" over moors, visible at a distance, but lost when we stand upon them.

The *cause* of the discolouration was the next point; and my examination satisfied me, that it was not, properly speaking, a diversion of the moraine, but that the particles of earth and sand, or disintegrated rock, which the winds and avalanches and water-runs spread over the entire breadth of the ice, *found a lodgement* in those portions of the glacier where the ice was most porous, and that consequently the " dirt bands" were merely *indices of a peculiarly porous veined structure traversing the mass of the glacier in these directions.* A most patient examination of the structure of the ice opposite to the Montanvert, satisfied me completely of the parallelism of the " veined structure" to the "dirt bands;" the former was the cause of the latter; and some more general cause, yet to be explained, caused the alternation of the porous veins at certain intervals along the glacier. This, then, tended to clear up a multitude of doubts respecting the real type of glacier structure in long or canal-shaped glaciers. That it was not merely trough-shaped was clear, but the direction and dip of the veins near the centre of the glacier was generally too confused to give a ready solution of its real structure. I now found that the veins appeared *generally* parallel to the moraines and sides of the glacier, only because the curves representing their real forms had branches which merged into parallelism, and that there really was a tendency in the direction of the veins on the two sides of the glacier to converge to a point in the centre. But the most difficult point to decide was, What is the form assumed

by the veins, where they meet in the centre, at the vertex of the curve?
After much attention, I found that the normal structure here (though
often obscured or annihilated) turned round and formed a loop exactly as in the oval-shaped glacier already described, the *direction* of
the structure being, for a short space, directly across the strata, and
dipping inwards at a considerable angle.   The ground-plan, transverse
section, and longitudinal section, (at the centre of the glacier,) of such
a structure would be the following :—

No. 1.   No. 2.   No. 3.

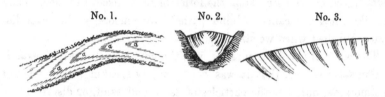

Opposite to the Montanvert, the dip inwards at *a a a* (that is,
towards the origin of the glacier) appeared to be 45°.   This is only

Twisted veins.

through a narrow space, and is often extremely
confused, but, whenever the structure appears
clearly, this is its position.   The ice is often contorted in the most fantastic manner, like limestone strata in the Alps, or the veins of knotty
wood.

Of course, after the discovery of these "dirt bands" below the Montanvert, it became an object to trace them throughout the glacier, mark
their variations, and compare them with the structure of the ice, so as
to ascertain that they rigorously corresponded ; lastly, to fix their
numbers, distances, and form.   Although at most times of the day I
could distinguish their position after once ascertaining their existence,
yet to see them well, or to count them throughout any extent of the
glacier, required an elevated position, and a peculiar effect of tempered
sunshine or moonlight.   In broad daylight, without clouds, only the
more conspicuous ones could be seen ; but it is not to be supposed from
this that there was anything illusory in their existence or position.   On
the contrary, both were so perfectly definite, that I have repeatedly

counted the bands visible from station L, (on the Charmoz, above the Montanvert,) all the way between the ice precipice at the Chapeau to the promontory of Trélaporte, which are exactly 18 in number as laid down in the map. The lower 10 bands (including 9 intervals) are contained between the right lines joining stations L and I, and L and F, and the distances of these lines are laid down on the map from actual survey. The mean intervals will be found by taking the distance along the axis of the glacier between the lines just mentioned, and dividing it by 9. That distance is 6400 feet, and consequently the average interval is 711 feet. But the intervals are not all alike ; indeed, they differed sensibly to the eye. The difference, however, for this part of the glacier, is probably not a tenth part of the mean for any one interval. The distance between the vertices of the two dirt bands immediately opposite to station D was found trigonometrically to be 667 feet.

The ground plan of the ribboned or veined structure generally, and of those porous veins in particular constituting the " dirt bands," may be pretty correctly judged of from the map ; but for a complete understanding of the structure, and its modifications, the following remarks are essential, which will be made plain by a reference to the following sections, taken from eye sketches made on the spot, and to the ground plan, which is here repeated, and which shows the lines of section. Thus,

a b	corresponds to section	No.	I.
H d e	——	——	No. II.
f G	——	——	No. III.
F A	——	——	No. IV.
D D	——	——	No. V.

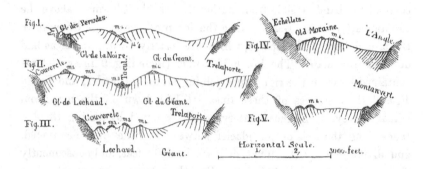

1. Opposite to the Montanvert, and beyond Les Echellets, the curved loops extend *across the entire glacier*. They are single, and therefore *cut* the medial moraine, though at a very slight angle. The structure of the ice to the east of the medial moraine is nearly parallel to the length of the glacier. It is also nearly vertical; but the whole trough-shaped structure, accommodating itself to the irregular form of the glacier, leans over towards the Angle as if tilted up by the promontory of Les Echellets, which is really the case, as shown in the sections Nos. IV. and V.

2. The position of the vertices of the curves of structure inclines towards the left bank of the glacier, as we approach the promontory of Trélaporte; and about that portion of the glacier we begin to distinguish a separation in the structure of the two confluent glaciers, which do not appear to be there fully consolidated. The Glacier du Géant has its own system of curves, and the Glacier de Léchaud its system, as shown in the map. From about the position where the *dislocation* of the moraines is marked on the maps near the Moulins, up to the promontory of the Tacul, the great medial moraine of the two glaciers, marks, as it were, a common vertical wall, formed by the mutual pressure of the ice streams, and throughout all that space the vertical structure of the ice follows precisely the direction of the moraine. On either side it begins to incline into the trough of its own glacier, as shown in the sections Nos. III. and II. After the glaciers have thoroughly amalgamated, the structure of the more powerful glacier (du Géant) predominates, and absorbs the other.

3. We have seen that the Frontal Dip, that is, the dip of the veined structure *inwards* throughout the very narrow space in which its direction is *transverse* to the glacier, or near the centre of the ice stream, is about 45°. This dip certainly increases as we ascend, exactly as I have shown, (see Appendix, No. I.,) and shall show, (Chap. X.,) that it does in those glaciers where, the ice being less confined, the frontal dip is a well marked angular phenomenon, as in the Glacier of the Rhone, of La Brenva, and at the lower extremities of many other glaciers. Now, just above Trélaporte, on the Glacier du Géant, though the frontal dip is undistinguishable, yet the curvature of the structural planes is perfectly clear, and likewise the occurrence of the dirt bands, which are here more rounded, and not so excessively drawn out as at the Montanvert. But, if we pursue the Glacier du Géant higher up, as opposite K, the *transverse structure in the centre of the glacier is perfectly distinct, and the frontal dip is vertical.* This is an important fact, and conformable to what I have observed on the Glacier of the Rhone.

4. The Glacier du Géant has a single or simple structure between Trélaporte and the Tacul, and for some way higher up. The system of curves, formed by the structural planes intersecting the surface of the ice, have their vertices near the centre of the glacier, and become parallel to its length near the banks, *cutting* the Moraine of La Noire, and stretching quite from side to side. But as we advance higher up, and approach the Aiguille de la Noire, which separates the great mass of the glacier from the small glacier descending from the range of Les Périades, we perceive a tendency to a double structure, as at the union of the Glaciers de Léchaud and Géant. (See Section No. I.) I am unable to state the exact number of dirt bands between the foot of the ice cascade opposite La Noire and the corner of Trélaporte. Under a favourable light they may perfectly well be counted, and I recollect doing so once, but the number was not noted, as I intended to make the observation more scrupulously another time, but was prevented by the fall of snow in September. Indeed, it is for but a very few weeks of the year that this part of the glacier is tolerably free of snow. My belief is, however, that these bands are not only more uniformly curved,

(as has already been said,) but are *compressed*, or more numerous in the same space. This appeared to me to depend partly upon the smallness of the declivity of the glacier.

5. If we follow the Glacier de Léchaud from the *Moulins*, we have, in the first instance, as has been said, the vertical stratification accompanying the medial moraine up to the Tacul. There are two medial moraines on the Glacier de Léchaud itself; one coming from the Jardin and the other from the Aiguille de Léchaud. The ice between the latter and the Couvercle is the ice of the Glacier du Taléfre ; that between the same moraine and the Tacul belongs to the Glacier de Léchaud, descending from the Grande Jorasse. Now, this moraine (*de l' Aiguille de Léchaud*) divides the separate structures belonging to these two ice streams, whilst the structure of the ice derived from the Taléfre cuts the moraine of the Jardin at an angle, and forms only a single system of curves. Both of these systems die out about the same time, after a complete union has been effected with the Glacier du Géant. I have not particularly noticed the dirt bands on the glacier de Léchaud, but I have carefully examined its structural planes, and traced them quite up to their disappearance, which takes place a little below station E, where the glacier is without any trace of structure. The structure commences a little below the junction of the steep glacier, descending from the foot of the Capucin du Tacul, and it is manifestly augmented, and becomes general after the confluence of the Glacier du Taléfre. I have often observed (and believe it to be a general rule,) that *where a glacier is contracted and jostled by its union with others*, if not violently crevassed, *there the structure comes out best*. The structure is rather elongated here, and not so transverse as in the glacier du Géant.

6. The structure of the ice of the Taléfre, forming the north-eastern portion of the Glacier de Léchaud, is remarkably well brought out, and instructive. At the Pierre Platte C, it is beautifully shown ; and here I first distinctly remarked, that the structure is not always parallel to a medial moraine, as I had at one time supposed. It evidently cuts the moraine of the Jardin, as already mentioned. This part of the glacier is steep, and its surface convex. It has very much the charac-

ter of a glacier poured out into a valley, as it really is, being derived from the stupendous ice cascade which falls from the basin of the Taléfre. The forms of the veined structure are more rounded than in most other parts of the Mer de Glace: I mean, that the superficial curves do not come to a sharp point, but have more of a circular sweep, and a well-defined transverse course, and a frontal dip inwards of 63°. But one of the most interesting points connected with this ice stream is the sudden change of structure which it undergoes at the foot of the ice fall descending from the Taléfre. The structure of the ice throughout the fall is more distinctly striated in a vertical direction, and parallel to the sides of the glacier, than I recollect to have observed in any glacier so violently crevassed and dislocated. The moraines are faintly perceptible by dirty stripes during the fall. But when the shattered ice is collected, and remoulded, upon reaching the foot of the precipice, by the pressure of the Glacier de Léchaud, a most remarkable and sudden change takes place. The ice, from fragmentary and fissured, becomes compact and swollen into a convex form, produced, no doubt, by the lateral pressure to which it is now subjected, and which it struggles to overcome. *Within the space of a few hundred feet, the transverse structure becomes developed*, the former longitudinal structure at right angles having disappeared in the interval, and the wave-like forms of the structure swell out more and more as the glacier is urged down the steep slope towards station C, with the Pierre Platte. The convexity of this part of the glacier will be perceived from section IV ; and as the glacier is swollen and pressed onwards, the crevasses in this part radiate, as from a centre, or in directions perpendicular to the lines of structure, exactly as I have described in the Glacier of the Rhone. These facts, which I have verified in many other glaciers, conclusively show, that *the structure is developed during the progress of the ice downwards*—is subject to the variations which its momentary conditions of constraint impress—and that it has not the slightest reference to the snow beds of the névé, or to any primitive conformation whatever.

7. When we trace the structure up to the icy basin of the Taléfre, we perceive the origin of the linear vertical structure of the ice which

accompanies it in its fall. The ice, near the moraines of the Jardin, is distinctly ribboned in a vertical direction parallel to those moraines; and this structure, so far as I have been able to observe it at the most favourable season, when this glacier is tolerably free of snow, spreads itself upwards, moulding itself by the forms of the rocky basin which confines it, nearly as represented in the map. The directions remind one irresistibly of the lines of floating matter upon a current of water converging towards a narrow outlet. The direction of the crevasses above the outlet, or icy cascade, is still perpendicular to the direction of the structure, and therefore their lines of fracture are *convex upwards*. Higher up the Glacier du Taléfre, as the structure of the ice becomes more snowy, and less crystalline, the ribboned appearance vanishes altogether at the surface, although it is probably continued at a greater depth.

Such are the facts which I have been able to observe most carefully with respect to the arrangement and distribution of this remarkable structure over a glacier of great size, and variety of surface. It will be found to represent very well the normal type of all glaciers, as we shall afterwards have occasion to illustrate by examples. In the meantime, I will say a few words respecting the accidents of crevasses, and then endeavour to explain the views which the study of the Mer de Glace suggested with respect to the cause of the veined appearance.

Perhaps the most usual and general rule for defining the direction of crevasses, when a glacier is not violently dislocated by moving over excessively steep or irregular surfaces, is, that they tend to a direction perpendicular to the structure; since, however, a rent once determined is often prolonged, irrespective of the immediately producing cause, such crevasses may, throughout their length, cut the structure at different angles, which they often do. Some of the crevasses of the Mer de Glace are probably 2000 feet long. I carefully examined a crevasse near the Montanvert, extending from the medial moraine quite to the western side; and in the higher parts of glaciers, as towards the Col du Géant, crevasses extend, by communication with one another, to far greater distances.

It has been stated by some authors, that crevasses are generally in lines transverse the glacier, and convex downwards;* but I doubt this being generally true. It is by no means so easy as it appears, to ascertain the general ground plan of a system of crevasses, for nothing is commoner, in viewing a glacier from a height, and seeing one system of crevasses, than to lose sight altogether of another set which cross the former. This is the case opposite to the Montanvert, where there are two distinct systems of crevasses, equally inclined to the axis of the glacier, and forming an angle on its surface of 65° with one another, so that each set deviates 32½° from a line transverse to the glacier. In turning round the promontory of Trélaporte, a series of fan-shaped crevasses succeed one another, as already remarked. It is extremely curious to observe the hyperbolic " dirt bands" maintaining their position amongst that confusion. Higher up, the crevasses become transverse, and less numerous.

When the glacier makes a rather abrupt turn, as between the Echellets and the Angle, it appeared to me that the crevasses of the higher glacier are stopped up by the pressure of the ice where it is *reflected* from the rock, and a new set open, corresponding to the new direction of motion. It is this interference of a current of water and its *reflection* from a promontory, which *breaks* the surface of a river into foam; and something of the same kind may be perceived, if I mistake not, between the Angle and the Montanvert. The old crevasses are sealed up, and new ones formed, cutting them across, which produce the tumultuous looking hillocks in that part of the ice.

But still more important are the circumstances attending the formation and change of crevasses during different seasons. Beyond the general admission that crevasses result from a glacier being pushed

---

* M. Agassiz, however, has maintained the contrary opinion, and supposes (in consequence of the crevasses being convex towards the origin of the glacier,) that the glacier moves faster at the sides than in the centre. Such are the differences of opinion upon what appear even the most simple and elementary points, in this subject.—*Etudes*, p. 167.

over a surface presenting great irregularities, which irregularities break the semi-rigid mass over them, little or nothing has been agreed upon by authors as to their origin. That crevasses form with a sudden noise, and are at first mere cracks into which the blade of a knife would scarcely enter, is beyond a doubt. But the fact for which I was least prepared, but which my long residence on the Mer de Glace last summer convinced me of, is this, that these crevasses, if not entirely renewed every year, are so at least in a great degree; that they are formed during spring, summer, and autumn, by which time the face of the glacier is in some respects entirely changed—much more so indeed during a few months than it ever is from one year to another,—so that a traveller may revisit a glacier from year to year, and think that he recognises localities on the ice, he may map the fissures and accidents, and seem to discover them afresh, but they are only the ghosts of his departed friends,—forms, which unlike a wave which moves on whilst the substance which moulds it is still, remain planted amidst motion, as if anchored in the icy cataract. This fact has formerly been insisted on, but what I wish now to make plain is the certain fact that the crevasses are in a good measure formed afresh every season.

When I traversed the glacier in a great many directions in the end of June, I had ample means of judging of its state from the obstacles which were opposed to a passage over it ; I had also an opportunity of noticing the width of the crevasses, their regularity, and the sharpness and verticality with which they generally terminated at the surface of the glacier. In July and August, during many excursions in the same directions, the change was most conspicuous, and especially in the higher parts of the glacier between the stations G, B, and H, where the snow had only recently uncovered the ice at my first visit. There the crevasses had increased to such a degree in number and breadth that the glacier seemed unlike what it was, and a space which I had formerly considered as almost sufficiently even for measuring a base line upon the ice, was now traversed by clefts. Even at the Montanvert the crevasses were visibly wider, and the whole texture of the ice more shaken.

But it was in the month of September that the change was most perceptible in the lower part of the glacier. I have already adverted to the loss of surface, and to the general subsidence of the whole mass of the glacier. The several stations where I made my regular observations on the ice had of course their topography and peculiarities firmly fixed on my memory, and there the change of feature, within a few weeks, was such as to render them scarcely recognisable. Great cavities or clefts were entirely soldered up,—others had encroached on their solid partitions so as to unite with independent ones; precipices had become gentle inclined planes ;—the landmarks of great stones were lost —they had tumbled into crevasses, or been so tossed over as to seem no longer the same: but the general character at this season was a subduing of all the angular rugged character of the ice in spring. The fissures, though wide, were many of them choked, their walls melted, and their edges deformed. The mid-day sun shines *along* the glacier, hence, (the fissures running generally from east to west,) the southern wall of ice was shaded, the northern exposed to the sun. Thus, it happened that in the month of September the northern edges of the

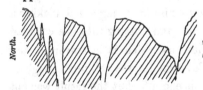

crevasses were nearly all degraded in the manner represented opposite, and the eminences falling into the hollows rendered the passage of the glacier much easier than it had been some weeks before. This occurred also in the higher parts :—above Trelaporte I observed crevasses similarly deformed, and at the same time *closed*, so that a mere crack now stood in the place of the open cleft.

It may here be proper to say one word about a system of crevices, of small dimension, which appear to traverse the ice of glaciers, and about which much has been said which is unimportant, and much has been supposed which is untrue.

We have already observed that glacier ice is eminently fragile ;— hence the facility of making steps with a hatchet, by which means alone many otherwise inaccessible summits are gained. This fragility depends upon the ice being traversed by an infinity of capillary fissures —generally invisible—but which become distinctly seen near the walls and moraines of glaciers, and wherever the ice is exposed to sudden alternations of temperature, by being in contact with rocks or stones. There the glacier consists of a congeries of tightly wedged polyhedrons, of the most irregular figures, often three inches or more in length, and of which a bunch may be held connectedly together, until, by melting, they become disengaged, and fall asunder. But, whilst the pieces remain thus connected, the fissures impart to the mass a certain rude flexibility within small limits, and they undoubtedly permit the free infiltration of surface water to great depths in the ice. These crevices and the granules which they separate, have been particularly described, and their existence insisted on by Scheuchzer, Hugi, De Charpentier, and Agassiz; and this has been described as the peculiar structure of the ice, while the veins of cleavage, or ribboned structure, remained unnoticed; it is, however, entirely subordinate to, and superinduced upon the latter, as I may later have occasion to show. Its existence near moraines and fixed rocks is too obvious to be doubted; but I was for some time sceptical as to its pervading the glacier generally. When I had the pleasure of visiting the Glacier of the Aar with M. Agassiz in 1841, I communicated my doubts to him, and suggested making a hole in the most compact part of the ice, and putting into it a coloured liquid, which might inject the crevices by which it is traversed, and thus demonstrate their existence. M. Agassiz was obliging enough to sacrifice two bottles of red wine to this enquiry, but the result was not entirely satisfactory to me, as, though the wine certainly escaped, it left no traces of its passage. I therefore resolved to perform the experiment more carefully in 1842, and took with me several portable colouring matters. To these, by the advice of M. Regnault, of Paris, I added some cakes of lithographic ink, which not being soluble, but only suspended in water, might, he ingeniously suggested, adhere to the capillary fissures, and indicate them more plainly. Holes about a foot square were made, to a small depth, in the most compact part of the ice, near the Montanvert, in the evening, when the superficial wet was least, and the black* and red dyes, very concentrated, were poured into

---

* The black was poured in in the morning, some hours before the final examination.

them, to the extent of some pints. I will state the result obtained the next forenoon in the words in which I noted it at the time :—

" With an axe I carefully cut the ice round the cup of ice in which the madder infusion had been put last evening, and also round another similar one, in which dissolved [diffused] lithographic ink had been placed this morning. Though much colouring matter yet remained in each, much was effectually and visibly infiltrated into the ice beneath and around ; the small solid colour-particles being visibly confined in the air cavities from which no visible capillary fissures extended, and [from which] they could not be removed by ordinary washing. This ice is seemingly compact ; it does not exhibit obvious traces of capillary fissures, and mere immersion in a coloured fluid produces no true infil-tration—the adhering colour may be immediately washed off ; whilst, where the ice is exposed to the air, it is fissured into the grains so often mentioned, and which may be immediately infiltrated with wine, ink, or any fluid. But this experiment shows that these do exist, and unite the air-cells, or many of them, though unperceived ; even the undis-solved fibres of madder and grains of lamp-black, had penetrated to con-siderable distances."

I therefore freely admit—what I formerly doubted—that a glacier in summer is penetrated to a great depth by water, which saturates all its pores. I am equally satisfied that during summer this water never freezes, and in winter only partially. Hence a glacier is not a mass of solid ice, but a compound of ice and water, more or less yielding, ac-cording to its state of wetness or infiltration.

It is now time to consider how these facts may be combined with those of the observed velocity of the glacier in different parts, in order to account for its veined structure.

Exactly a fortnight after observing the hyperbolic dirt bands oppo-site the Montanvert, I walked on the 7th August to the same spot, and I then obtained an insight into the cause of the phenomenon and of glacier motion generally, which I have no doubt is in substance the true one. The forms of the superficial curves before me recalled almost involuntarily the idea of *fluid motion ;*—they resembled perfectly the lines into which the froth or scum on the surface of a viscous fluid would form themselves if that fluid were propelled along an inclined trough or basin. The cause of such a form is evidently the greater rapidity of the centre than of the sides, a rapidity which, in the case of a viscous fluid, is occasioned by the less adhesion between its particles than between the fluid and the vessel in which it is contained ; and in *any* fluid a similar effect would arise from the friction of the banks or sides. Then the reflection naturally occurred—it is not only *probable* that such would be the motion of a semi-fluid or pasty mass placed in the conditions of the glacier, but it is *certain*, from my own experiments already detailed, that the actual motion *is* such as we have supposed it might be ; it *does* move faster in the centre than at the sides ; it is no hypothesis to say that the glacier moves as a viscous or pasty mass would move— we know that opposite the Montanvert the motion of the ice at the centre of the glacier is two-fifths greater than at even a very sensible distance from the bank. A glacier may, therefore, really be in its structure and formation, like what I had compared it to in 1841,—" A pailful of thickish mortar" poured out,* and the wrinkles on the surface of the one and of the other may have more than a vague analogy. But I carried my theory much farther. I considered that a semi-rigid mass, like a glacier, which has no pretension to be called a fluid in the common sense of the word, if it do not, (as it certainly does not,) move in all its parts parallel to itself, there must be a *solution of continuity* between the adjacent particles of ice to enable the middle to move faster than the sides. Imagine the surface of a glacier to be divided into a number of stripes parallel to its length, and adjoining but not cohering. If it be ascertained that each stripe nearer the centre moves faster than its neighbour nearer the side, the stripes will move past one another parallel to their length, the central stripes gaining upon the lateral ones. If we attempt to give such a varying motion to the parts of a flat stiff body, as a long sheet of paper, we cannot effect it without tearing the paper by rents parallel to its length, or the direction of movement. Now, such must be the case with a mass of ice which does not move with a uniform velocity in its transverse section, but where every line of particles has the velocity proper to its position in

---

* *Edinburgh Philosophical Journal*, January 1842.

the ice stream. The ice will, therefore, be rent by innumerable fissures, whose general direction will be parallel to its motion, and these fissures becoming filled with water and ultimately frozen during winter, will produce the appearance of bands traversing the general mass of the ice having a different texture.

It appeared to me not improbable that the recurrence of beds of more or less porous structure, to which we have seen that the phenomenon of " dirt bands" is due, might depend in some way upon the season of their first consolidation in the higher glacier, and that this character being preserved, with some modification, throughout their future course, would cause the recurrence of these porous bands at annual intervals, so that they may represent the " annual rings" of growth of a glacier, and the intervals between them its annual progress at any point. It is certain, that if their original formation in the higher glacier could be satisfactorily accounted for, the very elongated form of the curves towards its lower end would be exactly such as the difference of velocity in the central and lateral parts of the glacier would produce ; and the gentle curvature of the planes of structure in the higher parts of the glacier confirms such a view. It must be owned, however, that there are several difficulties which require to be removed as to the recurrence of these porous beds ; and especially it must not be forgotten that we have already shown in this chapter that the structure is not in any connection with the stratification of the névé.

But the theory of the cause of the ribboned structure goes much farther. We have hitherto spoken only of the influence of the *sides* of a canal upon fluid or viscous motion, but the *bottom* has also its influence. It cannot, I think, be doubted, after what has been stated, that the motion of the ice is more rapid at the *surface* than at the *bottom*, for the very same reason, that it is more rapid in the centre than at the side. The friction of the bottom must retard it ; and, be it remembered, that the less plastic the matter, the farther from the sides or bottom will the influence of friction extend. In a limpid fluid the friction of the containing vessel will extend but a little way ; in the case of a solid body sliding down its bed, the friction affects equally the motion of the whole mass ; in a viscid or plastic mass, it is partially transmitted throughout.   This fact is well known to all writers upon hydraulics who have attempted to assign the proportion which the superficial velocity bears to that at the bottom of a running stream, and the difference is greater in proportion as the stream is shallower.

In consequence of this it is *conceivable* that a glacier might remain permanently frozen to its bed, and yet move on ; that is, the strata might move over one another, and the highest most rapidly.   I am far from saying that this is actually the case ; for I believe that in most cases the glacier is detached from its bed by the natural warmth of the earth ;—yet, where the friction is so enormous, as when ice rubs over a channel of rough rock,—and when, what is far more important and unanswerable, a glacier, accumulating in an icy basin, flows out by a narrower aperture from that by which it entered, motion, as a *solid* body, is out of the question—it can only move by its parts yielding —in other words, it is plastic.   I therefore understand, that in every case where the friction of the icy particles against one another is on any account less than the friction of the ice against the rocky walls, then the ice will move fastest at its centre, and the motion of the centre due to gravity will draw after it the lateral and more sluggish parts—which justifies M. Elie de Beaumont's acute observation, that a glacier does not so much resemble a body thrust or pushed forward, as one dragged down and pulled.

It may appear that the explanation now hastily given of the varying rate of motion of the different portions of the ice, diminishing from the centre to the side and from the surface to the bottom, could only produce a series of trough-shaped fissures, of which the section should everywhere resemble fig. 2 of page 164, but that the circumstance of the frontal dip shown in fig. 3 would be wholly wanting.   A close attention to mechanical principles will, however, show that this is not the case, and that the forms will be such as has been already explained. Not to prolong an already tedious chapter, we will, however, postpone this general consideration to the recapitulation of the theory with which it is proposed to close this volume, when we shall also have an opportunity of supporting it by experimental analogies.

CHAPTER IX.

THE TOUR OF MONT BLANC—CHAMOUNI TO COURMAYEUR.

GLACIER DES BOSSONS—ITS CHIEF PHENOMENA—ROUTE TO THE MONTAN-VERT—GLACIER DE TACCONAY—ROCHES MONTONNEES AT PONT PELISSIER—BATHS OF ST. GERVAIS—ORIGIN OF THE BLOCKS OF THE VAL MONTJOIE—NANTBOURANT—COL DU BONHOMME—COL DE LA SEIGNE—ALLEE BLANCHE—COURMAYEUR.

WHAT is called the *Tour* or circuit of Mont Blanc, is an easy journey round its base, beginning and ending at Chamouni. It is familiarly described in many works, and well deserves all the praise which can be bestowed upon the admirable and varied scenery through which it leads us. To those who look at matters more closely, it offers great interest, because it gives an opportunity of examining in succession every one of the valleys and ravines which take their origin in the chain of Mont Blanc, and which are usually in part or entirely filled with glaciers. I shall suppose the traveller starting from Chamouni so as to cross Mont Blanc at its western shoulder called the Col du Bonhomme, where he comes amongst valleys which pour their streams into the Isère, and thence to the Rhone; turning next to the eastward, and crossing the Col de la Seigne, he enters the Allée Blanche, a valley of singular grandeur on the southern side of Mont Blanc, and parallel to that of Chamouni. Here the river Doire (*Dora Baltea*) takes its origin, which, joining the Po below Ivrea, goes to swell the waters of the Adriatic. Courmayeur, a Piedmontese watering place, is situated on the Doire, immediately behind the chain of Mont Blanc.   The map, facing page 1, contains the route which we are now considering.

The first object of importance after leaving Chamouni is the Glacier des Bossons,* (the *patois* of Buissons, as it is spelt by De Saussure,) of which the exquisite purity is known to all travellers.   I shall not stop to describe the phenomena of its *Aiguilles* of ice, and its greenish blue crevasses, so familiarly known, but I will point out shortly what seems most worthy of remark, especially in connection with the theory of glaciers.

1. The Glacier of Bossons belongs to what may be called (see page 30) the *intermediate* kind of glacier, that which, taking its origin at a great elevation, pours itself down in a confused mass into a valley at a low level, where it spreads itself out as far as the principle of the equality of waste and supply (page 19) will permit.   This glacier has brought down beside and beneath it a great mass of debris of the rocks of Mont Blanc, (including serpentines of doubtful origin, but most likely from the foot of the Aiguille du Midi,) and these have formed a steep embankment, projecting into the valley, upon whose top the glacier rests.   This gives to it a very remarkable appearance, especially as seen from Les Ouches, farther down the valley, where the fir woods conceal the origin of the glacier, and the lower part, thrust forward as it were from out of the side of the hill, stands forth like an island of crystal in the bottom of the valley.   This part of the glacier is very nearly flat, and it is there easily crossed.   Quite at its termination it falls over the slope of its moraine, and forms deep chasms and lofty pinnacles.

2. The Glacier of Bossons, like most of those in the same neighbourhood, attained, in 1820, its greatest extent in recent times, when the moraines advanced over cultivated fields, very near to the Hameau des Bossons.   The traces of this progress are very visible. One enormous block has rolled out from amongst its neighbours on the eastern side of the glacier, and has mowed down a path for itself,

---

* Before crossing from the right to the left bank of the Arve, some fine springs are passed at the foot of the Breven : they are called *Eaux de Gailland*.   The temperature on the 27th August 1832 was 44°6 Fahrenheit.

through the wood, on that side, and there it lies on a slope surrounded by trees, exactly like the moraines of the Chaumont, or of Monthey, (Chap. III.)

3. The glacier of Bossons has no medial moraine. It descends (as De Saussure has remarked) in an unbroken continuity of ice from the very summit of Mont Blanc. Its great feeder is the Grand Plateau, and almost the only rocks which break its passage are the *Grands Mulêts*, the first stage on the ascent of Mont Blanc. The *detritus* of these is, however, too inconsiderable to afford any medial moraine, especially as the glacier is one of the most precipitous, for its extent, in the Alps.

4. The structure of this glacier is generally homogeneous, and almost snowy, or at least opaque white, with little green or blue tinge, except near its edges, where it is most icy. The *veins*, or bands, are distinct near the sides, and fall towards the centre in the usual manner. They are not formed in this glacier by a simple alternation of parallel layers, but the icy bands have all the appearance of *posterior* infiltration, occasioned by fissures thinning off both ways, and filled with frozen water. The icy cascade above seems to have little or no structure. The structure is gradually developed as the glacier consolidates and moves more horizontally, but it is never perfect, owing apparently to the shortness of its course, and its great breadth compared to its length. The tendency, however, is evidently towards the usual type of such glaciers, the structural veins bending round in a loop, as seen on the surface, and with a frontal dip diminishing as the glacier approaches its termination, where the bands are more distinct, and indeed well defined, inclining altogether forwards and parallel to the soil on which the ice rests.

5. The paucity of moraines, and the very slightly developed structure near the centre of the ice, occasion the extraordinary purity of the Glacier des Bossons, in which it has a remarkable analogy with that of Rosenlaui, in the canton of Berne, which has a somewhat similar course. We have seen that it is the veined structure which intercepts and retains the sand of the moraines. Now, in the case before us, where the glacier is in contact with the lateral

moraines, we perceive fragments of stone and earthy matters intermixed with the ice to a considerable thickness, and evidently following the direction of its cleavage.   These are, no doubt, the
earthy beds of which De Charpentier speaks,* and which he distinguishes from true stratification, but of which, nevertheless, he gives
a very unsatisfactory account, supposing that they arise from debris
which had fallen into crevasses, and which had afterwards become parallel to the sides of the glacier, or its line of contact with the moraine,
by some process which he does not explain.   The real explanation,
upon the theory of these veins which I have given in the last Chapter,
appears to be, that they are due to the fissures developed near the
edge of the glacier, where its friction is greatest, and the velocity
of its layers most unequal, and, owing to this inequality, the faster
moving parts of the ice drag along with them some of the particles
of the moraine with which they have become soiled.   In these parts,
the icy structure is perfect, owing to the complete thaw which the near
contact of the warm ground produces, for the lower level of the glacier
of Bossons is unusually deep in the valley, not probably more than
3300 feet above the sea, or at least 5000 feet below perpetual snow.

6. The Glacier of Bossons, then, by showing the exact manner in
which an almost homogeneous mass of opaque white ice begins to have
a structure developed 10,000 feet below its origin, by the formation of
fissures into which water being infiltrated assumes the appearance of
bluish veins, which finally present the usual forms of glacier structure,
is highly illustrative of the views formerly explained.   I must add,
that the peculiar phenomena of *dirt bands* on a great scale described in
page 162 are not here wanting, although from the dazzling whiteness of
the ice they may very easily be overlooked.   They are best seen in
cloudy weather, when two or three of great breadth may be easily seen
traversing the lower end or *snout* of the glacier where it dies away in
the valley.

The Glacier of Bossons is bounded on the east by a steep grassy hill,

* Essai, p. 75.

which rises to the foot of the Aiguille du Midi, where it is surmounted by the Glacier des Pélerins. A very interesting and by no means dangerous excursion may be made in this direction from the Glacier des Bossons to the Montanvert, or the reverse. Above the Chalet of Para (on the slope just mentioned, and the last habitation passed on the ascent of Mont Blanc,) is a grassy height which may be from 7000 to 8000 feet above the sea, and whence a most interesting view is obtained of the highest part of the Glacier des Bossons, the Grands Mulêts rocks, the Grand Plateau, and, indeed, the whole course of the route to the summit of Mont Blanc. From thence the Glacier des Pélerins is crossed (where De Saussure met with one of the narrowest escapes of his life,*) to the Sommite des Croix, another green hill-top which offers a magnificent view; and continuing nearly on the same level, avoiding or crossing with precaution the glaciers of Blaitière and Grépon, the ridge of the Charmoz is gained, along which the descent upon the Montanvert is easy.

The western side of the Glacier des Bossons is bounded by the Montagne de la Côte, a very narrow and steep ridge of rock, covered, however, by many pines, which separates the glacier just named from that of Tacconay, which descends immediately to the westward, and has a common origin with it amidst the snows of Mont Blanc. Naturally enough the earlier attempts to ascend Mont Blanc were made by the Montagne de la Côte, but it has been found on the whole easier to traverse the glacier. It was by the Montagne de la Côte that De Saussure ascended, and he slept on the summit the first night. The Glacier of Tacconay is remarkable from this circumstance, that it appears to have diminished notably in modern times, whilst that of Bossons has either increased or perhaps remained stationary. The modern glacier of Tacconay has but small moraines, whilst the ground below, and indeed the whole neighbouring valley in the direction of Les Ouches, is strewed with immense fragments of the granite of Mont Blanc, which it seems impossible to doubt having been transported by this glacier when it

---

* *Voyages*, § 675.

formerly attained a greater bulk, and crossing the Arve, deposited these blocks on its farther bank, where the river takes a sudden turn to enter the valley of Servoz. Limestone occurs on both sides of the Arve, in the neighbourhood of Les Ouches, and is connected with the great secondary chain to the north of the Breven. Farther down, however, it is succeeded by a nondescript quartoze rock, forming the ridge between Servoz and St. Gervais. Between Les Ouches and Pont Pelissier, this rock is furrowed and polished in the most characteristic manner of the glacier action of the Alps, in a direction parallel to the length of the valley, and which it is impossible for one moment to doubt being due to the abrasion of some heavy superincumbent rubbing body. These forms may be compared to those produced in ductile plaster by the wooden mould with which the workman finishes a cornice. They extend to some height on the western slope where I first noticed them in descending from the Col du Forclaz. The whole of this part of the valley scarcely contains one angular fixed rock—all are smoothed and polished. Near Pont Pelissier, and on the western side of the Arve, are several hillocks presenting precisely the phenomena of *roches montonnées*, and that their forms are due to glacier action, is rendered the more probable from the occurrence of blocks amongst them, one of which, of immense size and angular shape, seems poised on the very top of one of these bee-hive-like summits; such phenomena have been called by De Charpentier *blocs perchés*, and it is impossible to see a better example than the one I have just mentioned. It is truly surprising that in the minute mineralogical description which De Saussure gives of this route * he makes no allusion to these phenomena. This is one example amongst many how obvious facts may escape the most experienced and assiduous observer, for De Saussure must have passed through this valley dozens, if not hundreds of times.

Some miles below Servoz, the valley of the Arve is joined by the Vallée de Montjoie on the left, transversed by the rapid and cheerful stream of the Bon Nant, which forms a remarkably pretty and

---

* *Voyages*, tom i., chap. 6.

well-known cascade immediately behind the Baths of St. Gervais. These baths are situated in a deep and picturesque ravine, a little below the village of the same name, whose gay and neat appearance at a distance, with its fantastic spire, decorated, like most of the churches of the province of Faucigny, with burnished tin plate, gives a sparkling character to the landscape. The mineral springs of St. Gervais issue from alluvium, through the floor of a subterranean gallery. The three hottest vary in temperature from 104° to 106° Fahrenheit. They contain iron and sulphur. Like most thermal springs, they issue near the union of different rocks. The valley on one side being composed of slate, quartz rock, and conglomerate, and on the other of limestone, limestone shale, and thick beds of gypsum from which copious springs rise, with a temperature of 51°, at no great distance from the others. Several excursions of interest may be made from St. Gervais, which we will not stop to particularize; the views are very striking, although the higher Alps are concealed; but the limestone range of the Aiguille de Varens which rises above St. Martin, is singularly picturesque in its outline and detail. What interested me most, however, in my last visit to St. Gervais, was the discovery of what I cannot doubt to be numerous and extensive Moraines in its neighbourhood, although the nearest modern glacier is some hours' walk distant.

It is to be observed, in the first place, that the valley is choked, as it were, in its lower part, by a mass of debris, through which the river has worked its way below the village of St. Gervais. The rock, where it appears, is usually slaty limestone; but the surface of the soil is every here and there strewed with blocks of granite, some of them insulated and of great size, at other times accumulated in ridge-like mounds along the face of the slopes, exactly like moraines. Amongst the woods on the western side of the valley, not far from the baths, I found blocks of from thirty to forty feet in length, composed of well characterized protogine or granite of the chain of Mont Blanc. An extensive and well marked moraine stretches along the face of the hill in the direction of Sallenches, and on the slope fronting the valley of the Arve, where it is almost inconceivable that a torrent could have been embayed, so as to deposit its blocks, supposing it could have moved such immense ones. They lie high above the open plain, and in a regular ridge, exactly like that figured on page 17, from the *Mer de Glace* of Chamouni. The ridge just mentioned is partly grassy, and partly covered with small trees, but there is ample evidence of its composition being similar to that of a moraine.

The most direct route from Chamouni to St. Gervais is not by Servoz, but across the Col de la Forclaz, which rises immediately above the village of St. Gervais. For a great height on this path, angular granite blocks are strewed about.

The Col de la Forclaz is a gorge, and therefore offers no view from the summit. The Col de Bellevue or Col de Voza, which crosses the chain of Mont Lacha, somewhat higher up, and communicates between the village of Les Ouches, in the valley of Chamouni, and that of Bionnay in the Val Montjoie, which commands the prospect of Chamouni and Mont Blanc, is, therefore, deservedly more frequented. It also gives an opportunity of inspecting the Glacier of Bionassay, which descends in a north-western direction from near the summit of Mont Blanc, and approaches near the Chalets of the same name. The Pavillon de Bellevue on the Col is nearly 7000 feet above the sea,* and yet erratic blocks are strewed all around. Not only is it inconceivable that a torrent should have passed over a hill like this, fit to carry great blocks of granite, but *the erratics of the Col mix insensibly with the modern moraine of the Glacier of Bionassay beneath, so that it is impossible to say where the erratic phenomenon ends, and where the glacial phenomenon begins.* This is an argument very striking on the spot, in favour of the glacial theory of erratics, and these very blocks of protogine may be traced, I believe, without any intermission, down to the Baths of St. Gervais, and perhaps to Sallenches. There are three, if not four, distinct glaciers, which occupy the higher parts of valleys communicating with that of Montjoie: Bionassay, (already mentioned,) Miage, (to be distinguished from that of the same name in the Allée Blanche,) and Trelatête, which descends opposite to the Chalets of Nantbourant. All of these

---

* 6939 English feet. See Decandolle, *Hypsométrie.*

transport numerous primitive blocks, and sometimes deposit them upon insulated summits near the openings of the respective valleys. From Contamines (where there is an indifferent inn) to Nôtre Dame de la Gorge, (a chapel and mission-house, without a village,) the scenery is cheerful and pretty. There the defile narrows, and the steep rocks of gneiss on either hand, between which the stream struggles, are picturesquely clothed with larch and pines; and here, as is almost universal in valleys containing erratics, the surface of the rock is worn, rounded, and cut by long smooth furrows, which resemble those produced by glaciers. The torrent is passed by a bridge immediately above a fine waterfall, and we find ourselves in an upland pastoral country, but still pleasingly diversified by wood. A main branch of the Bon Nant descends a narrow rough gorge from the Glacier of Trelatête. We are now at Nantbourant, where travellers, making the tour of Mont Blanc, usually pass the night.

Nantbourant is about seven hours walk from Chaumoni. The Col du Bonhomme is between two and three hours farther. The way lies chiefly over upland pastures, not unmixed with good trees; but the higher part is bare rock, with patches of snow. The upper portion of the valley is composed of secondary limestone, containing Belemnites, and presents no granite blocks. But though the little plain of Les Barmes is covered with vast calcareous fragments fallen from the cliffs above, these do not extend (so far as I have observed) into the valley beneath; and the numerous primitive blocks already mentioned cease entirely above Nantbourant, that is, they commence with the Glacier of Trelatête; thus showing, that the transporting cause of these erratics had its origin, not in the natural prolongation of the valley, (at the Col du Bonhomme,) but in the highest tributary which contains a glacier.

The passage of the Bonhomme is one of the most dreary in the Alps; and in bad weather it is dreaded by the guides. The strong west wind spends itself upon this great outlier of the chain of Mont Blanc, and raises the snow into fearful eddies, called *tourmentes* in the French, and *guxen* in the German Alps, which are justly feared by those who have been exposed to them. Here two English travellers lost their lives some years since. Their last entry is still to be found in the traveller's book at Nantbourant. I have crossed the Col du Bonhomme three times, and on one of these occasions, having merely a porter with me, who did not know the way, we got bewildered in fog amongst the rocks, from which we were only extricated by my referring to the map and compass, instead of following the directions of my companion. When the summit is gained, a wide view is seen over the valleys of the Tarentaise; and the traveller naturally thinks of descending immediately by a path right before him. Let him, however, beware of this, for it will lead him into the valley of Beaufort, which is most likely not his intended route. If going to Courmayeur, he follows an ill traced path on his left, over black shale, (or snow during part of the season,) which conducts him nearly on a level, after a quarter of an hour's walk, to a point somewhat higher than the last, which is called La Croix du Bonhomme, and which, on my last journey, I found to be 8195 feet above the sea. The view from thence is striking, although Mont Blanc is concealed. The mountains of the Upper Isère, stretching away towards the Mont Cenis, are fully in view; and conspicuous amongst these is the Aiguille de la Vanoise, a snow-clad pyramidal summit between Moutier and Lans-le-Bourg, and which is undeniably one of the most elegant mountains in the Alps.

Immediately before the spectator is the very deep valley of Bonneval, which takes its rise at the foot of the Col de la Seigne, and which, turning sharply at the Chalets of Chapiu, (whose position may be seen at an immense depth below,) forms a very wild and uninhabited gorge, extending nearly to Bourg St. Maurice, in the valley of the Isère. By this route the traveller reaches the pass of the little St. Bernard, which he may traverse to the Val d'Aoste. If, on the other hand, he wish to reach Courmayeur directly, he may either descend from the Croix du Bonhomme to Chapiu, and ascend to the Hameau du Glacier at the head of the valley, or he may cross the Col des Fours, which conducts him by a shorter, but rougher road; or, finally, he may scramble along the rocks by an intermediate path, without descending so low as Chapiu. The passage of the Col des Fours is still more savage than that of the Bonhomme, and it is considerably higher. I shall long remember an hour spent here in magnetic and barometric observations, in August 1832, amongst perpetual snow, and exposed to a biting wind. It is about 850 feet higher than the Col du Bonhomme. The middle path just alluded to is, in some respects, interesting. Instead of descending the steep pastures of Chapiu, we follow an obscure track amongst the rocks towards the east ; and after traversing for some distance the limestone strata rising towards the north, of which the main chain is here formed, we come to a mass of granite, rising from the valley, *and overlying them* at a considerable angle. Near the same point, there is a magnificent view of Mont Blanc and the adjacent mountains, seen above the Col de la Seigne, which appears just in front. It presents the whole range, from the Grande Jorasse on the east, to the summit called Aiguille du Glacier on the west, from which the vast glacier descends which occupies the head of the valley of Bonneval.

At the Chalets of Motet, or Glacier, the traveller will probably make as short a stay as possible, and will then proceed to ascend the Col de la Seigne, which (as has been said) separates the tributary streams of the Rhone from those of the Po. The ascent is very easy, but tedious. The summit is 8422 feet above the sea, by my observations, and was fortified (as I was informed) when the French army endeavoured to force this pass. From the top, the extent of the Allée Blanche is well seen, with the great masses of the chain of Mont Blanc, which bound it on the left. Mont Blanc itself presents a singular appearance in this direction, and would not be easily recognised by those who know it only in a northern or eastern direction. The western and southern faces are very steep, although not so absolutely precipitous as they would appear to be, when viewed in front. The former falls abruptly towards the Glacier du Miage ; the latter, in the direction of the Cramont, or into the Allée Blanche itself. The bottom of the valley is here not more than 4000 English feet above the sea ; consequently, this colossal mountain rises above it at a very short horizontal distance, and no less than 11,700 feet of vertical height, which, though not an unbroken precipice, is composed entirely of steep and savage rock, upon which the snow cannot lie for any extent. Its aspect is, therefore, far more imminent and imposing than on the side of Chamouni, where the eye is greatly deceived as to the actual *distance* of the top, and consequently as to its height. But here the details rather aid the perspective, and when seen in profile from the Col de la Seigne, the stupendous buttresses, by which the mountain is supported, and especially one prodigious Aiguille of granite, called Mont Péteret, come out in relief, although, when a front view is taken from Courmayeur, or its neighbourhood, these pinnacles, thousands of feet in height, are lost against the towering mass behind, which then seems to rise like a wall. I am unable to state the exact line of junction of the limestone with the central mass of granite. I apprehend, however, that it runs from some way to the north of the Col de la Seigne (which is calcareous) to the Cime des Fours, and so down nearly to Nantbourant, leaving the Aiguille du Glacier, and the greater part of the Glacier de Trelatête, within the primitive boundary. To the east the limit is, in a good measure, determined by the direction of the Allée Blanche, which separates, for some distance, the granite from the limestone. Two conspicuous summits, however, which appear near the foreground of the view, a little higher than the Col de la Seigne, are the Pyramides Calcaires de l'Allée Blanche of De Saussure. They are upon the left hand in descending. It is a walk of nearly five hours from the top of the Col to Courmayeur, during which we traverse the whole length of the Allée Blanche.* It is there met by another parallel valley, which opens exactly opposite to it, and forms, as it were, the prolongation of the Allée Blanche, for about five hours farther. This is called the Val Ferret, and terminates at the Col Ferret.

The chief glaciers of the Allée Blanche (on the north side) are the following:—1. The Glacier de l'Estellette; 2. The Glacier de l'Allée Blanche; 3. The Glacier de Miage; 4. The Glacier de la Brenva. The second and third of these have formed barriers across the valley, by moraines, so as to have occasioned lakes from the interruption of the course of the river. That formed by the Glacier de l'Allée Blanche is nearly filled up by alluvial matter; but an extensive flat attests its former

---

* De Saussure states that below the Glacier de Miage the name of Allée Blanche is exchanged for that of Val de Veni.

existence, together with the extensive barricade of debris, through which the river now tumbles in a foaming rapid. The moraine of the Glacier de Miage is, perhaps, the most extraordinary in the whole Alps, and has given rise to the Lac de Combal, which will be especially described in the next chapter. Below the Moraine of Miage, which occupies the valley for a great space, are some chalets, and then a level, fertile plain, whilst the valley widens, and becomes more romantic and less savage. Trees appear on both sides, especially on the right, where the forest is very fine, and clothes all the northern slope of a remarkable hill, with a conical summit, called the Mont Chetif, or Pain de Sucre, which is composed of granite, although separated from the great chain by secondary rocks. The paths through these woods are amongst the most beautiful and striking with which I am acquainted. That leading to Courmayeur, after attaining some height above the torrent, proceeds nearly on a level, until, emerging from the trees, we come into full view of the majestic Glacier de la Brenva which, formed in a hollow to the east of Mont Blanc, pours its mass into the valley which it has, in a good measure, filled up with its moraine, forming a kind of bridge, which it has pushed before it, and on which it bestrides obliquely the Allée Blanche, abutting against its opposite side, at the foot of the Mont Chetif. Its appearance and phenomena will also be described in the next chapter. A chapel, dedicated to Nôtre Dame de la Guérison, stands on the right hand side of the way, exactly opposite to the ice; and another steep descent conducts us again to the bank of the river, which here turns abruptly, after its confluence with the stream of the Val Ferret, into a ravine, cutting the range of the Pain de Sucre. The united streams are passed by a wooden bridge at the Baths of la Saxe, and twenty minutes more brings the traveller to the beautifully situated village of Courmayeur, after a laborious walk of eleven hours from Nantbourant.

The Glacier de Miage and its Moraine.

a. Pyramides Calcaires.   b. Glacier de l'Allée Blanche.
c. Glacier de Miage.   d. Col de la Seigne.

CHAPTER X.

THE GLACIERS OF MIAGE AND LA BRENVA.

THE ASCENT OF THE ALLEE BLANCHE——MORAINE OF MIAGE——ITS HEIGHT AND
EXTENT——CHAMOIS——TRIBUTARY GLACIERS——THEIR STRUCTURE AND FORMS
OF UNION WITH THE PRINCIPAL ONE——SCENE OF DESOLATION ON A MORAINE
——LA BRENVA——ITS REMARKABLE STRUCTURE——A SUPERIMPOSED GLACIER
——INTERESTING CONTACT OF THE ICE WITH THE ROCK BENEATH——INCREASE
OF THE GLACIER OF LA BRENVA IN 1818——A TRADITION.

" I am acquainted with only one other scene in the world which can pretend to rival, in
natural magnificence, the Glacier de Miage ; I mean Niagara."

BASIL HALL.

COURMAYEUR would be worth a visit, if it were only for the purpose of
examining in detail the Glaciers of the Allée Blanche. But this excursion is rarely made.   Travellers are usually content with what they see
of them in descending from the Col de la Seigne, and there are but few
guides who have ever traversed either of these glaciers.   A short day
is sufficient for visiting the Glacier of La Brenva, but it is a laborious

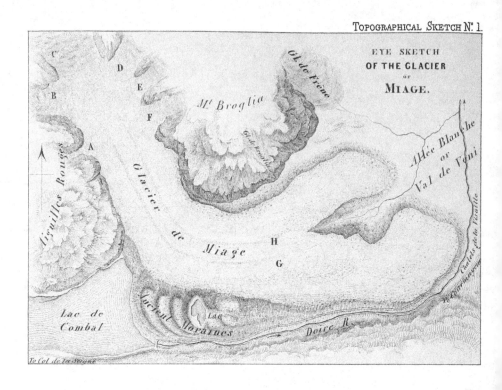

EYE SKETCH
OF THE GLACIER
OF
MIAGE.

C

D

B

E

Ch. de Frène

F

Mt Broglia

Glacier de Mouttes

Aiguilles Rouges

A

Glacier de Miage

Allée Blanche
or
Val de Veni

Chalets de la Visaille

H

H

G

G

To Courmayeur

Lac de
Combal

Ancient Moraines

Lac

Doire R.

To Col de la Seigne

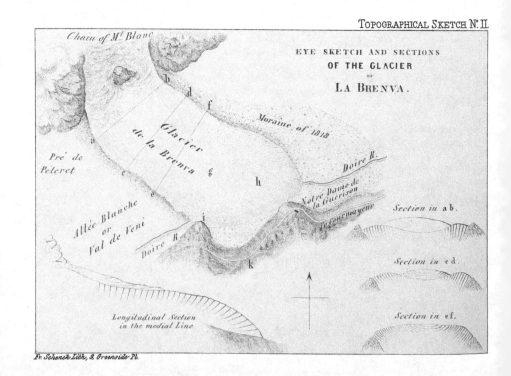

Chain of Mt Blanc

EYE SKETCH AND SECTIONS
OF THE GLACIER
OF
LA BRENVA.

b

d

f

Moraine of 1818

a

Glacier de la Brenva

g

Pré de
Peteret

c

Doire R.

e

h

Notre Dame de
la Guerison

Section in a b.

Allée Blanche
or
Val de Veni

i

Doire R.

To Courmayeur

Section in c d.

k

Section in e f.

Longitudinal Section
in the medial Line

Fr. Schenck Lith., 9. Greenside Pl.

day's work fully to examine the Glacier de Miage. I shall begin with the latter.

I had twice before passed the Lac de Combal, and the moraine of the glacier which I have described as pushed out into the valley which it occupies for *several miles in length*, nearly a mile in breadth, and several hundred feet in depth. I had no small curiosity to see the chasm in the mountains whence this mass of debris had been derived, and to examine the glacier which had been and still continues to be so powerful an agent of degradation and transport. Accordingly, on the 15th July 1842, I left Courmayeur at half past five A. M., on foot, and reached the lower extremity of the moraine at the Chalets of La Visaille in about two hours. The Doire there struggles through the narrow ravine left between the moraine and the foot of the calcareous hills on the south side. The path keeps the side of the moraine, and is every year more or less injured by the falls of rubbish. In this ravine on the south side is a deep hole in the gypsum rock which occurs there, in which my guide Antoine Proment assured me that chamois frequently pass the night, and their young are sometimes taken alive. This surprised me, and I was inclined to doubt it, but we actually saw traces of them on a patch of snow within a short distance. In three hours from Courmayeur I reached the Lac de Combal, where the Doire issues from it, (see the Topographical Sketch, No. I.) A dam has been formed so as to secure its regulated discharge, and to prevent accidents. This lake, as has been already said, is formed entirely by the moraine of the glacier, which is here shot out from the side ravine, and occupies the entire breadth of the valley. The moraine consists of two parts, the old and the new. It is the old which bounds the lake; the new moraine rises to a greater height, and sweeps more gently round, until it becomes parallel to the length of the valley. The old moraines are still fortified by the low walls with slits for musquetry, probably erected by the Piedmontese troops in 1794. It is strange to see this application of the artificial looking mounds which the glacier has raised, and which bear no slight resemblance to a series of gigantic outworks of an extensive fortification. It is the outermost of these ridges which is so occupied. The arrangement of the others is abundantly singular, forming

N

a series of four semilunar curves with their convexity up the valley, as shown in the ground-plan which is taken from a careful sketch made upon the spot. A small lake is formed behind these moraines, which is farther enclosed by other convex, though less perfect moraines beyond, of which the greater part are now grass-grown. I am by no means satisfied as to the way in which these successive ridges of debris were deposited by the glacier. They may either have been frontal moraines, or the contents of vast fissures which were deposited as the glacier melted. Something of the latter kind I have since observed to have taken place in the recent retirement of the Glacier de Lys, in the valley of Gressouay, near Monte Rosa. But as I cannot give any certain explanation, I will not dwell upon it.

Having observed the barometer at the level of the lake, I proceeded to ascend the modern moraine which is higher than would readily be believed from mere inspection, and when I had gained the top and commanded a view of the Glacier de Miage, I observed the barometer again, and found the vertical height of the moraine (besides what is below the level of the lake) to be 395 feet.* Here I found the veined structure of the ice distinct, parallel to the length of the glacier, but dipping inwards at an angle of 70°.

The Glacier de Miage, as I have said, is here pretty level; it is shot out as it were from a narrow valley which works its way back into the very entrails of the great chain, so that the head of the valley is considerably to the north-west of the summit of Mont Blanc, which here presents inaccessible escarpments. The valley is almost straight, and the sides parallel, without subdividing itself into considerable branches. The ice is shoved along this uniform canal, and receives a few tributaries from either side, which descend with great steepness. One which I remarked on the right bank of the glacier, at a spot marked A on the map, descends at an angle, which, so far as I could ascertain it without being on its surface, was inclined 50°, and which is the steepest *unbroken* surface of ice I have ever seen. It descended a nar-

* The height of the Lac de Combal is, by my observation, 2091 feet above Courmayeur, or 6302 feet above the sea.

row couloir from the Aiguilles Rouges (called Mont Suc by de Saussure,) from a great height. The narrowness of the valley makes it
like an unfinished excavation intended to have cut the chain of Mont
Blanc in two, and struck me with surprise, although I was somewhat prepared for it after viewing the prodigious mass of solid matter
which the glacier has poured out into the valley. It may be cited
as a most striking instance of excavation by the ceaseless action of
seemingly trifling causes. The continual fall of fragments detached
from the neighbouring summits, loads the glacier with debris which it
bears incessantly down from the head of the valley, and discharges
into the Allée Blanche; and as we judge of the size of a quarry
from viewing its rubbish heaps, so here we have the mould and the
cast, the die and the relief, the matter transported and the spot of
its excavation.

I traversed the glacier in several directions with a view to examine
its structure, and whilst standing on the moraine I saw a female chamois and her calf cross the glacier, within a very short distance, towards the Aiguilles Rouges, nearly opposite the couloir marked A.
They were followed by eight full-grown chamois, which I could watch
all at once. They were tame, and stopped frequently to look about
them without apparent alarm, and took gently up the hill. They are
almost never hunted here. Near this part of the glacier, and also on
the face of the Aiguilles Rouges, at a very considerable height, a mine
of lead and silver was worked for some years. It was a strangely wild
position for the hope of gain to allure any speculators to establish themselves. After the ore had been excavated and brought down the face
of the cliffs, it had to be carried on men's backs for several miles over
the ice before even a mule track was reached.

Two principal medial moraines occupy the centre of the glacier, and,
as usual, their magnitude becomes apparently greater the farther the
glacier descends, owing to their exposure, by the melting of the ice,
as I have elsewhere explained. The materials of these moraines are
rather remarkable, and have been minutely described by De Saussure,
who is the only author I have met with who describes this glacier.
He particularizes a beautiful granitello, composed of crystallized fel-

spar and schorl ; amianthus, of the fine short kind like delicate fur,
mixed with quartz, and which occurs in all the cabinets of the minerals
of Mont Blanc; several kinds of serpentine, and (what I have not
seen,) carbonate of lime crystallized with quartz.*

The tributary glaciers of the Miage, are, as already said, very steep,
and sometimes pour their icy flood down unbroken, at other times they
descend in *avalanches* upon the main glacier, and become gradually and
completely amalgamated with it.   This is in the higher part, where the
descending masses are rather of compact snow than ice.   In this sense
it is perfectly true, as stated by De Saussure, that the glaciers are
partly fed by avalanches, a position which has been too flatly contradicted.†   Such is the feeder marked B on the map.   After three
hours' walk upon the ice, I reached a considerable height upon the
north-western tributary of the glacier, which was in this part covered
with snow, and, indeed, passed into the state of névé.   I took the
height of the barometer, and found the elevation above the sea,
the highest which I attained upon this glacier, to be 8051 feet.
I then crossed the head of the glacier, which was here wet with
wide water runs, and remarkably free from crevasses, and very carefully examined the structure of the tributary glaciers, which fall into the
principal glacier from the precipices of Mont Blanc.   These afforded
valuable studies of the manner of development of glacier structure, a
subject which at the time particularly engaged my attention, and which
I have since abundantly confirmed in similar cases.   Each tributary is,
in the *first* instance, amorphous, without any apparent structure, and
confusedly thrown together in fragments, as it descends a steep and very
uneven slope.   As it approaches the foot of the steep, it accumulates
upon itself in the *second* stage of the process, and becomes a consoli-

---

* I may mention that carbonate of lime is found, though very rarely, in the granites in the
very heart of the chain ; as at Les Courtes, near the Jardin, on the Glacier du Taléfre.   An
enormous price was asked last summer at Chamouni for a large crystallized specimen of this
kind.

† By De Charpentier and Agassiz.

dated glacier, in which a wavelike structure is developed, with convex arcs on the surface, directed downwards, and the bands which form these arcs, dipping inwards, and approaching horizontality, as the glacier approaches the level of the other. But after it has done so, and the tributary glacier no longer falls forwards, but has its advancing motion resisted by the ice of the main stream, against which it is laterally forced, the structure planes become steeper, and they gradually assimilate themselves, in the *third* stage, to those of the main glacier, becoming erect, and repressing the others. This arrangement is shown in section in fig. 1, and in ground plan in fig. 2. These figures were drawn on the spot from the tributaries marked D, E, and F. It was from the examination of these that I first drew the conclusion, which I have since found to be quite general, that the structure of glaciers is developed from time to time, according to the conditions of the ice, that the structure of one part is not necessarily any modification of the structure of another, and that it is in vain to attempt to trace the stratification of the névé into the vertical bands of the middle glacier, or these into the conchoidal surfaces near the lower extremity. The appearance of fig. 2 occurs when the tributary is of

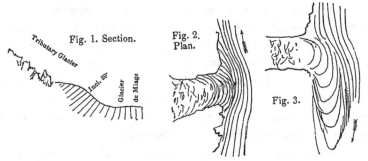

Fig. 1. Section.   Fig. 2. Plan.

Fig. 3.

insignificant dimensions, compared to the primary glacier; its structure is immediately overpowered, and becomes subject to the law of the preponderating one. But where the two streams are comparable in magnitude, the inosculation is more gradual, and the structure is a more complete mixture of the two. Such a condition is shown in fig. 3, which is the condition (for example) of the Glacier du Taléfre uniting with the Glacier de Léchaud.

In general, the structure of the glacier, whilst it is bounded by the chain of Mont Blanc, is well developed, both near the medial moraines, and near the sides, in nearly vertical planes, parallel to the length of the glacier.

Near the promontory of the Montagne de la Broglia, round which the glacier sweeps, so as to turn sharply into the Allée Blanche, the whole structure inclines *from* that promontory exactly as I have described in the case of the Mer de Glace, rounding the promontory of Les Echellets, and as is figured in the glacier section, No. IV., page 166. The rocks are smoothed by the action of the glacier at several points on both sides.

I was not satisfied with having traversed the upper and more level part of the glacier in its whole extent, but I resolved to follow the surface of the ice as far as possible, after it spreads itself abroad in the Allce Blanche, in order to examine its wonderful moraines, and if possible to trace its structure. From a distance, this appears not to be very difficult, for the surface is not steep, its mean inclination in its middle part being about $5\frac{1}{2}°$. Its immense extent, however, deceives the eye as to its inequalities, and I scarcely ever remember to have had a more laborious or rougher walk than the traverse of the lower part of the Glacier de Miage, which I followed down its centre to the spot where, as will be seen by the eye-sketch, it divides into two branches. This icy torrent, as spread out into the Allée Blanche, appeared to me to be $3\frac{1}{4}$ miles long and $1\frac{1}{2}$ wide; but I am aware of the uncertainty of these measures. After struggling for a long time amongst fissures and moraines, I at length mounted a heap of blocks higher than the rest, and surveyed at leisure the wonderful scene of desolation, which might compare with that of chaos, around me. The fissures were numerous and large, not regular, like those of the Mer de Glace, traversing the glacier laterally, but so uneven, and at such angles, as often to leave nothing like a plain surface to the ice, but a series of unformed ridges, like the heaving of a sluggish mass struggling with intestine commotion, and tossing about over its surface, as if in sport, the stupendous blocks of granite which half choke its crevasses, and to which the traveller is often glad to cling when the glacier itself yields him no farther passage. It is then that he surveys with astonishment the strange law of the ice-world, that stones always falling seem never to be absorbed—that, like the fable of Sisyphus reversed, the lumbering mass, ever falling, never arrives at the bottom, but seems urged by an unseen force still to ride on the highest pinnacles of the rugged surface. But let the pedestrian beware how he trusts to these huge masses, or considers them as stable. Yonder huge rock, which seems "fixed as Snowdon," and which interrupts his path along a narrow ridge of ice, having a gulf on either hand, is so nicely poised, "obsequious to the gentlest touch," that the fall of a pebble, or the pressure of a passing foot, will shove it into one or other abyss, and, the chances are, may carry him along with it. Let him beware, too, how he treads on that gravelly bank, which seems to offer a rough and sure footing, for underneath there is sure to be the most pellucid ice; and a light footstep there, which might not disturb a rocking stone, is pregnant with danger. All is on the eve of motion. Let him sit awhile, as I did, on the moraine of Miage, and watch the silent energy of the ice and the sun. No animal ever passes, but yet the stillness of death is not there; the ice is cracking and straining onwards—the gravel slides over the bed to which it was frozen during the night, but now lubricated by the effect of sunshine. The fine sand detached loosens the gravel which it supported, the gravel the little fragments, and the little fragments the great, till, after some preliminary noise, the thunder of clashing rocks is heard, which settle into the bottom of some crevasse, and all is again still. In walking over ordinary rugged ground or rocks, the presumption is, that the masses have become shaken into the position of stable equilibrium—that is, that if a block be moveable, it will tend to roll back to its former position. But, on the glacier, the conditions are exactly reversed, and the consequences are proportionably more serious.

I had the satisfaction of perceiving, as I descended the glacier, that where it spread laterally, and at the same time fell forward, the structure of the Glacier of the Rhone, and those of similar type, (see p. 30,) developed itself; the vertical bands bent round in front of the descending glacier, and dipped inwards at the point marked H, at an angle of 65°, and at G, at an angle of 45°. Here the fissures become lon-

gitudinal or radiating, and the ice still more difficult to traverse.
Nearly the whole surface is covered with the moraine. The extreme
difficulty of finding a path, prevented me from ascertaining its
structure right and left, where it divides into two branches ; but I
have no doubt that in this, as in other similar cases, the divided
streams have each the usual structure of a single glacier.

The bifurcation of the glacier does not appear to be the result of any
fixed obstacle in the valley itself, which interrupts its progress. It is
occasioned solely by the prodigious accumulations of the medial moraines, which, having for ages discharged their contents in front of the
glacier, at length accumulated a mound in the centre which parted the
ice in two with less resistance than would have been required to shove
the prodigious mass forward. Arrived at the point of separation, I
looked from the edge of the glacier into a hollow or ravine several hundred feet deep, having very steep sides, composed entirely of the most
massive blocks which the glacier has brought down, and which are
piled in vast confusion. Down these I scrambled with some labour,
and found at the bottom, not the natural soil of the valley, but apparently the surface of an older moraine, which had spread wider, though
not to so great a height. A stream struggles from amongst the
blocks, and waters a small valley containing some stunted larches and
alders, almost surrounded by the two arms of the glacier, whose moraines nearly meet below, but the two streams do not again coalesce.*
Into this wild enclosure a few sheep are annually driven. I then crossed
the torrent, descending from the Glacier of Frène, which falls from
the chain of Mont Blanc, but little below the moraine of Miage, and
returned to Courmayeur by a pleasant path through the Châlets of
Frène on the same side of the valley.

The Glacier of La Brenva may rank amongst the most accessible in
the Alps. It descends more prominently into the lower valleys than

---

* The lowest part of the modern moraine is 5483 feet above the sea, or 1819 feet below
the Lac de Combal.

THE GLACIER OF LA BRENVA IN THE ALLÉE BLANCHE, FROM ENTRÈVES.

almost any with which I am acquainted, and may be very completely seen from a convenient mule-road which traverses the Allée Blanche, at a distance of less than three miles from the village of Courmayeur. I have already mentioned the extreme beauty of the ride through the pine-woods which clothe the northern face of the Mont Chetif, from which the stupendous chain of Alps may be surveyed like a theatrical scene, and amongst the trees beneath the dazzling white of the glacier presents itself, supported on the bridge of rubbish by means of which it crosses the valley, and presents itself to our close inspection.

Two circumstances are especially worthy of note in this magnificent glacier: its veined structure, and the remarkable changes of dimension which it has lately undergone. As I am not aware of any author who has traversed this glacier, or who has described either of these interesting facts, I will devote a short space to them.

The Glacier of La Brenva consists (as De Saussure so far very well described it) of two distinct parts; first, the rugged and fissured portion, which is quite inaccessible, and which descends a ravine, having its origin very near the summit of Mont Blanc, exactly behind the Monts Maudits; and the inferior, or gently sloping portion, which traverses the valley, as already said, upon a mound or embankment formed by itself, and beneath which the river Doire at present makes its way The middle of the craggy descent is interrupted by a great prominence of rock, over which the descending ice falls in avalanches, and is so completely pulverized, as to be reduced almost to a snowy condition, in which it lies on the surface of the consolidated glacier, and goes through the same changes as in the transformation from névé into ice in ordinary glaciers. It is, indeed, a little parasitic glacier, cradled in the ice of the old one. The Topographical Sketch, No. II., is only intended to explain the sections of the ice, showing its structure, and to give a general idea of the position of the glacier; but has no pretensions to exact topography, not having been sketched with a view to publication. I ascended the glacier to a little way above the line a b, which I found to be 1717 feet above Courmayeur, or at least 1500 above the point where the Doire issues from beneath it. The mean slope of the glacier seen from this point, looking downwards, is 12°. The first section shows the superposition of the powdery ice upon the glacier, which last is there traversed by almost vertical bands, well developed, and which I traced towards the centre, until lost beneath the other.

The cause of the sudden twist in its direction, which, as will be seen by the sketch, the glacier takes when it issues from between the rocks, is probably this, — The glacier, when it descended from the mountain, gradually accumulated an enormous moraine exactly in front of it, so that a valley was formed between the rock and the moraine on either hand, down which the glacier might naturally pass. It took to the left in the direction of the valley beneath, and the old moraine formed one of its barriers. There is no doubt, that in this and similar cases, the vast moraine on which the glacier rides is hollow at its centre, and the debris on either hand form a sustaining wall.

At $c\ d$, the parasitic glacier assumes a tolerable structure, and it is there clearly of the cup-shaped, or conchoidal form, described in page 30 as belonging to the Glacier of the Rhone. This observation is important, as showing, that the arrangement of the structural bands in glaciers of the second order, is independent of the annual layers of the névé, (see page 32.) The superimposed glacier is here *radially* fissured like the Glacier of the Rhone.

Between $c\ d$ and $e\ f$, the parasitic glacier gradually disappears; the vertical bands of the fundamental glacier may there be seen to turn round, so as to present their edges across the glacier, and to dip inwards at a considerable angle. (See figures, page 164.) In the centre of the line $e\ f$, the frontal dip inwards (see the longitudinal section attached to the Topographical Sketch) is 65°. Here the glacier is flattest. Its mean slope is 8°. I traversed carefully the breadth of the glacier; the sides being supported only by the moraines, the dip no longer approaches verticality. The structure planes at the sides dip inwards at 55°, as shown in the transverse section along $e\ f$.

The next stage of the surface (marked g in the plan) inclines 14°. The frontal dip inwards of the structure planes is here 30°. Opposite the chapel, at the point marked h, where the glacier is steep, but still

Drawn from Nature by Professor Forbes. J. Haghe lith.

GLACIER OF LA BRENVA, SHEWING THE STRUCTURE OF THE ICE.

is at a considerable height above its lower termination, the frontal dip inwards is 19°; and immediately above the vault of the torrent, the inward dip of the structure is only 5°. The four sections on the lithographed plan fully illustrate, it is hoped, the geometrical structure of this glacier, together with the superimposed one. It will be found to agree most accurately with what I have described as the normal type of glacier structure, and especially with the description given of the Glacier of the Rhone, which that of La Brenva very much resembles in some respects, in my earliest paper on the subject, reprinted in the Appendix.

The alternation of bluish-green and greenish-white bands, which compose this structure, gives to this glacier a most beautiful appearance, as seen from the mule road. An attempt has been made in Plate V to give some idea of this most characteristic display, and which is better seen here than in any other glacier whatever with which I am acquainted. The sketch was taken by myself from the point marked *k* in the map, in July 1842.

When the ice of the glacier abuts against the foot of Mont Chetif, at the promontory marked *i* on the map, it is violently forced forward, as if it would make its way up the face of the hill. Here the contact of the ice and soil is very well seen; and my friend M. le Chanoine Carrel of Aoste, with whom I walked several times in this neighbourhood, and who took an interest in such questions, discovered a point of contact between the limestone and a protuberant mass of ice which admitted of easy removal, thus showing the immediate action of the ice and rock. Having taken a man furnished with a strong axe, we proceeded together to the spot. The soil near the ice appeared to have been but recently exposed by the summer's melting of the ice. It was chiefly composed of clayey debris from the blue limestone. At the point marked by M. Carrel a piece of fixed rock opposed the ice, and was still partly covered by a protuberance of the glacier, which we speedily but gently cut away with the hatchet. The ice removed, a layer of fine mud covered the rock, not composed, however, alone of the clayey limestone mud, but of sharp sand, derived from the granitic moraines of the glacier, and brought down with it from the opposite

side of the valley.   Upon examining the face of the ice removed from
contact with the rock, we found it *set* all over with sharp angular
fragments, from the size of grains of sand to that of a cherry, or
larger, of the same species of rock, and which were so firmly fixed in
the ice as to demonstrate the impossibility of such a surface being
forcibly urged forward without sawing and tearing any comparatively
soft body which might be below it.   Accordingly, it was not difficult
to discover in the limestone the very grooves and scratches which were
in the act of being made at the time by the pressure of the ice and its contained fragments of stone.   By washing the surface of the limestone we
found it delicately smoothed, and at the same time furrowed in the direction in which the glacier was moving, that is, against the slope of the hill.
We succeeded in detaching some fragments of the rock with hammers,
having even the sharp sand adhering to it, which I afterwards secured
with gum-water, in order to illustrate the exact condition of a rock
subjected to glacier action.   It would be impossible to catch nature
more completely in the fact than in the observation just stated.   I
afterwards returned with a skilful mason, who, with much labour, succeeded in detaching several specimens of the striated and polished surface.*   Not only was the limestone friable, but the cleavage being perpendicular to the surface, rendered it impossible to obtain a slab of any
extent.

On the path leading to Courmayeur—a few minutes walk below where
the glacier now ends—are some admirable specimens of ancient polished
and striated surfaces of the same limestone, which it seems impossible
to doubt were produced by the ice at a former period.

So far as we can judge from the view which De Saussure has given†
of the Glacier of La Brenva, and which he states was drawn in 1767,
we must infer that the glacier was then greatly less extensive than at
present.   It seems almost certain that at that time the Doire did not
pass *under* the glacier at all, but in front of it.   He likewise mentions

---

* One of these specimens is deposited in the museum of the Royal Society of Edinburgh.
† *Voyages,* Tom. ii., Plate III.

the chapel to which I have referred as exactly opposite to the glacier, and which is indicated in the map under the name of Notre Dame de la Guérison. It is also called Chapelle de Berrier. De Saussure speaks of it as in ruins in his time, having been allowed to go to decay on account of the superstitions to which it gave rise.* It appears, however, to have been rebuilt, and was again reduced to ruins, under much more remarkable circumstances. Its position relatively to the glacier at the present time will best be judged of from Plate IV., which gives a view of it as seen from near Entrèves, looking up the course of the Doire from below. The position of the modern chapel will be observed on a rock at a great height above the glacier, on the left hand, near an aged larch tree. The height of this rock is about 300 feet. Looking at that view, it will scarcely be believed that the glacier attained, only twenty-four years ago, so enormous a size as to have risen up to the level of that rock, (which is of limestone,) and to have worked with such tremendous force upon the promontory on which the old chapel stood, built upon the rock itself, not fifty yards from the present one, as actually to have heaved both rock and building to such a degree as to fill both with fissures, and to cause the latter to be removed by authority, as in a dangerous state.

The notoriety and recent occurrence of these facts, makes it now easy to establish them beyond a doubt; and I have thought it well to do so on account of their great interest. That a series of comparatively cold seasons should have produced so enormous an increase in the unmelted portion of a glacier, is a fact of the highest importance to any speculations as to the circumstances under which glaciers might be enormously more extended than at present. So far was there from being any *marked* change of climate at the period when this and many other glaciers were undergoing an enormous enlargement, that, for the five years preceding 1818, when the glacier of la Brenva attained its greatest size, the mean temperature at Geneva was 7°. 61 Réaumur, whilst the mean of the last forty years has been

---

* *Voyages*, § 855.

7°. 75,*—a difference of not *one-third* of a degree of Fahrenheit. This difference is so insignificant, that it is most likely that the increase of the glaciers at that time, depended rather upon an increased fall of snow, than upon any change of temperature.

The height of the ice was such in 1818, that the glacier rose up against the opposing wall of rock, until it covered the path, as Captain Hall attests ;† and I was assured by eye-witnesses, that the hermitage connected with the chapel was supplied with water from a conduit, which descended from the ice of the glacier, which then had a higher level.

I obtained from the Syndic of Courmayeur a certificate, in the following terms, of the fact being entered in the archives of the commune :—
" Je sous-signé, Syndic de Courmayeur, déclare après la vérification sur les régistres des archives du présent lieu, que la chapelle de Berrier à coté du glacier de la Brenva à été écroulée en 1818, dans l' endroit où elle étoit batie anciennement, par l'accroissement du dit glacier, qui étoit monté au niveau de la dite chapelle ; que la Nôtre Dame a été transportée dans l'Eglise de cette Commune où elle y resta pendant deux ou trois ans environ, et que la dite Chapelle fut rebatie dans l'endroit où elle est maintenant en 1821–22."

I have examined various other documents put into my hands by the Curé of Courmayeur, including the builder's report upon the state of the chapel, which leave not the slightest doubt of the extent and cause of the damage.‡   Indeed, the force by which the strata of limestone, forming the promontory, have been dislocated and rent asunder, is abundantly evident by inspection.

Tradition relates that the glacier in former times did not occupy the bottom of the valley, which was then covered with meadows and fields. My guide imparted to me the following story, which I give as I received it :—

---

* Dove, Temperaturvertheilung auf der Erde, p. 26.

† Patchwork, vol. i. p. 108.

‡ He says, " Je l'ai trouvé écroulée par la force du glacier, d,ou il resulte de toute nécessité de la rétablir, puisqu'il n'existe que les ruines."

On St. Margaret's day, the 15th July, no one knows in what year, the inhabitants of the village of St. Jean de Pertus, which was then overhung by the Glacier of La Brenva, instead of keeping the *fête*, pursued their worldly occupations:—the hay is dry, they said ; the weather is fine; let us secure it. But the sacrilege was soon punished. Next day the glacier descended in a moment, and swallowed up the village with its inhabitants. My guide added, in proof of the existence of this buried hamlet, that a person now living at Courmayeur, having gone, when a child of seven years old, with many others, for devotion, to the Chapel of Berrier, overlooking the glacier, heard the chaunting of vespers from under the ice, and saw a procession come out and return; but the vision was only seen by the child, for when he called the attention of the others to it, they beheld and heard nothing.

CHAPTER XI.

ENVIRONS OF COURMAYEUR—GEOLOGY.

MINERAL SPRINGS OF COURMAYEUR AND ST. DIDIER—REMARKABLE RELATIONS
OF LIMESTONE AND GRANITE IN THE VAL FERRET—MONTAGNE DE LA SAXE
—CROIX DE LA BERNADA AND MONT CHETIF—SYMMETRY OF THE GEOLOGY
ON EITHER SIDE OF THE ALPS—ASCENT OF THE CRAMONT—OBSERVATIONS
ON SOLAR RADIATON.

COURMAYEUR, by twenty-four corresponding barometrical observations
which I have made, is 876.5 metres, or 2776 English feet above
Geneva, and therefore 4211 above the sea. It is the highest considerable village in the great valley of Aosta, which takes its origin in the
Allée Blanche and Val Ferret, at the southern foot of Mont Blanc,
and merges into the valley of the Po at Ivrea. It is frequented by
the Piedmontese in considerable numbers every summer, both on
account of the mineral springs in its neighbourhood, and for the sake
of the exquisite freshness of its climate. A more complete contrast, than
between the walks of Courmayeur and the streets of Turin, in the
month of July, it is hardly possible to conceive.

All who have visited this place, under favourable circumstances,
agree in considering its position one of the finest in the Alps. No less
than six routes diverge from it,—the road to Aosta; that of the Little
St. Bernard; the Allée Blanche; the Col du Géant; the Col Ferret;
and the Col de Serène, leading to the Great St. Bernard. I have
travelled over all of these but the last, and several of them more than
once. Consequently my visits to Courmayeur have been frequent; but it
was only in 1842 that I made any stay there. I devoted a fortnight to

explore its most interesting neighbourhood. At present, I shall only describe a few of the most prominent points chiefly connected with its geology.

The occurrence of mineral waters first strikes us. This is a phenomenon peculiarly interesting in a geological point of view, for it very generally happens that the appearance of mineral springs, especially if warm, indicates a great disturbance of the strata, and very generally the appearance of what are called *intrusive* rocks, such as granite. I have shown, for example, that in the Pyrenées, a district unparalleled perhaps for the multitude of its thermal springs, that these occur almost invariably at or near the contact of granite with stratified rocks.* The springs near Courmayeur have been described by De Saussure, and I have little to add respecting them. The waters of La Victoire and La Marguerite rise from alluvium, and are saline and purgative. The waters of La Saxe rise in the defile by which the Doire issues from the base of Mont Blanc, exactly at the junction of the limestone strata with a remarkable mass of granite presently to be mentioned. They are sulphureous, and are used both for baths and internally ; but the bathing establishment is rather mean. All the above springs are cold. Four miles below Courmayeur, at St. Didier, is another bathing house formerly much more frequented, and which is supplied by a hot spring which issues in the deep and picturesque ravine immediately adjoining, through which a torrent descends from the Little St. Bernard. The spring is conveyed partly through a subterraneous gallery. In 1839 when I visited the source, I found the temperature to be 95°.0 Fahr., or 28° Réaumur : De Saussure found it be 27°.5 R.

The relations of the limestone and granite in the neighbourhood of Courmayeur are very interesting and remarkable, and offer so striking an analogy to the phenomena of the same kind seen on the northern side of the Alps, that we cannot but regard them as important with respect to the formation of this chain. The Topographical Sketch, and Section No. III. are intended to illustrate these peculiarities. I

---

* *Philosophical Transactions,* 1836.

had observed on my former visits to Courmayeur that there were appearances of limestone dipping under the granite of Mont Blanc, or rather of the Grande Jorasse on the north side of the Val Ferret. This I was enabled fully to establish, on my last visit, at several points. I obtained an excellent section by passing the moraine of the Glacier of La Brenva to the west of Entrèves, and ascending the ravine, marked on the sketch, between that village and the glacier. There is there a complete superposition of gneiss to lias shale forming a precise counterpart to that described in page 66, as occurring under the Aiguille du Bochard at Chamouni, and forming a portion of the fan-shaped stratification exhibited in the section, and which had been so far anticipated by De Saussure and M. Necker. In the ravine now mentioned the junction may be traced for a long way towards the centre of the chain, the line of contact between the limestone and the overlying Protogine or Gneiss, being inclined in the higher part of the section 38° to the horizon, (dipping north-west,) and in the lower part of the section 50°. The strata are therefore bent at the junction, but at a little distance they have a pretty uniform north-west dip of 38°.

There is no difficulty in reaching the junction. The limestone shale is altered and crystalline near the contact. The gneiss is altered also. These phenomena bear the most striking analogy with those which I have seen in the Alps of Dauphiné, and which have been so well described by M. Elie de Beaumont. The junction may be traced nearly as far as the glacier of La Brenva, but not (I think) farther west. The Mont Frety, which lies immediately to the east of the ravine in question, is also of limestone, which dips under the granite of the Col du Géant, and a close examination would, I have no doubt, give proofs of the same thing all along the north side of the Val Ferret as far as the Col of that name, where the limestone becomes nearly vertical.

This analogy in the arrangement of the rocks on either side of the great chain is not the only one, for on either side of Mont Blanc is a secondary range also composed partly of granite. The Aiguilles Rouges (which however are not included in the section) are granitic, although separated from the main chain by the limestones of the valley of Chamouni, and the Mont Chetif, and part of the opposing Montagne de la

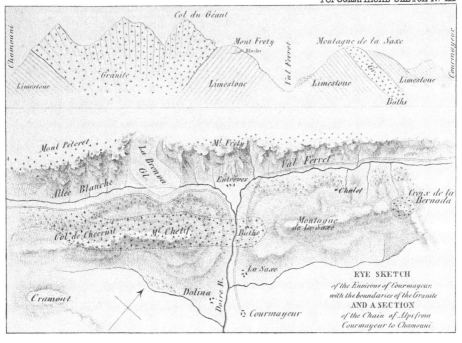

Topographical Sketch Nº III.

Chamouni — Col du Géant — Mont Frety — *Blachs* — Val Ferret — Montagne de la Saxe — Courmayeur

Granite — Limestone — Limestone — Limestone — Limestone — Baths

Mont Peteret — Mᵗ Frety — La Brenva Gᵗ — Entreves — Val Ferret

Allée Blanche — Chalet — Croix de la Bernada

Col de Checruit — Mᵗ Chetif — Baths — Montagne de la Saxe

Cramont — Dolina — Doire R. — La Saxe — Courmayeur

EYE SKETCH
of the Environs of Courmayeur,
with the boundaries of the Granite
AND A SECTION
of the Chain of Alps from
Courmayeur to Chamouni

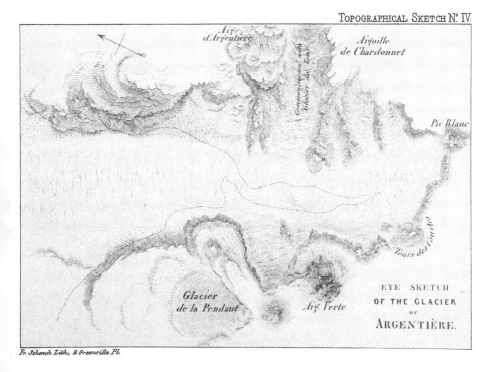

Aig. d'Argentière — Aiguille de Chardonnet — Communicates with Vallorie du Tour — Pic Blanc

Tours des Courtes

Glacier de la Pendant — Aig. Verte

EYE SKETCH
OF THE GLACIER
OF
ARGENTIÈRE.

Fr. Schenck Lith., 9. Greenside Pl.

Saxe near Courmayeur are in like manner granitic. The form of the latter mass, as shown in the section, is a great tabular body of imperfect granite, greenish, slaty, and containing an excess of quartz, with limestone above and below, very nearly in the manner in which the greenstone of Salisbury Crags, near Edinburgh, is interposed between the sandstones. Both the granite and limestone *rise towards Mont Blanc*, consequently, the limestones on the two sides of the Val Ferret rise towards the axis of that valley,—a very remarkable arrangement. The tabular mass of Mont Chetif is cut through by the Doire at the baths of La Saxe, where there is an excellent section: the granite is then lost under the Montagne de la Saxe to the eastward, which is chiefly composed of limestone which envelopes the granite, and is also covered with herbage. I had however remarked a summit parallel to the axis of Mont Blanc, on the eastern part of the ridge of this hill, which I suspected to be granite, and having made an excursion on purpose, I found my conjecture to be confirmed. This summit is called La Croix de la Bernada ; it may be easily reached either from the Val Ferret, or from the little valley of La Saxe. Farther east the granite is again lost under the limestone. The general dip of the limestone mountains farther from the main chain is towards the south-east.

In returning from the Croix de la Bernada by the Val Ferret, I observed a very remarkable accumulation of debris of granite which occupies the bottom of the valley to a great depth, and which has been evidently cut in two by the river, the deposit being of Alpine boulders resembling a moraine, which lie heaped upon the north side of the Montagne de la Saxe, as shown in the section already referred to. The existence of this moraine, if we may so call it, taken in connection with the deposit of similar blocks upon the face of the limestone outlier of the great chain called Mont Frety, and which will be more particularly mentioned in the next chapter, certainly appears to favour the conclusion that the glaciers, such as those of Entrèves and Mont Frety, which have now retreated towards the Alpine summits, once filled the entire space below, and transported these *debris*. They are deposited close to the sudden turn of the river between the Val Ferret and the baths of La Saxe.

I made another excursion towards the Mont Chetif to determine the relations of the granite in that quarter. I ascended the little valley above the village of Dolina, marked in the sketch behind the Mont Chetif, until I reached a col or passage which leads into the Allée Blanche, and which commands a magnificent view of the range of Mont Blanc. This is called the Col de Checruit. I had here an opportunity of examining the granite of the ridge on which I stood, and of seeing it disappear to the westward under the limestone, which it has greatly altered just at the Col. It is impossible to trace the connection of the granite of Mont Chetif with that of Mont Blanc, owing to the mass of debris and verdure with which the north slope is covered. I apprehend, however, that there is an undoubted connection between the granite of Mont Péteret and that of Mont Chetif, and that it crosses the valley in that place. The last exposed limestone is seen (as observed in the last chapter) on the south side of the valley just opposite to the Glacier of La Brenva.

From the Col de Checruit, I saw very distinctly the dip of the limestone of Mont Frety, under the granite of the Col du Géant, which I afterwards confirmed on the spot. The descent into the Allée Blanche, through some of the finest pine forests in the Alps, is a most interesting walk. Every one has noticed how rarely fine trees are to be seen in almost any part of the Alps. The forests on the north side of Mont Chetif are an exception, and whilst those in the valley of Courmayeur and La Thuille, are very generally in a dying state, from some cause which seems not to be understood,—these are flourishing. Several encampments of charcoal burners are met with during the descent; and the latter part of the walk may be performed along a conduit of water through the wood, from which, at intervals, the noblest views of the unequalled range of mountains and glaciers beyond, and in both directions, may be obtained. The path of the Allée Blanche being reached, I returned to Courmayeur by La Saxe.

De Saussure mentions the granite of La Saxe, though he does not advert to the peculiarity of its position, as respects the great chain. He notices, however, what he calls, " La superposition monstreuse des roches primitives sur les secondaires,"* at La Saxe. In the haste and exhaustion with which he descended from the Col du Géant,† he probably omitted to examine the rocks of Mont Frety. M. Sismonda, the able geologist of Turin, mentions‡ the superposition of granite to limestone at Pra Secco, beneath the Grande Jorasse, where I noticed it in 1841. But the remarkable symmetry of the chain on both sides, has not, so far as I am aware, been hitherto remarked.

The ascent of the Cramont is one of the best known excursions near Courmayeur. The great object is to command the complete view of the southern precipices of Mont Blanc and the adjoining chain. Its elevation is considerable, being, according to my observations, 4932 feet above Courmayeur, and by contemporaneous observations at Aosta, I find it to be 9081 English feet (2768 metres) above the sea. The route usually followed is, to descend the valley of the Doire as far as Pré St. Didier, and to ascend the Cramont by its southern slope, although that mountain lies nearly due west of Courmayeur. It is, however, extremely precipitous on all sides, except the south. On the present occasion, I walked down to St. Didier in the evening, in company with M. Carrel, whom I have already mentioned, and having gone to bed for a few hours, we started by starlight, in a beautiful morning, at half-past three A. M., so as to gain the summit early. The first stage of the journey is on the mule-path of the Little St. Bernard, which rapidly ascends the ravine whence the hot spring issues, as already mentioned. On this road is one of the grandest bursts of scenery in the Alps,—that, namely, which is enjoyed in descending from La Thuille, at the instant that the Aiguille du Géant, the Grande Jorasse, and the whole of the eastern chain of Mont Blanc come first into view. The road is soon after left ; and a long but easy path, through meadows, brings the traveller insensibly above the level of the adjoining hills. At length, the highest irrigation is passed, and a full hour's

----

* *Voyages*, § 881.　　　　　　　† *Voyages*, § 2034.
‡ Memoria sui Terreni stratificati delle Alpi, di Angelo Sismonda, p. 12.

ascent remains, over the short turf, by which the top of the Cramont may easily be reached in four hours from St. Didier. I was so fully imbued with De Saussure's enthusiastic picture of the grandeur of the station, that I was a little disappointed to find it, not only equalled in height by some others in the neighbourhood, but overtopped by one, also of limestone, which stands between the Cramont and the Allée Blanche, effectually preventing the eye from diving into its depths, and thus measuring Mont Blanc at once from top to bottom, as is the case in the view from the Breven, above the valley of Chamouni. This interfering summit, which I cannot help thinking has been mistaken by some topographers for the Cramont described by De Saussure, lies nearly west from the Cramont, and at the head of the valley whose streamlet passes Dolina. See the Topographical Sketch, No. III. It is, in fact, the prolongation of the Mont Chetif and Col de Checruit, and separates that valley from the Allée Blanche. The ascent is obviously easy and direct, much more so than that of the Cramont; the height is greater: it is nearer Mont Blanc, and commands completely the Allée Blanche and its glaciers. On all these accounts, I do not doubt that this hill is worth ascending, although it appears to be unknown to tourists, and even to natives, for I could not learn its name.

The Cramont is part of the limestone group, whose strata dip southwards, and the northern face being composed of the broken edges, is extremely abrupt. A ragged cliff extends for a long way, without any great variation of height.

M. Carrel, myself, and my guide Antoine Proment, had carried to the summit a considerable collection of meteorological instruments; for my intention was to spend the entire day upon the top, in order to observe the force of solar radiation. It is a familiar fact to mountaineers, that the sun's rays have an intensity and energy at great heights, which they entirely want on the plains. At first, this might be supposed imaginary, or to result from the reflection of the heat by the snow. On a station like the Cramont, where there is no permanent snow, this error is avoided; and no one who has compared the effect of a single day's exposure amongst the Alps, in discolouring the hands and face, with that of the hottest weather at Paris or Marseilles, will be disposed to question the former assertion. The difference admits of being shown instrumentally, by means of the valuable apparatus, called an actinometer, invented by Sir John Herschel, and I was provided with two of these instruments on the present occasion. My object was, in completion of some experiments made in former years, in other parts of the Alps, to ascertain the varying solar force at different hours of the day, at a height and at a season of the year in which the sun's rays travel through the atmosphere with least resistance.* I had, accordingly, brought these instruments on purpose from England, and I sought this hill in the month of July, soon after the solstice, for no other purpose. But such experiments are attended with numberless chances of disappointment. The day, though fine and bright, was by no means so cloudless as to warrant any conclusions from the experiments, which I continued every hour from 8 A. M. to 5 P. M., the whole of which time I spent upon the summit of the mountain. I had, therefore, abundance of time to survey the magnificent panorama by which I was surrounded; and having brought up a very good telescope by Tulley, of $2\frac{1}{4}$ inches aperture, with a tripod stand, I could inspect minutely the forms and details, both of the nearer and more distant objects,—Mont Blanc, with its glaciers; the pass of the Col du Géant, exactly opposite to me, on which, with the glass, I could discover almost every step, and every difficulty of the road; and to the eastwards, the summits of Mont Cervin and Monte Rosa especially engaged my attention.

As it was now late, I proposed to Proment (M. Carrel had left us early) to descend to Courmayeur by the rocks. He had not before done it; but we found little difficulty in discovering a most direct and not dangerous passage of the cliff, which is here at least 1500 feet high. Observing the limit of the larch in the valley of Courmayeur to be remarkably well defined, I took the level of it, which I found to be 7200 feet above the sea. From this point, the walk to Courmayeur was easy and pleasant, and remarkably direct.

---

* See a paper on this subject in the Philosophical Transactions for 1842, being the Bakerian Lecture for that year. See also Appendix, No. III.

CHAPTER XII.

THE PASSAGE OF THE COL DU GEANT.

PASSES OF THE CHAIN OF MONT BLANC—HISTORY OF THIS PASS—PRELIMINARY
OBSTACLES—DEPARTURE FROM COURMAYEUR—ASCENT OF MONT FRETY—
EXPERIMENT ON THE COMPARATIVE INTENSITY OF MOONLIGHT, TWILIGHT,
AND THAT OF A TOTAL ECLIPSE—GRANITE AND GRANITE BLOCKS OF MONT
FRETY—ARRIVAL ON THE COL—THE VIEW—HISTORY OF DE SAUSSURE'S
SOJOURN—AND OF HIS OBSERVATIONS—THE DESCENT—DIFFICULTIES OF
THE GLACIER—FOLLOW THE TRACK OF A CHAMOIS—REACH THE MER DE
GLACE—MONTANVERT.

> And followed where the flying chamois leaps
> Across the dark blue rifts, the unfathom'd glacier deeps.
> HEMANS.

THE chain of which Mont Blanc forms the culminating point has
a very peculiar structure, and is connected in a remarkable manner
with the great chain of Alps. One would hardly guess from the common maps, that Mont Blanc, and its adjacent tributaries, form a kind
of oval group rather than a portion of a line of mountain continuous
from the Mediterranean to the Tyrol, such as the Alps are usually
represented. In length this group extends from the Col du Bonhomme,
on the confines of the Tarentaise, to the Mont Catogne, in the valley
of St. Branchier, above Martigny, a distance of thirty English miles
in a north-east and south-west direction, whilst its breadth at right
angles to the former, from Chamouni to Courmayeur, is only thirteen
English miles. Now to perform these thirteen miles, a tedious journey
of two days (one of them of nearly 12 hours' walking,) is necessary,
because this chain or group, being, generally speaking, impassable,
must be gone round.

To avoid so great a circuit, the Col du Géant offers the shortest passage from the one valley to the other. It forms the crest of the chain, where the western branch of the Mer de Glace takes its rise ; and, notwithstanding its immense height, it would probably be frequented but for the dangers of the glacier on its northern side. A tradition, common to this, and many other passes of the Alps, states, that formerly the glacier was less formidable, and that communication was not unfrequent between Chamouni and Courmayeur.* This has not occurred, however, within some centuries from the present time. The passage of the Col du Géant appears to have been reckoned impracticable as late as 1781. M. Bourrit, writing in that year, and speaking of the aspect of that branch of the Mer de Glace of Chamouni called the Glacier du Tacul, says, with respect to the crevasses :—" Elles sont si effroyables qu'elles font désespérer de retrouver jamais la route qui conduisait à la Val d'Aoste."† De Saussure, in the second volume of his travels, speaking of the Glacier du Tacul, does not say one word of this historical passage of the Alps, though he seems to have thought it just possible that the summit of Mont Blanc might be gained in this direction ;‡ and, in the fourth volume, written some years later, when about to give an account of his memorable residence on the Col du Géant, he speaks of " la route nouvellement découverte,"‖ from Chamouni to Courmayeur. This was in 1788.

There is said to be a passage which has been effected from the Glacier de Miage, which penetrates very deeply indeed on the south side of the chain of Mont Blanc, to the valley of Contamines, by the glacier also bearing the name of Miage, on the north side ; but I have no accurate information of its accomplishment, and the appearance of the head of the glacier on the south side gives little encouragement to the attempt.

One other passage of the chain has, however, been made, and that is by the Glacier of Le Tour, near the Col de Balme, descending by the Glacier of Salena into the Val Ferret. This was discovered a few

---

* BOURRIT, *Voyages*, I., 72.   † Ib., I., 106.
‡ S 629.   ‖ § 2025.

years since by a guide of Chamouni, named Meunier. It cannot be very long, and is probably not very dangerous.

Such are the only known passes of this wild country.

I was induced to undertake the passage of the Col du Géant, chiefly for two reasons; in the first place, from a desire which I had long entertained to visit a spot rendered memorable by De Saussure's extraordinary residence, and admirable observations; and, secondly, having occasion, on other grounds, to visit Courmayeur, and to return to Chamouni, I preferred any alternative to that of experiencing once more the tedium of either of the circuits, by the Cols de Bonhomme and La Seigne, on the one hand, or the Cols de Ferret and Balme on the other. I had already traversed the former three times on foot in different years; and, though I had passed the latter only once, I wished to avoid the repetition of so long and dull a route.

Accordingly, having reached Courmayeur, in the beginning of July 1842, by the Col de Bonhomme, in order to go to Turin to see the total eclipse of the sun, my resolution was taken to return by the Col du Géant.

The guides of Courmayeur were, with one exception, unacquainted with the passage. I therefore wrote to Chamouni about the middle of July, desiring my old guide, Jean Marie Couttet, who knew the passage well, to come by the Col de Bonhomme, on the 19th, to be ready to return by the Géant on the 20th. I had previously ascertained that my guide of Courmayeur, Antoine Proment, would consent to undertake the passage with a single competent guide of Chamouni, for I had seen so much of the uselessness and inconvenience of numerous guides on such expeditions, that I resolved to take two only. Another item of expense and trouble was saved at the suggestion of Proment. Hitherto the passage had, in every instance, been effected in two days. In starting from Chamouni, the Tacul was the place of the first night's bivouac; and, in the one or two passages which have been made from the side of Courmayeur, the travellers had slept, or at least *lain* on the exposed and almost precipitous face on the southern ascent, which offers no spot at all adapted for the most indifferent night's quarters. Proment suggested passing the Col without any halt, as the first part of the way, being without danger, might be performed in the dark. I determined, accordingly, to leave Courmayeur in the night, and to reach Col soon after sunrise, or at least before the morning was far advanced.

Couttet arrived a day before his time, and the day of his arrival was also the last of fine weather, which had continued almost without interruption for a month. The south wind began to blow, the dew point rose, fogs covered the range of the Cramont, and formed a belt along the chain of Mont Blanc, and it was but too evident that the weather was deranged for some days. The provisions were ready, the guides astir, and I was called at midnight of the 19th, to consult upon the state of the weather ; when it was unanimously agreed to be unfit for such an expedition. A repetition of the same occurrences took place for several successive days and nights. I was immoveably fixed in my purpose to return by no other route, and as resolute not to attempt the Pass but with the finest weather. Proment, who was at home, bore the tantalizing delay philosophically enough, but Couttet fretted himself into such a state of impatience, that I believed he would have left me, and returned to Chamouni. Sometimes he urged me to depart, whatever might be the weather ; but, when the hour of midnight came, and the council was called, his better sense warned him not to make so rash an attempt ; then he tried to induce me to give up the plan, and return by the Bonhomme ;—anything to avoid the ennui of Courmayeur. But I was inflexible. The 20th, 21st, and 22d July were spent thus. On the evening of the latter day, the weather gave a promise of mending, whilst the snow which had fallen on the Col, and even a great deal lower, gave the prospect of some inconvenience from the cold, and increased difficulty in passing the glacier. Couttet put these prominently before me, as the last temptation to abandon my project ; but, finding me resolute, he made up his mind for departure that night, good or bad.

I was called a little after midnight, between the 22d and 23d July, and to my inexpressible satisfaction, I beheld a magnificent calm night, illuminated by a moon just full. I had sent off by an opportunity some days before my heavier luggage, so that my pacquet was soon

made. I carried, as usual, my barometer, hammer, compass, and telescope; one guide took my little knapsack, and the other a similar one
containing provisions. I took some soup before departing; and we
were detained, and my temper a little ruffled, by the stale imposition
of a supplementary bill, containing items left out by inadvertence in
the regular account paid the night before, which was presented to me
at one o'clock in the morning, when remonstrance and appeal were alike
unavailing. Travellers who undertake expeditions beyond the common run of excursions, cannot be too much put upon their guard against
the systematic extortion of innkeepers, seconded by the love of indulgence of their guides. The better way would be to let the guides pay
for themselves in every case.

Being fairly on foot at 30 minutes past 1 A.M. of the 23d July, my
ill-humour was soon dissipated by the exquisite beauty of the scene
which the valley of Courmayeur presented. The full moon was riding
at its highest noon in a cloudless sky—the air calm and slightly fresh,
blowing very gently down the valley. The village and neighbourhood
lay, of course, in all the stillness of the dead of night; and as I headed
our little caravan, and walked musingly up the familiar road which led
to the Allée Blanche and the foot of Mont Blanc,—that vast wall of
mountain, crowned with its eternal glaciers, seemed to raise itself aloft,
and to close in the narrow and half shaded valley of Courmayeur, verdant with all the luxuriance of summer, and smelling freshly after the
lately fallen rain. Of all the views in the Alps, few, if any, can, to
my mind, be compared with the majesty of this, and seen at such a
moment, and with the pleasing excitement of thinking, that within a
few hours I hoped to be standing on the very icy battlements which
now rose so proudly and so inaccessibly, it may be believed that I had
never before regarded it with so much complacency.

Having left the baths of La Saxe on our right, we crossed the stream
descending from the Val Ferret, and skirting the village of Entrèves
under the guidance of Proment, who knew the bye-paths through the
fields, we gained, after about an hour of pleasant walking, the woods of
larch which clothe the south-eastern foot of the Mont Frety, as the
pasture-mountain is called, above which the Col du Géant stands.

The Mont Frety may be ascended either on its eastern or western side; both are steep and rugged, but not difficult. Some of the trees are of considerable size, and every now and then from between their trunks, I caught an admirable peep of the still scenery of the low country, bathed in moonlight, whilst, as we gradually but steadily ascended, our progress was measured by the successive hills or mountains which we left below our level: first, the Montagne de la Saxe—then the Pain de Sucre—finally, the Cramont itself sunk its head amongst more distant ranges of hills. Couttet had now taken the lead, and kept going steadily up hill at a very easy measured pace, but without the least intermission. In this way, admirable progress is made; the mind yields to the monotony of the exertion, and ceases to measure time, or to long for a remission of so moderate an effort. The footing being easy, no annoyance was felt from the want of full daylight, and the eye was left generally free to dwell on the objects around.

Two hours had passed from the time of starting before we emerged from the larch wood upon the bare slope of Mont Frety. Twilight was beginning to make evident progress in the serene sky above the Col Ferret. The moon was still high in the south-west, 20° or 25° above the horizon; and I was curious to notice the relative intensity of the moonlight and the dawn with reference to some experiments which I had made during the total eclipse of the sun a fortnight before. On that occasion, the light permitted me to distinguish small print with difficulty in the open air, and I think I could not have read writing. I compared it afterwards to the darkness in a clear evening one and a quarter to one and a half hours after sunset. The moonlight now was evidently incomparably brighter than the light of the eclipsed sun, and enabled me to read writing easily. As we ascended the slope with the increasing dawn on the right hand, and the setting moon on the left, I referred continually to a written paper in my hand, to mark the moment when it should appear equally legible by either. The difference of colour of the light caused some difficulty. It was the contrary of what we usually perceive: the moonlight seemed yellow and warm, the dawn was cold and grey. This was evidently no illusion, and arose from the quantity of blue rays reflected by the large surface of sky whence the twilight was derived. At 3 h. 30 m. A.M., I judged the two lights to be equal, and in a very few minutes the dawn had so manifestly gained upon the other, that it showed the method to be susceptible of some accuracy. Now, the summit of Mont Blanc was not touched by the sun until 4 h. 20 m. or 50 minutes later. This corresponds sufficiently well with my former estimate of the darkness of the total eclipse. It was very far less bright than the light of the full moon; as much less, in fact, as the dawn 80 or 90 minutes before sunrise (in the month of July) is than the dawn 50 minutes before sunrise, which is probably not much more than a fourth part.

This little experiment required no delay, and we kept always advancing. The Mont Frety projects considerably towards Courmayeur from the great chain, although, viewed from below, it seems an almost precipitous slope. There is a ravine on either hand, the highest portion of which contains a glacier—the Glacier du Mont Frety on the west, and the Glacier d'Entrèves on the right.* What may be called the summit of Mont Frety is a green pasturage, interspersed with enormous blocks. By frequent examination from below with a telescope, I had satisfied myself that the upper part was of granite, overlying strata of limestone, which dipped inwards at a considerable angle, and also that the blocks on the summit were granitic masses removed from some distance; both of these conjectures were confirmed by examination. The dimness of twilight permitted me only to ascertain generally the fact of the superposition of the granite to the limestone. As I approached the level of the scattered blocks of granite, I was struck by the peculiarity of their position. These enormous masses lie on an isolated ridge of very little extent, and on a steep declivity. There are ravines on either hand; precipices above, and the valley nearly 3000 feet below. The level at which they occur is very remarkably preserved; and without by any means vouching for the explanation, they seem to me not to have alighted on this promontory in the course of rolling down from the cliffs above, which is scarcely probable, but rather to have been deposited by the glaciers descending on either hand. If those glaciers formerly reached the valley beneath—which is not unlikely—they probably occasioned the remarkable deposit of

---

* These are the names given by De Saussure, § 2035.

boulders exactly opposite to Mont Frety, on the farthest or south side of the torrent of Val Ferret, described in the last chapter. The section in the Topographical Section, No. III., will give an idea of the combination of these remarkable phenomena, which contribute to render the environs of Courmayeur very interesting to the geologist. I have only to add, that the granite of the boulders on Mont Frety does not resemble the rock on which they lie, being more crystalline, and evidently derived from the neighbourhood of the Col du Géant. The blocks in the valley have the same character.

Having passed the sort of top or prominent flat of Mont Frety, and having now arrived at the foot of the final ascent after three hours of continuous walking without any pause, we halted by a spring to break our fast at 30 minutes past 4.

The sun was just about to rise, and this was the coldest period of the morning ; at the height which we had now reached the frost was pretty intense, and the herbage white and crisp. I breakfasted heartily on hard eggs and cold tea, of which I had brought a good store in a gourd. After a halt of about 20 minutes, we proceeded, the cold continuing sharp—the thermometer was 30°.

The ascent now began in earnest, and, before long, we had left all grassy slopes behind, and clambered upon the bare rock. This was at first precipitous, though not dangerous. I had so completely studied the route with the telescope from the Cramont,* that I should have had no difficulty in selecting, had it been necessary, the easiest path. There was but one point where it was necessary to touch the snow, and that but for a few steps. Keeping always along the ridge, we climbed patiently amongst the loose masses of rock, which it required some care not to overthrow upon one another. We were yet nearly 1000 feet below the top, where Couttet felt his breathing a little affected, though not distressingly so. This is a symptom very common, and depending much upon the state of health at the time. I scarcely felt it even at the top; but in 1841 I was distinctly incom-

---

* The vignette on the next page gives an imperfect representation of the ascent of the Col du Géant as seen from the Cramont. It is, however, somewhat deficient both in clearness and accuracy.

moded at a lower level on the ascent of the Jungfrau. The guides
say that it depends upon the state of the air ; and David Couttet has
assured me, that on some days, he and his brother have *simultaneously*
felt inconvenience from the action of the lungs at very moderate eleva-

tions. Continuing steadily to
mount, and invigorated rather
than incommoded by the sun's
rays, which now began to beat
upon us, we reached the summit with scarcely any halt at
20 minutes past 7 A.M., or in
5 hours 50 minutes from Courmayeur. The vertical elevation
is 7000 English feet, and it
never before occurred to me to
make a long ascent so nearly
in one right line. The point
at which we arrived, (marked *a*
in the sketch) is the very lowest
point of the chain, and is precisely at De Saussure's station.

The disagreeable feeling of cold had now entirely subsided. The
sun's rays had taken off the frosty chill, though, in consequence of
our increased height, the thermometer was only 29°, we established
ourselves, nevertheless, not uncomfortably, in a hollow of the rock
facing the south, where we could rest after this, the most toilsome,
though not the most difficult part of the day's work, and survey the
astonishing prospect which was spread out before us.

We were at a height of 11,140 feet above the sea. It is very rare
to be at this elevation at so early an hour as seven in the morning, and
still rarer to combine this essential for a distant prospect with such
magnificent weather as the day in question afforded. The atmosphere
was, perhaps, as the event proved, too clear for very permanently
fine weather,—not a cloud—not even a vapour was visible. The air
of this lofty region was in the most tranquil state. Range over
range of the Alps, to the east, south, and west, rose before us, with a

perfect definition up to the extreme limit which the actual horizon permitted us to see. Never in my life have I seen a distant mountain view in the perfection that I did this, and yet I have often been upon the alert to gain the summits before the hazy veil of day had spread itself.

Perhaps it enhanced my admiration of the scene, that a great part of the labyrinth of mountains were familiar in their forms to my eye, and that from having penetrated many of their recesses in different journeys, this wide glance filled my mind with a pleasing confusion of the images of grandeur and beauty which had been laboriously gathered during many pedestrian tours, whose course and bounds I now overlooked at a glance. To the eastward, the Mont Cervin, with its obelisk form, never to be mistaken, presented evidently the same outline as I had sketched last year, from a point diametrically opposed, near Zermatt;—close to it, on the left, rose another peak, which I conjectured and afterwards ascertained to be the Dent d'Erin.* A little to the right, most exquisitely defined in outline, yet with every detail delicately subdued by the undefinable blue of immense distance, was the whole mass of Monte Rosa, the rival of Mont Blanc, with its many heads of nearly equal height, whose geography I looked forward to exploring in the course of the summer. The hirsute and jagged rocks of the Valpelline and its neighbourhood formed the base out of which

the chain of Monte Rosa seemed to rise; and a little more to the right lay the indentation of the Val d'Aoste, well marked by the complete separation which it forms between the mountains just men-

Monte Rosa from the Col du Géant.

tioned and those which formed the middle group of the picture, the savage chain of Cogne to the south of Aoste. These mountains (which I had partly traversed in 1839) contain many summits of 11,000 and 12,000

---

* I cannot positively assert that the Mont Cervin is visible from the very Col. I rather think not, but I saw it as described from a little lower level. I verified my recognition of the mountains, on the spot, by the excellent reduced map of the Sardinian Government triangulation, connecting France with Italy.

feet high, scarcely known even by name, such as the Becca di
None, 11,738 English feet above the sea, which has been repeatedly
ascended by M. Carrel of Aoste, who even passed the night of the 7th
July there, in order to witness the solar eclipse:—the Montagne de
Cogne, the Grand Paradis, and the Aiguille de la Sassière, all stream-
ing with glaciers. These were flanked on the left by the stern grey
mountains of Champorcher, and on the right by the snowy wastes of
the Ruitor. Behind the last rose the vast mass of Mont Iseran, which
completely conceals the Alpine chain beyond, and of course the Monte
Viso, which I had hoped to have recognised. Hitherward from the
Ruitor the pass of the little St. Bernard carries the eye to the valley
of the Isère, whose whole course I had also followed up to its pa-
rent glaciers, in the year 1839. Then a fresh range of snowy mountains to the right, above which rises conspicuous the Aiguille de la
Vanoise, (between Moutiers and Lanslebourg,) a mountain which, for
elegance, vies with any in the whole chain. To the west, and beyond,
stood forth in clear perspective the yet more distant range of Mont
Thabor, separating the valleys of the Arc and the Durance; and Savoy
from France. There, a very well defined, though very distant group
of familiar forms reached my eye.

Profile of Mont Pelvoux.

It was the Mont Pelvoux in
Dauphiné, rising proudly from its
rugged basis of lofty hills, the
highest mountain between Mont
Blanc and the Mediterranean, and
of which I had laboriously made the circuit in 1841, in company with
Mr. Heath, by passing Cols themselves above 10,000 feet in height.
The adjacent mass of Les Grandes Rousses, sloping towards Grenoble,
closed this admirable panorama, which was thus cut short exactly
where it would have become uninteresting, by the colossal mass of Mont
Blanc, which, with its huge sentinel, the Mont Pétéret, (that vast
rocky Aiguille which guards it on the side of the Allée Blanche,)
stood forth in the closest proximity, and still at a height of 4600 feet
above us.

I will not stop to describe the appearance of the valleys immediately
beneath us, and of which the eye seized at once the ground plan from

the great height at which we stood. It is very rare, as I have observed, to find so long and uniform a slope, affording a clear view to the very bottom, near 8000 feet deep. The Allée Blanche, with its glaciers, its lake, and its torrents, all *in plano*, the peaks of the Mont Chetif, and even the Cramont, now completely subdued, the monotonous length of the Val Ferret, the hamlets of Courmayeur and La Saxe almost at our feet, and the meadows of St. Didier, green as an emerald, and set in a solid chasing of precipices, begirt with pines,—all these familiar objects scarcely withdrew my attention from the magnificence of the wide Alpine view beyond.

The barometer (one of Bunten's) had been set up on our arrival, and whilst admiring the scenery, a second and more substantial breakfast of cold fowl was proceeding with marked advantage to the prospects of the journey,—for our appetites were excellent. I scarcely tasted the wine, and not at all of the brandy which Couttet had plentifully provided and liberally partook of. We had yet many hours' walk in the heat of the day, over dry snow, where no drop of water is ever seen.

The barometer had been exposed for forty minutes in the shade, and was now carefully observed. It stands .08 millimetre lower than the corrected barometer at Geneva Observatory.

Col du Géant, 1842, 23d July, 8 h. 0 m. A.M.,	Barom. m.m.	Att. Ther. Cent.	Det. Ther. Fahr.
	507.9	+0.6	29.8°

The following had been the readings at Courmayeur (hotel de l'Ange, second floor) the previous day, during the whole of which the barometer had been steadily rising :—

Courmayeur, 22d July,	4 A.M.	657.5	18.0	
	10 ,,	659.4	18.3	61
	12 ,,	659.8	18.1	62
	4 P.M.	660.25	18.0	65
	8½ ,,	660.85	15.7	55
	12½ ,,	661.35	17.5	50

The corresponding height of the barometer at Geneva was,

729.85 m.m. at 0° cent. D. T. 17° 2 cent.

whence the height of the Col du Géant above Geneva is 9803 feet,* above the sea 11,146 feet. Above Courmayeur; by the previous observations, 6979 feet. The Col du Géant, by observations at the Montanvert, on arriving there, is 4841 feet above that station. This result we shall afterwards find to agree with the direct comparison with Geneva, and hence we are disposed to place the Col du Géant at 11,146 feet above the level of the sea. De Saussure determined it, trigonometrically, by reference to Chamouni, using the Aiguille du Midi as an intermediate point seen from both, and taking the barometrical height of Chamouni, he obtained for the Col du Géant 1747 toises, or 11,172 English feet. By his seventeen days' barometrical observations, compared with simultaneous ones at Chamouni, he obtained by the formula of Trembley, 16 toises less, reducing the height to 11,070 English feet. I have recalculated his simultaneous observations at the Col du Géant and Geneva, and have obtained so low a result as 11,028 feet.

The rock under which we breakfasted had supported the " Cabane" of De Saussure. I pleased myself with contemplating a board which yet remained of the materials of his habitation, and a very considerable quantity of straw, which lay under the stones which had formed its walls. The frosts of this elevation had preserved the straw in a pretty fresh state for half a century. There was also an empty bottle entire. This, indeed, had no claim to be so old, but it might be a relic of another illustrious guest,—M. Elie de Beaumont, the last traveller but one, who, seven years ago, had passed this wild spot.

De Saussure's habitation, as figured very intelligibly in the fourth volume of his work, consisted of a wretched stone hovel, six feet square, and two tents. Here this remarkable man passed sixteen days and nights, keeping, together with his son, M. Théodore de Saussure, (the only surviving sharer of the expedition,) almost perpetual watch upon the instruments which he had undertaken to observe. No system of connected physical observations, at a great height in the atmosphere,

---

* Calculated both by Baily's Tables and those of the French "*Annuaire.*"

has ever been undertaken which can compare with that of De Saussure. At any time such self-denial and perseverance would be admirable, but if we look to the small acquaintance which philosophers of sixty years ago had with the dangers of the higher Alps, and the consequently exaggerated colouring which was given to them, it must be pronounced heroic.

De Saussure and his son arrived at the Col du Géant on the 3d July 1788, accompanied by a number of guides and porters, who carried two tents, and the utensils required for a long residence, having slept by the Lake of the Tacul. On the 19th of the same month he descended on the side of Courmayeur, having remained seventeen days at this great elevation. It may be believed, that those guides who remained to share the wretched accommodations of this truly philosophical encampment, were not a little exhausted by the tedium of such prolonged hardships. De Saussure states, that he believes they secreted the provisions appropriated to the day of their descent, in order to render impossible a prolongation of their exile from the world. The astonishment of the country people on the side of Piedmont, whence the position of De Saussure's cabin is distinctly visible, it may be believed, was great; and it naturally showed itself in the form of superstition. It is still well remembered at Courmayeur, that that month of July, having been exceedingly dry, the report arose, that the sorcerers who had established themselves on the mountain had stopped the avenues of rain, and that it was gravely proposed to send a deputation, to dislodge them by force,—a task, probably, of some difficulty, for a few men could defend the Col du Géant against an army.

If we look to what was accomplished by these indefatigable observers, we shall find, that it was fully commensurate to the efforts made to attain it. Scarcely a point in the "Physique du Globe," which was not illustrated by their experiments. Geology, meteorology, and magnetism, were amongst the most conspicuous. I will pause a moment to state some of their leading results, which, as respect meteorology, are of permanent and, even now, almost of unique interest in the science. It were, indeed, to be desired, that the original registers, which are understood to be in the possession of the family, were published entire.

After mentioning the few observations which could be made on the plants and animals of this wild spot, and the rocks of which the Col is composed, the Meteorological Observations are next discussed.* These were conducted every two hours, from 4 A.M. to midnight, by the alternate care of M. de Saussure and his son. We extract the following from the simple history of their days, each so like another, as to make the time seem to pass with extreme rapidity:—" Vers les 10 heures du soir le vent se calmoit ; c'étoit l' heure où je laissais mon fils se coucher dans la cabane ; j'allais alors dans la tente de la boussole me blottir dans ma fourrure, avec une pierre chaude sous mes pieds, prendre des notes de ce que j'avais fait dans la journée. Je sortais par intervalle pour observer mes instrumens et le ciel, qui presque toujours était alors de la plus grande pureté. Ces deux heures me paroissaient extrêmement douces: j'allois ensuite me coucher dans la cabane sur mon petit matelas étendu à terre à côté de celui de mon fils ; et j'y trouvois un meilleur sommeil que dans mon lit de la plaine."†

The mean height of the barometer during 85 observations was 227.355 French lines.‡ At Chamouni the corresponding mean height was 300.638 lines, and at Geneva 323.668 lines, the temperatures of the air being 3°.630, 17°.288, and 19°.934 Réaumur, respectively. The temperature of the mercury of the barometer is not given. De Saussure clearly established,—at a period, too, when the diurnal variations of the barometer were little attended to,—that these oscillations are *reversed in their direction at great heights*, the barometer standing highest at 2 o'clock in the day, and lowest in the morning and evening.

His thermometric observations are not less interesting or original. His deduction of the law of decrease of temperature in the atmosphere is, probably, the best that we yet possess, 1° Réaumur for 100 toises of ascent. He shows, that a decreasing arithmetical progression satis-

---

* *Voyages*, § 2049.   † § 2032.

‡ Ingenuity never contrived a more perverse system of notation than the subdivisions of the barometer in the time of De Saussure, who gives his results in inches, lines, (or twelfths,) 16ths of these lines, and 1000ths of these 16ths. I have reduced them to lines and decimals.

fies the observations better than the harmonic law proposed by Euler;
he points out the importance of his conclusions to the theory of astronomical refractions; he insists on the diminishing range of daily and
annual temperature as we ascend, and observes, that this causes a corresponding daily and annual change in the rate of decrement with
height; and he shows that he had a clear idea of *space* possessing a
definite temperature at a distance from any planetary body. He considers, with much neatness and simplicity, the variations in the progress and extremes of daily temperature in the month of July at the
three stations of the Col du Géant, Chamouni, and Geneva. The
mean daily ranges were

4°.257 Réaumur,

10.092,

11.035,

or in the proportion of 2 to 5 nearly at the first and last stations.
The progress of the diurnal warmth is most rapid at the higher station,
for whilst the lowest temperature of the night occurred at all the stations at 4 A.M., the mean temperature of the day was already attained
at 6 A.M. at the Col, at Chamouni at 8, at Geneva only at 9 A.M.
These experiments are amongst the most definite and exact which we
yet possess on these subjects.[*]

On solar radiation the experiments of De Saussure were not so conclusive as on most other subjects. He employed undefended thermometers, exposed in the sun and shade, and generally not even black-
ened. Hence the difference of these was always trifling, and depended
fully as much on the force of the wind, (as he himself acutely notices,)
as upon any other circumstance. The effect of radiation from the surface of the snow, reducing its temperature below that of the surrounding air, he seems to have particularly noticed; and though he quotes
Dr. Wilson's paper on the subject, (§ 2054,) it may be inferred, that
he was not familiar with that curious observation at the time of his
own experiment.

---

[*] See a paper on the Diminution of Temperature with Height in the Atmosphere, and
on the Diurnal Curves.— Trans. xiv. 489.

This remark, however, seems to have led him to make some most interesting observations on the temperature of the interior mass of snow. He notices, that the hard crust of congealed snow on the Col du Géant extended to the depth of only some inches, and that below that, down to 12 feet, the temperature was continually 0° Réaumur, or the freezing point. The following passage, in which De Saussure reasons respecting the progress of the winter's cold into masses of snow and ice, compared to that in common soils, is so important to the modern theories of glaciers, and is, I think, so just, that I will quote it entire :—" La croute gelée," says he, " qui recouvre les neiges, est sans doute plus épaisse en hiver qu' en été ; je ne crois cependant pas qu'elle ait plus de dix pieds d'épaisseur, et je suis persuadé, qu'au delà de cette profondeur les neiges demeurent tendres, et comme en été, au terme de la congélation. En effet si l'on adopte le principe que j'ai posé dans l'article précédente que la différence entre la température des plaines et celle des hautes montagnes n' est en hiver que les deux tiers de ce qu' elle est en été, on verra que, puisque la température moyenne du Col du Géant, n' est en été que de 15 degrés plus froide que celle de Genève, elle ne le sera que de 10 en hiver. Ainsi comme nos plus grands froids n' excèdent guères 15 degrés au dessous de zero, ceux du Col n' excéderoient guères 25, et ceux de la cime du Mont Blanc 30 ou 31 ; ce qui est un peu moins que les plus grands froids de St. Pétersbourg. Or, puisqu' à la baie de Hudson, dont le climat est beaucoup plus froid que celui de St. Pétersbourg, la terre ne gèle qu' à la profondeur de 16 pieds anglais, environ 15 pieds de France, on ne s' écartera pas beaucoup de la vérité en supposant que, sur les hautes cimes des Alpes, la neige ne gèle en hiver qu' à 10 pieds de profondeur ; surtout si l'on considère que la neige se laisse pénétrer par le froid plus difficilement que la terre."* These views will be found to be in accordance with those which have lately been brought forward to illustrate the Theory of Glaciers.

On the electricity of the atmosphere, De Saussure made many observations on the Col du Géant, of which it may be said, that the imperfections were those of every observation of the kind, and that even at the present day, it would be difficult to suggest very material improvements. He found the diurnal variations similar to those at the same season in the plains, showing that variation of temperature merely, is not the cause of the dissimilar phenomena presented at different seasons.

A very interesting chapter refers to experiments on evaporation, and the dryness of the air, which, though tinged by the erroneous views on Hygrometry then prevalent, present several results of value. The rate of evaporation was determined by the ingenious device of exposing a moistened cloth on a stretching frame, whose loss of weight, in a given time, was determined by means of a nice balance. He thus ascertained, by direct experiment, " that other things being the same, with respect to temperature and dryness, a diminution of about one-third in the density of the air, doubles the amount of evaporation."[*]

Besides these, we have observations of great interest upon clouds, the formation of hail, an elaborate series of experiments upon the blue colour of the sky, with the cyanometer invented by himself, on falling stars, on the colour of shadows, on the transparency of the air, on the scintillation of stars, and on the duration of twilight. He observed a sensible twilight when the sun was 45° below the horizon, instead of 18°, as is usually reckoned in the plains. Pictet concluded,[†] that this reflected light was derived from an elevation in the atmosphere of 121 leagues, where the air must be inconceivably rare, if indeed it exist at all. It seems so much more natural to suppose, as Arago has done, that the light of twilight has undergone several successive reflections, from comparatively dense air, that one wonders that so probable an opinion was not earlier held. De Saussure likewise made use of the influence of light in facilitating certain chemical operations, as a measure of the intensity of light at the Col du Géant, compared to the level of Geneva.

---

[*] § 2063. [†] De Saussure, *Voyages*, § 2090, *note*.

Besides all these varied subjects of inquiry, we find that De Saussure devoted particular attention to the phenomena of magnetism on the Col du Géant. Indeed, it was one of his chief objects, as was shown by the extreme pains which he bestowed on the arrangement and observation of his magnetic apparatus. Seven times was the pedestal of his variation instrument constructed before it presented sufficient stability to afford consistent results, and it is not easy to appreciate the zeal which, in such trying circumstances, returned so often to the fulfilment of its object. He found the diurnal variations to subsist at this height as at Geneva and Chamouni, and to have generally the same direction. Their magnitude did not appear to be considerably altered. He was also probably the first person who attempted to inquire, whether the terrestrial magnetic intensity is sensibly diminished at these great heights. The observations made at Chamouni and the Col du Géant, at nearly the same temperature, agree very closely, and do not seem to warrant the supposition towards which De Saussure seems to lean (though with his usual caution), that the diminution was very apparent.[*]

In reviewing thus hastily the results of the memorable journey of De Saussure, we cannot but be struck with the completeness of a plan of observation in terrestrial physics, to which it would be difficult, even at the present day, to make any considerable addition, except as to *methods*. Himself on the borders of fifty, and with the assistance only of his son, at the age of eighteen, he filled actively the part of geologist, naturalist, and *physician*, during seventeen days and nights, at a height which, but a few years before, was believed to be inaccessible in Europe,[†] and where it might well have been doubted whether human life could continue to be supported. Whilst the ascent of Mont Blanc has ever been considered De Saussure's most·popular claim to his deserved reputation, the annals of science will register the residence on the Col du Géant as the more striking, as well as more useful achievement.

---

[*] § 2103. See also a paper by the author, Edin. Transactions, vol. xiv. p. 22.

[†] " Environ 180 toises plus haut que la cime du Buet, qui passait il y a quelques anneés pour la sommité accessible la plus élevée des Alpes."—*Voyages*, § 2032.

I left the Col to descend its northern side towards Chamouni at 8 A.M. A few steps brought me to the edge of the glacier, which may be considered as the head of the Mer de Glace in this direction. The view, though very grand, wants the effect of distance which the southern panorama presented. The summit of Mont Blanc is perfectly distinct; but it appears close at hand, and its elevation, though still 4600 feet above the spectator, loses somewhat of its grandeur from its apparent proximity. The chain of *Aiguilles*, which separates this branch of the Mer de Glace (or Glacier du Géant or du Tacul) from the valley of Chamouni, completely bounds the view to the north, and yet does not rise to a great height above the eye. The row of their summits, exactly in the reversed order from that in which they are seen from Chamouni, is, however, abundantly striking, commencing with the Aiguille du Midi on the left, succeeded by the Aiguilles de Blaitière, Grepon, and Charmoz. The great tooth-like form of the Aiguille du Géant, belonging to the chain on which we stood, rose imposingly on the right, supported by a mass which completely cut off any view in the easterly direction. The comparatively small summits of the Aiguilles Marbrées, figured by Saussure, occupied the foreground in that direction. But perhaps the most striking part of the northern prospect was the dazzling mass of glacier upon whose surface we were now to walk for some hours, which occupied the basin to the depth of several thousand feet beneath us, intermixed with craggy pinnacles, which here and there connected themselves with the rocks on either hand, or stood out as islets amidst the breadth of unbroken white.

On rising from breakfast on the Col, we had taken the precaution to tie ourselves together with two strong new cords which Couttet had provided; and as he took the lead, I being in the centre, and Proment behind, about 10 feet apart, we had soon occasion to test their utility. The snow had fallen to a considerable depth during the late stormy days, and added considerably to the difficulty of detecting hidden chasms in the ice; almost the first step that Couttet took upon the glacier, he sunk up to his middle in a hole. By dint of reasonable precaution in sounding with a staff, even so trifling an accident was not repeated, and we passed safely over the beautiful snow beds, sloping at first gently towards the north. The map of the Mer de Glace gives a tolerably correct idea of the serrated ridges of granite peaks which break the monotony of the scene. The first which we passed on our left is called *La Tour Ronde.* This is connected with the main ridge of Alps, a little to the westward of the Cabin of De Saussure, where it terminates in a remarkably shaped hill, called Le Flambeau. It must be observed, however, that there are two rocks of this name, and which resemble one another extremely. The one marked on the Map 2^d Flambeau, is still farther west, and forms part of a transversal, and apparently inaccessible, ridge, which stretches quite across from the Glacier of La Brenva on the south to that of Bossons on the north, forming the mass of the Monts Maudits. These appear effectually to cut off access to the summit of Mont Blanc on this side, nor does De Saussure hint at the possibility of ascending it from hence. The western, or Second Flambeau, is a summit conspicuous from several points, whence it could hardly be expected to be seen, as, for instance, from the Col de Balme.

The glacier here, enclosed between La Tour Ronde and the Aiguille du Géant, is very broad, but it is only one of the tributaries which aliment this branch of the Mer de Glace—another descends from between the first and second Flambeau by the foot of a promontory called Le Capucin, (see the Map,) owing to the fantastical forms which the granitic obelisks here assume, and one of which has the rude outline of a human figure. Another and very large ice-flow descends from the Aiguille du Midi, and is more precipitous and broken; it breaks against a small rock called Le Rognon, nearly opposite to the Aiguille Noire, and which is surrounded entirely by the glacier. It was up this glacier that Col. Beaufoy first, and afterwards M. Romilly of Geneva, ascended the Aiguille du Midi, at least up to the foot of the last rocky summit, which I believe is inaccessible.

We continued to descend with precaution, though without any inconvenience, excepting from the sun, which was now high and brilliant, and its light reflected with more intensity than I had ever felt it from the *facettes* of the highly crystallised and fresh snow by which we were surrounded. I began to think that the passage was to be effected without any difficulty worth mentioning, until we arrived at the part of the valley where the three tributary glaciers already mentioned begin to unite, and are together squeezed through the comparatively narrow passage between the Aiguille Noire on the right, and the rock which I have marked *petit Rognon* on the left. It is difficult to say, whether the ascent or descent of such a glacier is more arduous; but in descending, one is at least more taken by surprise; the eye wanders over the wilds of ice sloping forwards, and in which the most terrific chasms and rents are hidden like the ditch in a *ha-ha* fence. The crevasses of the glacier gradually widened; the uniting streams from different quarters met and justled, sometimes tossing high their icy waves, at others leaving yawning vacuities. The slope, at first gradual, and covered continually with snow, became steeper, and as we risked less from hidden rents, the multitude and length of the open ones caused us to make considerable circuits.

But the slope ended at last almost in a precipice. At the point where the glacier is narrowest it is also steepest, and the descending ice is torn piece-meal in its effort to extricate itself from the strait. Almost in a moment, we found ourselves amidst toppling crags and vertical precipices of ice, and divided from the Mer de Glace beneath by a chaos of fissures of seemingly impassable depth and width, and without order or number. Our embarrassment was still farther increased by the very small distance to which it was possible to command by the eye the details of the labyrinth through which we must pass. The most promising track might end in inextricable difficulties, and the most difficult might chance ultimately to be the only safe one.

The spectacle gave us pause. We had made for the north-western side of the glacier, near the foot of the Petit Rognon, hoping to get down near the side of the rocks, although not upon them. But when we neared this part of the glacier, even Couttet shook his head, and proposed rather to attempt the old passage by the foot of the Aiguille Noire, where De Saussure left his ladder,—a passage avoided by the guides on account of the steep icy slopes it presents, and the great danger which is run from the fragments of stone which, during the heat of the day, are discharged, and roll down from the rocks above. These stones are amongst the most dangerous accidents of glacier travels. A stone, even if seen before hand, may fall in a direction from which the traveller, engaged amidst the perils of crevasses, or on the precarious footing of a narrow ledge of rock, cannot possibly withdraw in time to avoid it. And seldom do they come alone. Like an avalanche, they gain others during their descent. Urged with the velocity acquired in half rolling, half bounding down a precipitous slope of a thousand feet high, they strike fire by collision with their neighbours—are split perhaps into a thousand shivers, and detach by the blow a still greater mass; which, once discharged, thunders with an explosive roar upon the glacier beneath, accompanied by clouds of dust or smoke, produced in the collision. I have sometimes been exposed to these dry avalanches; they are amongst the most terrible of the ammunition with which the genius of these mountain solitudes repels the approach of curious man.* Their course is marked on the rocks, and they are most studiously avoided by every prudent guide.

It was, however, in the direction of La Noire that it was thought that we might pass; and we accordingly crossed the glacier to inspect the passage. But there, barriers still more insurmountable appeared. One prodigious chasm stretched *quite across the glacier ;* and the width of this chasm was not less than 500 feet. It terminated opposite to the precipices of the Aiguille Noire in one vast *enfoncement* of ice bounded on the hither side by precipices not less terrible. A glance convinced every one that here, at least, there was not a chance of passing, unprovided as we were with long ropes or ladders. Nothing remained but to resume the track we had at first abandoned; for the whole centre of the glacier was completely cut off from the lower world by this stupendous cleft. Here the experience of Couttet stood us in good stead, and his presence of mind inspired me with perfect confi-

---

* At saxum quoties ingenti ponderis ictu
Excutitur, qualis rupes quam vertice montis
Abscidit, impulsu ventorum adjuta, vetustas,
Frangit cuncta ruens : nec tantum corpora pressa
Exanimat ; totos cum sanguine dissipat artus.

LUCAN, *Phar. III.*, 465.

dence, so that we soon set about ascertaining, by a method of trial and error, whether any passage could be forced amongst the labyrinth of smaller crevasses on the northern side of the glacier. A chamois, whose track we had followed earlier, seemed here to have been as much baffled as ourselves, for he had made so many crossings back and forward upon the glacier, and had been so often forced to return upon his steps, that we lost the track for a time. This animal is exceedingly timorous upon a glacier covered with snow, since the form of the foot prevents it from offering almost any resistance when hidden rents are to be crossed. We had accordingly passed earlier in many places where the chamois had not ventured; but the case was now different on the hard ice. He took leaps upon which we dared not venture; and as we were never sure of not being obliged to retrace every step we made, we took good care never to make a descending leap which might cut off our retreat. Many a time we were obliged to return, and many a weary circuit was to be made in order to recommence again; but we seldom failed ultimately to recover the chamois track, which is the safest guide in such situations. The excitement was highly pleasing. The extrication from our dilemma was like playing a complicated game, and the difficulty of the steps was forgotten in the interest of observing whether any progress had been gained; for now we were obliged to descend into the bosom of the glacier, and to select its most jagged and pulverized parts, in order to cross the crevasses where they had become choked by the decay and subsidence of their walls. Thus hampered by our icy prison, we only emerged occasionally so as to catch a glimpse of what lay beyond, and to estimate our slow and devious progress. At length, by great skill on the part of Couttet, and patience on the part of all of us, (for we remained inseparably tied together all this time) by clambering down one side of a chasm, up another, and round a third, hewing our steps,* and holding on one by one with the rope, we gradually extri-

---

* A geological hammer sharpened at one end is nearly as good an implement for this purpose as a hatchet. For this reason, amongst others, I generally wore it. A person so provided, if he falls uninjured into a crevasse, possesses the most essential means of extrication.

cated ourselves from a chaos which at first sight appeared absolutely impenetrable, and that without any very dangerous positions.

Whilst we were in the middle of this confusion and difficulty, I could not help remarking how totally unserviceable any addition to the number of guides would have been. On saying as much to Couttet, he replied, " ils ne seraient bons que pour faire peur les uns aux autres," which was perfectly true. At length, having been for some hours engaged in these toils, we saw a comparatively clear field before us, the glacier became more level and compact, the crevasses were knit, and though no trace of life or habitation, nor the most stunted tree, was within any part of our horizon, the familiar localities of the Mer de Glace were apparent, the Tacul with the branching glacier, the Couvercle, the Jardin, the Charmoz, and the Moine. Here we halted about one o'clock, for we had now reached *water*, always a joyful sight to those who have been long wandering over snow fields. We drank of it freely, and the guides added fresh libations of brandy, which caused them to complain of intolerable thirst and heat of the head all the rest of the way to the Montanvert, which, by confining myself to cold tea, and a very little wine with water, I entirely escaped.

As I have not described this branch of the Mer de Glace, above the Tacul, I will here add the very few words which it requires.

The Aiguille Noire, on the south, and the Aiguilles de Blaitière and Grepon, on the north, here bound the Glacier of Tacul (or Géant.) The former gives rise to a pretty extensive lateral glacier, which descends from the foot of the Aiguille du Géant, and the Mont Mallet. I distinguish these two, as it will be seen is done on the map. But the Aiguille du Géant is itself sometimes called Mont Mallet, on the south side of the Alps. What I have termed Mont Mallet, on the authority of the guides of Chamouni, is a very remarkable peak, a little to the north-east of the Aiguille du Géant ; and, so far as I can judge from a single altitude with the theodolite, somewhat higher, as indeed I had suspected, from observing both in different positions. The Geant appears to be 13,099 feet above the sea. Mont Mallet 13,068. The glacier descending from them is very convex and copious ; and, by its union with the others, tends to consolidate the whole. It is from the

Aiguille Noire, (probably so called from having formerly been visited in search of smoky quartz crystals), that the fourth moraine of the Mer de Glace, mentioned in a former chapter, descends. This moraine offers a feature similar to that of the Glaciers du Taléfre and de Léchaud, namely, that it is at first imperceptible, or nearly so, and increases in distinctness and mass as we descend the glacier. It is several miles below its origin, namely, near the "Moulins," that it is best developed. This very singular fact admits of no contest, but the mode of explanation varies. Some have supposed that it arises from the rejection of the stones through the matter of the ice, which presupposes that the fragments have been mixed up with or engaged in the solid ice. I believe that it arises from a very simple cause. Where two glaciers do not unite at exactly the same level, (the most common case), or even where, the level being the same, the one vastly preponderates, the lower or smaller glacier flows or forces itself some way under the upper or greater, and thus the fragments of rock borne by each to the point of union, are naturally carried inwards at the sloping junction, where they lie for a time buried, as in figure 1, page 166, which represents the section of the glacier at this place, until the thaw or waste of the surface brings them gradually to light. This is attempted to be represented on the map, and it is one of the most striking features of these accumulations. I must add, that, at the foot of the icy precipice opposite to La Noire, I found rocks and sand appearing on the surface in a way not very easy to comprehend. They were probably, or almost certainly, derived from the Petit Rognon, but by what mechanism they were brought to light I am unable satisfactorily to decide. As soon as the glacier becomes compact and moderately fissured, the veined structure of the ice makes its appearance, and continues the whole way down the Mer de Glace, as has been already particularly described.

The Glacier du Géant, below La Noire, is of great and nearly uniform width. I have, on the present and other occasions, traversed it in various directions. It is little fissured, and consequently great watercourses are formed, which pursue their way along the surface of the glacier, of which the inequalities are sometimes very considerable, so that the water at last finds an exit through some great funnel, or vertical opening in the ice ; and here and there it stands in pools to a great depth. About half-way from La Noire to the Tacul, there is a rocky promontory on the right bank of the glacier, marked K in the map, which was one of my points of observation, and opposite to it is an offset from the range of the Aiguilles of Chamouni, on the left, which forms a series of very fantastical summits, one of which might deserve a peculiar name, and is figured in the map as the *Aiguille de Blaitière derrière*.

Truncated glaciers of the second order festoon the wild enclosures of the valley on both sides. Those on the left are nearly continuous, and may, I believe, be traversed, so as to reach the shoulder of the Charmoz, or station G*, from the upper part of the Glacier du Géant, an experiment which I was prevented from trying by premature bad weather.

From the Aiguille de la Noire it seems but a step to the foot of the Tacul, but the elevation is considerable, the glacier very wide, and I was surprised at the distance which separated me from the regions with which I was then familiar. I must not omit to add, that the view in descending the Glacier du Géant is admirable. The picturesque mass of the Aiguilles du Moine and Dru, terminating in the enormous elevation of the Aiguille Verte, forms a group of singular majesty, which cannot be so well appreciated from any other point. The basin of the Glacier du Taléfre is likewise exposed, and the triangular rock of the Jardin stands forth in form and dimensions very apparent.

We all felt an exuberant cheerfulness at being relieved from our embarrassments, and ran cheerfully down the magnificent glacier, leaping crevasses which at another moment we would rather have avoided. Soon on the platform at the confluence with the Glacier de Léchaud, all was plain and direct; and I reached the Montanvert at a quarter before four P.M., without fatigue, headach, or lassitude. Here I remained, intending to spend some weeks. My guides, having finished their brandy, descended to Chamouni, where their arrival created, I was told, some astonishment, as no one had before crossed the Col du Géant in a single day, and as it was supposed that the fresh snow must at any rate have rendered the attempt impracticable. I slept that night somewhat sounder and longer than usual, but rose next morning with a freshness and elasticity to which the inhabitant of the plains is a stranger. A threatening of inflammation of the eyes confined me partly to the house, but it fortunately subsided: I felt at first a slight shortness of breathing on ascending a hill, but that also disappeared the second day. My guides, as I afterwards learned, entirely lost the skin off their faces. The barometer on my arrival was—

	m.m.	A.T.	D.T.
Montanvert, 1842. July 23. 3 h. 45 m. p.m.	610.8	15.8 C.	51 F.
5 15 ,,	610.2	11.4	51

This, compared with the observation of the Col du Géant, gives 4841 feet for its height above the Montanvert, or 11,144 above the sea.

CHAPTER XIII.

FROM COURMAYEUR TO CHAMOUNI, BY THE COL FERRET AND COL DE BALME.

PIEDMONTESE VAL FERRET—GLACIER OF TRIOLET—VIEW FROM THE COL—
SWISS VAL FERRET—MARTIGNY TO CHAMOUNI—GLACIER OF TRIENT—COL
DE BALME—GLACIER OF ARGENTIERE.

IN order to complete our narrative of the tour, or circuit of Mont Blanc, I proceed to describe shortly the route by the Col Ferret across the great chain of Alps, and that from Martigny to Chamouni by the Col de Balme, and those glaciers of the valley of Chamouni which have not as yet been enumerated. The former part of the route I performed in 1841, in company with Mr. Heath; I have three times visited the Col de Balme, in different years.

The passage of the Col Ferret is tedious, and perhaps less interesting than most others in the Alps; travellers usually, and perhaps wisely, prefer the longer round, by Aoste and the Great St. Bernard, which offers greater variety. This route, however, completes the closer inspection of the great chain of Mont Blanc, which is very completely separated, both geographically and geologically, by the Col Ferret, from the mountains of which Mont Velan forms the culminating point. After having ascended the Piedmontese Val Ferret (the prolongation of the Allée Blanche,) and descended the Swiss Val Ferret to Orsières; and having, either by Martigny or otherwise, reached the Col de Balme, and thus passed into the valley of Chamouni, the circuit of Mont Blanc and its chain is complete. Unless by passing difficult or dangerous glaciers, as in the case of the Col du Géant, this extensive chain may be considered as impracticable, or nearly so, in its whole length.

The ascent of the Val Ferret from Courmayeur seems monotonous after the more varied grandeur of the Allée Blanche and Val de Veni:—for here, though there are numerous glaciers on the left hand, they do not descend completely into the valley except near the head of it, and the mural precipices of the Jorasses, which separates this valley from the tributaries of the Mer de Glace of Chamouni, although magnificent at a distance, rise here so completely overhead as to conceal their own elevation, and the magnificent summits by which they are crowned. As the secondary mountains on the right hand—forming the prolongation of the Montagne de la Saxe, or Mont du Pré—offer nothing of interest beyond what has been already mentioned in a former chapter, I shall merely enumerate the glaciers which descend from the primary chain so far as I was able to ascertain their names from native guides. I am aware that the guides of Chamouni differ a little in their nomenclature. Eastwards from the Glacier of La Brenva, we have first the Glacier of Mont Frety, and then that of Entrêves with the Mont Frety between. From the Aiguille du Géant descends the Glacier de Rochefort, and between it and the Grande Jorasse the Glacier de la Grande Jorasse.

The next in order is the Glacier de Triolet, which, as already mentioned in the fifth chapter, is nearly opposite to the head of the Glacier de Léchaud, and descends from a summit called by the Chamouni guides, " Montagne des Eboulements." The event to which the name refers took place, I believe, in 1728, though I failed in obtaining at Courmayeur any *authentic* documentary evidence respecting it. According to a small printed work, which was shown to me, the avalanche, or sudden descent of the whole glacier, took place on the night of the 15–16th August in that year, and completely overwhelmed the chalets of Pré de Bar, which were situated exactly in front of it, destroying of course the inmates and cattle. The modern chalets of Pré de Bar are higher up on the southern side of the valley. They are very filthy.

Beyond the glacier just named is the Mont Ru, which separates it from the Glacier of Mondolent, the highest in the valley. This one appears to have greatly retreated of late years.

There are two passages of the Col Ferret, the Petit Ferret, which is a foot path, and the horse road, which is more circuitous. It is five hours' walk from Courmayeur to the Col. The path of the Petit Ferret is close to the junction of the limestone and granite. The former is nearly vertical, rising against the latter at an angle of at least 70°. The junction is well marked, and the limestone is a tabular slate. Indeed, the chief interest of this route consists in the closeness with which the geological boundary is followed. Behind the Grande Jorasse, at a point called Pra Sec, two hours from Courmayeur, is a junction and apparent superposition of granite to limestone, which I noticed in 1841, and again from a distance in 1842. On neither occasion had I any doubt that the limestone actually dipped under the granite as, in the interval of the two observations, I had established that it does farther west. De Saussure, however, who ascended to the junction, maintains that the strata rise towards the granite (§ 871) although he seems to admit that farther west both the granite and limestone dips inwards; but he never asserts the superposition distinctly.

The view from the Col Ferret, looking back, is certainly one of the finest which I have seen. The prodigious outworks which sustain the mass of Mont Blanc on the southern side are more conspicuous here than from any other point, especially the Mont Péteret which stands out like a majestic Gothic pinnacle. From hence, as from the Col de la Seigne, we see how far this side of the chain is from being an absolute precipice as it appears when viewed in front, as from the Cramont. The descent of the Swiss Val Ferret to Orsières, offers no great interest, and it is of most tedious length. On the right hand is seen the passage of the Col de Fenêtres leading to the Great St. Bernard, by which the produce of the valley, and especially fire-wood, the property of the convent, is conveyed with the aid of mules.

Several glaciers are passed on the left; since, however, the side of the valley is exceedingly steep, several of these are only seen peeping over the precipices. One of them has evidently descended formerly into the valley, and has deposited in it an immense transversal moraine which now stands alone ;—the glacier having retreated into the upland ravine. It is commonly supposed to be from these glaciers that the vast granite masses descended which are still found on all the neighbouring slopes at a great height above the valleys, the Blocks of Monthey and those upon the Jura. The Aiguilles to the east of Mont Blanc are indeed the only ones in this district capable of yielding rocks of the kind in question, and the secondary mountains adjoining Orsières are strewed with masses, having evidently a common origin with those in the valley of the Rhone. These were well known to De Saussure,* and accurately described by his correspondent M. Murrith, but they form one of the especial grounds of the theory of Venetz and De Charpentier and have been more particularly described by the latter.

I shall not dwell upon the descent of the Dranse to Martigny, or the circumstances of the debacle of the Val de Bagnes, to which I shall shortly again recur ; but I proceed to describe a journey which I took from Martigny to Chamouni, in September 1842, in which, avoiding as much as possible the common route, I visited the Glaciers of Trient and Argentière. The Glacier of Trient may be reached from Orsières by crossing the Mont Catogne, or from Martigny by the Col de la Forclaz. In the latter case, the village of Trient being passed, instead of turning to the right in ascending the valley, which would lead to the Col de Balme, I followed the eastern side of the glacier stream, and after a rough walk, (having missed the path,) I arrived at a group of châlets. The glacier is then well seen ; it descends into a kind of basin, apparently inaccessible in its higher parts, from granitic pinnacles which divide this valley from the Val Ferret. Of these the most conspicuous is a fine point on the right hand, looking towards the head of the glacier ; it was named to me Salena ; and is no doubt also at the head of the glacier so called, whilst at the same time it separates

* *Voyages*, § 1022.

the Glacier of Trient from that of Le Tour. I think it most likely that this is the Pointe d'Ornex, seen from Orsières.*

The lower end of the Glacier de Trient is about an hour's walk above the village of the same name. It is a well spread out glacier, with few ramifications, and a rather attenuated front; it somewhat resembles in contour the Glacier of the Rhone, or that of La Brenva, but it communicates more directly with the higher slopes. An inspection of the structure proved it to be quite normal; so much so, indeed, that I could have accurately predicted it before hand, by seeing merely the external form of the ice. Suffice it to say, that it corresponds generally to the structure figured on page 30. The crevasses in the lower part are also *radial*, as in every glacier of this order, (see the full lines marked *a* on the figure, page 29.) In its middle or mean portion, the glacier is, as usual, most readily traversed, and here very easily. I crossed over, making observations in different directions, and observing especially the character of the granite blocks which come down the western moraine from the summit of Salena just mentioned. These blocks are remarkably chafed and rounded, no doubt from the friction they have experienced between the ice and rocks; but neither in this or in any other case have I perceived an approach to *polish* on glacier-moved blocks, which cannot (I think) for a moment be confounded with those *smooth* pebbles and boulders plentifully found in the diluvium of all countries, and composing many of those gravel heaps which have been styled moraines. The nature of the granite, or protogine, appeared to me accurately to resemble that of the blocks of Monthey, and those on the Jura. Supposing them to have been derived from the Pointe d'Ornex, they may either have descended the Glacier de Trient when it filled the valley of the Tête Noire, and joined that of the Rhone below Salvent, or, (as is more probable, from the distribution of the blocks,) followed the exterior of the chain by St. Branchier and Martigny.

The highest chalets on the eastern side, named Lali, are somewhat

---

* See more on this subject in the next Chapter.

higher than where I crossed the glacier, and I reached the western bank under the chalets of Chazettes, which are close to a ravine which contains a stream from a glacier, which fills its higher part, and which descends from the ridge of the Aiguille du Tour. Finding nothing more particularly worth exploring, I proceeded to look for the path which, I had been informed, led directly to the Col de Balme, without descending to Trient. It was, I was told, above the precipices which bound the valley of Trient to a great height on its western side. Although I met with no one here to give me information, I succeeded in discovering the path, which is a bold and romantic one, and crosses the mountain by which the Col de Balme is separated from the Glacier de Trient, at a great height on its precipitous eastern side. In the course of this walk, I obtained a more correct idea of the chain to which the Dent du Midi of Bex belongs, than I before had. Instead of being an insulated pyramid, or a pair of summits, as it appears from most points, it belongs to a jagged ridge, which is very elevated, and which extends from east to west, including great fields of snow, and glaciers of the second order. I arrived early at the little inn upon the Col de Balme, and slept there.

Next morning I left the Col de Balme at six, with fine weather, intending to explore the Glacier of Argentière. I had long had a great curiosity to visit this glacier, because, though so near Chamouni, it is very little known ; and still more, because on all the models it is represented like an unbroken, perfectly uniform, nearly level canal, extending to the very axis of the Alps ; and I was anxious, if possible, to determine its boundaries as respected the barriers of the Glacier du Taléfre, to which I understood it to be contiguous. It is a glacier little known to the guides of Chamouni ; but a few of whom frequent it for the sake of the crystals, with which it is said to abound ; but the length of the way is so great, and the snow lies so long and so deep upon the higher parts, which are sheltered from the sun by their northern exposure, that it is an expedition only to be attempted (I mean for the search of minerals) in the finest weather, and at a late season of the year, when the boundary of the snow is highest. But as the days are then short, it is necessary to sleep out, and this is no pleasant task in so very wild and remote a spot. So far as the report of the guides may be believed as to the locality of the minerals, (a matter on which the current information is little to be believed,) the Glacier d'Argentière is the richest field in the chain of Mont Blanc ; and specimens of red Fluor Spar and smoky Quartz,—the most expensive in the cabinets of Chamouni,—are understood to have been brought from thence, often at imminent peril to those who secured them.

I have said that few of the professed guides have been on the higher part of the Glacier d'Argentière. The makers of the two best models of this part of the Alps have admitted to me, that they took their design of its locality from the perspective view on the Buet, which looks right up it. De Saussure, I believe, only mentions it once ; * and as he speaks of having visited it and the Glacier des Bois in early spring, it is certain that he can only have examined its lowest part. It is unnoticed, or all but unnoticed, by Ebel and by Pictet.

Understanding from the innkeeper on the Col de Balme—himself a good mountaineer—that the Glacier d'Argentière presented no unusual difficulties, I contented myself with taking along with me the man who usually accompanied me, although he was also unacquainted with the way. As we knew that we must again ascend, we unwillingly went down the great depth which separates the Col de Balme from the foot of the Glacier of Le Tour. I then regretted that I had not taken the guide of the Col de Balme, who offered to conduct me by a little known route across the upper part of the Glacier of Le Tour, and to descend upon that of Argentière, near the Aiguille, which bears its name. But I was anxious to see the glacier in all its length, and not to come upon it in the middle. The Glacier of Le Tour has considerably shrunk in its dimensions of late years, as well as that of Trient. Beyond the village of Le Tour, which I left on the right, a sharp ascent led me through extensive pastures, up to about the level whence we had started, and keeping along about that line,

---

* *Voyages,* § 740.

we there came in sight of the Glacier of Argentière, at a great depth below us. I did not descend, however, but kept along the face of the hill, represented in the upper left hand corner of the Topographical Sketch No. IV., so as not to lose the height we had gained. The path became smaller,—then a mere sheep track,—and that again was sub-divided. The mountain face became precipitous, and in some places went sheer down to the glacier. As my guide, or rather companion, was somewhat nervous on untried excursions,—rather, perhaps, from a caution characteristic of the Savoyard character, in getting himself into trouble by bringing a traveller into danger, than from any want of personal courage,—I took the lead both on this occasion and on the previous day, and fortunately extricated myself satisfactorily from the precipices, which, when seen in the afternoon from the opposite side of the glacier, were of a sufficiently dangerous kind, and had we attempted a passage either higher or lower, we must have failed. The precipices passed, a long and fatiguing slope of debris was to be crossed, and then a vast lateral moraine of the glacier, covering a great surface with huge blocks, which, however, afforded solid and comparatively easy footing, after what we had passed. Amongst these blocks I was astonished to observe some sheep, which must have been driven across the nearly pathless rocks which I had traversed.

Nearly opposite this moraine, which is marked on the Sketch, the glacier is tolerably flat, and might be traversed from side to side; but being precipitous both above and below, I continued along the moraine until I came to the foot of the rocks descending immediately from the Aiguille d'Argentière to the glacier. There I made for the ice, having had, rather to my surprise, a fatiguing walk of four hours from the Col de Balme, before setting foot upon the glacier.

The Aiguilles of Argentière and of Chardonnet separate the glaciers of Le Tour and Argentière, and between these Aiguilles there descends a steep tributary glacier to the level of the latter. On the ridge connected with the Aiguille de Chardonnet there is a remarkable instance of a glacier of the second order, which appears to be rapidly disappearing. It is marked *a* on the Topographical Sketch No. IV. Its for-

mer boundary is indicated by the whiteness of the rock where it has
been beneath the ice, of which there is now scarcely a trace.

On the Glacier of Argentière there is only one medial moraine of
any extent which comes from the higher part of the Glacier, on the
left in ascending. There are two lateral glaciers also on the left,
which appear to communicate with the Glacier of Le Tour. Having
gained the ice, I proceeded without difficulty, for on the higher part
it is not much crevassed, and the higher we ascend the more level it
becomes. The Aiguille Verte rises in great majesty on the right, and
from its rugged sides some short glaciers descend to meet that of Argentière. I walked on, having reached the névé, or perpetual snow, until
I had left the Aiguille Verte quite behind me, and was now within a
short distance of the head of the glacier, that is to say, not much exceeding an hour's walk. The surface is even, and the whole topography is easily seized. The direction of the glacier, which up to the
Aiguille Verte had been S. 25° E., now became S. 50° E. This bend
in the direction corresponds to the basin of the Glacier du Taléfre,
which is only separated, as has been said, from the higher part of the
Glacier d'Argentière by the range of the Tours des Courtes, which
appears to be of small thickness, and is one continued precipice on its
north-eastern side. I can only guess at the height of the upper part
of the Glacier d'Argentière, as I was provided with an imperfect instrument. It is, no doubt, more than 8000 feet above the sea. The
extremity of the view is terminated by a snowy peak, which I believe
is probably that marked [A] on the large map of the Mer de Glace,
and which was also visible at the Jardin,—perhaps the Mondolent.

The structure of this glacier is very confused. The vertical linear
bands are, of course, visible throughout up to the névé; but it would
be difficult to trace the curves. The middle and lower part is excessively crevassed; and the extremity near Argentière has very much
shrunk of late years.

After a careful examination of the higher part I returned by the
western side, under the Aiguille Verte, and gained the bank somewhat below the tributary glacier on that side. There is a small
snowy peak to the north of the Aiguille Verte, which is connected

with it by a ridge dividing the ice which falls in the direction of the Mer de Glace and in that of Argentière. From the same peak descends a small glacier on the north side, called Glacier de la Pendant, or de l'Oignon, which, judging from the polished rocks below, appears to have been formerly more extensive. From the highest Châlets there is a path to the village of Argentière, and another less easily found, which descends near Lavanchi. Both pass through fine fir wood. From thence the village or Prieuré of Chamouni, is soon reached.

CHAPTER XIV.

JOURNEY FROM CHAMOUNI TO VAL PELLINE, BY THE VAL DE BAGNES AND COL DE FENETRES.

TRACES OF ANCIENT GLACIERS FROM LES MONTETS TO THE TETE NOIRE—ARRIVAL AT THE GREAT ST. BERNARD—FIND M. STUDER—RETURN TO ORSIERES—THE VAL DE BAGNES—CHABLE—THE INHABITANTS—GLACIER OF GETROZ—THE DEBACLE OF 1818—CHALETS OF TOREMBEC—ECONOMY OF CHALETS, AND MANNERS OF THE INMATES—GLACIER OF CHERMONTANE —COL DE FENETRES—VIEW INTO ITALY—VALLEY OF OLLOMONT—GOITRES —ARRIVAL AT VALPELLINE.

BEFORE going to Chamouni in June 1842, I had visited my friend M. Studer, Professor of Geology at Berne. We then agreed, that a plan which had been vaguely discussed between us the year before—of visiting the neighbourhood of Monte Rosa, and the almost unexplored valleys to the westward—should, if possible, be accomplished in company that summer. M. Studer visited me on the 1st August, at the Montanvert, and we then fixed the 12th of that month for a *rendezvous* at the Convent of the great St. Bernard, he, in the meanwhile, making an excursion into the Tarentaise, whilst I remained pursuing my survey of the Mer de Glace, and determining its motion. Accordingly, on the 11th, I left Chamouni, having engaged an active young man, (not a professed guide,) of the neighbourhood, named Victor Tairraz, to accompany me on the expedition, and to carry my haversack and instruments. M. Studer and myself had already decided on taking one man a-piece as a personal attendant, and to secure guides from time to time, to assist in carrying the provisions,—which he was well aware would be requisite, from having in 1841 visited the valley of Erin, and seen the almost total deprivation which there exists of the commoner commodities of life.

I had proposed crossing the chain of Mont Blanc, by the Glacier of

Le Tour, to the valley of Orsières, a pass which has already been alluded to ; but I was prevented, partly from the difficulties and endless formalities always made by the guides of Chamouni, when any unusual expedition is contemplated, with a view of enhancing their services—and partly from a trifling accident to my foot, which yet occasioned me some concern, with the prospect of a prolonged and difficult expedition before me. I therefore rode to Martigny by the Tête Noire, a route with which I was already pretty well acquainted, but which offered me new subjects of remark and speculation connected with the ancient extension of glaciers. I observed the distinct prolongation of the ancient moraine of the Glacier d'Argentière towards the pass leading by Les Montets from the valley of Chamouni into that of Valorsine. This moraine seemed to me not less clear in its origin and details than that of the Glacier des Bois at Tines ; and the low ridge of rock separating the two valleys is strongly marked by glacier action, which has also deposited a number of granite boulders on the summit of the pass. The whole valley of the Tête Noire shows, from time to time, proofs of having formerly been filled with moving ice, and between the cascade of La Barbarine and the little inn of Tête Noire, I observed the celebrated Valorsine puddingstone rock, which is exceedingly hard, beautifully fluted and polished, at a great height above the bed of the torrent.

I slept at Martigny, and next day proceeded in company with other travellers as far as Liddes, in a char, whence we walked to the convent, where I had the great satisfaction of finding that M. Studer had arrived only half an hour before from the southern side of the Alps, together with his tried and faithful attendant, Siegfrid Klaus, a peasant of the Oberland, who, for twenty summers, has followed the indefatigable Professor of Berne in his geological rambles, and has rendered himself a deserved favourite and friend, by his experience, hardihood, simplicity, and that peculiar patience and fertility in expedients which characterizes the best guides of German Switzerland, together with an honest warmth, and even playfulness, which is less commonly united with it.

Our greetings were hearty when we met around the hospitable fire, which, even in August, is the chiefest luxury in the domicile of the worthy Fathers of the great St. Bernard. The evening was partly spent in discussing our plans, to which the priests lent an interested ear. One of them, the Chanoine L'Eglise, almost volunteered to accompany us on a part of our journey, but unavoidable engagements in the convent prevented it; however, he kindly gave us letters, which proved of service.

The next morning, at eight o'clock, I found water to boil at 199.08 Fahr., the convent barometer being at 576.1 millimetres, unusually high in this position. Accordingly, the Fathers predicted favourable weather for our expedition.*

We walked leisurely down to Orsières by the same road as I had ascended the previous day, for we had decided upon commencing our journey by ascending the valley of Bagnes, which separates at St. Branchier, a little below Orsières, from the valley of Entremont leading to the Great St. Bernard. I was struck with the extremely small interest of the Swiss side of the St. Bernard Pass. It was ten years, within a few days, since I had last visited it, but I well remembered the tedium of that interminable descent to Martigny. All the higher part is bare and wild, without either grandeur or variety,—of course I mean in comparison with other Alpine passes.

At Orsières we introduced ourselves to M. Biselx, formerly Prior of the Convent, and now Curé of Orsières, a man known in the scientific world by his zeal and acquirements, an intimate friend of M. de Charpentier, and partaking his views on glacier theories. Our introduction was easy, and the evening passed pleasantly in his society. Indeed, we had a marked proof both of his skill and experience; for learning that M. Studer's syphon barometer was injured by having taken air, and considering the interesting results which it might afford on our present excursion, he begged to be allowed to boil the mercury in the tube, a critical and disagreeable operation, as every one knows, but which he most effectually accomplished on the spot with his own

---

* I determined the geographical position of the Great St. Bernard, as I did that of Chamouni in 1832, and found it to be

Lat. 45° 50' 60'' N. Long. 7° 4' 45'' E. of Greenwich.

hands over a charcoal stove in the kitchen of the inn ; he then bade us a hearty farewell.

At Orsières, we made a considerable provision of food for our journey, for we were immediately to leave the beaten track. A guide was engaged to go as far as Chable, the principal village of the Val de Bagnes, where M. Studer had already been the preceding year, and had made an acquaintance who might be useful in procuring us a person acquainted with the higher parts of the valley, and with the Col de Fenêtres leading into Italy.

At length, all preliminaries being settled, we left Orsières, in a beautiful morning. The view towards the chain of Mont Blanc was particularly fine, as seen by the early sunlight. The landlord of the Hôtel des Alpes particularly pointed out to us a conspicuous granite peak, which he called Pointe d'Ornex, and which he assured us was known by no other name in these parts. This must, therefore, un-doubtedly be the same as Van Buch referred to in his paper in the *Berlin Memoirs* on the distribution of erratic blocks, and to the neighbourhood of which he referred the origin of the Pierre à Bot and other masses of granite on the Jura range. The Mont Catogne, a conspicuous hill between Orsières and St. Branchier on the left, is composed partly of granite, but its eastern face, which is very steep, presents a vast triangular *revêtement* of limestone, which here, as elsewhere, rises against the primitive rock, which, as we have seen, bounds the Val Ferret in its whole extent. On the face of this limestone slope lies one of those vast masses of transported granite described by M. de Charpentier, under the name of *blocs perchés*, which afford so strong an evidence in favour of his theory of glacier extension. This vast mass may be distinctly seen, notwithstanding its distance and height, from Orsières, on a steep part of the rock, free from the trees which nearly surround it. Its position is exceedingly remarkable, for it seems impossible to conceive a block of that size deposited by the mere force of water at such a height above the bed of the valley.

Our party now amounted to five, of whom the three guides were all considerably laden, for, besides personal effects, and some instruments, we carried a provision of rice, bread, and meat, intended for three days. M.

Studer's barometer was the only instrument for measuring heights which we could at the time depend upon, but I had a portable sympiesometer, by Adie, constructed on purpose for this journey, but whose indications required a special correction difficult to determine, and one of those very convenient Russian furnaces, made by Stevenson of Edinburgh, which proved an invaluable adjunct for melting snow, for making tea, and at the same time for ascertaining the temperature of boiling water by a thermometer, which I had adapted to it, reading from 185° to 213° Fahr., and on which a fiftieth of a degree was capable of estimation. This is the only instrument which I have found capable of resisting sufficiently the influence of wind and cold to produce boiling water even from snow, in almost any situation, and it replaced the barometer usefully, on several occasions, as will be seen.* Our appearance was sufficiently remarkable to attract the attention of the passers by, of whom, at this early hour, there were a number on their way, to spend the day at Orsières, as it happened to be a great festival in this and the neighbouring valleys,—the eve of the Assumption of the Virgin. The day, as I have said, was splendid, and promised to be very warm; but our course, as far as Chable, lay almost entirely on the shady side of the valley of Bagnes, which we entered by turning abruptly to our right, before entering the village of St. Branchier, an hour's walk below Orsières.

The path, which was scarcely traced on the left bank of the rapid and impetuous Dranse, passed through woods and meadows, and the whole scene was refreshing and peaceful in the highest degree, and seemed to augur success to an excursion so happily commenced. Chable is a considerable village, very pleasantly situated in a tolerably open space, into which the valley enlarges itself, near the foot of the Pierre à Voie, a conspicuous summit, which separates this valley from that of the Rhone, and not far from which a path leads from Chable to Riddes,

---

* An account of the method used for calculating heights from the temperature of boiling water will be found in the *Edinburgh Transactions*, vol. xv., part 3. I have found that the temperature of the boiling point falls 1° Fahr. for 550 feet of ascent, *uniformly* for all heights.

on the Simplon road. The neighbourhood is very fertile, covered with fruit trees and meadows, and studded with several villages ; at this season it has a peculiarly cheerful and thriving aspect. As we approached the village (having joined the great road) we were struck by the appearance of the peasantry, and by the great numbers who had met together on occasion of the festival. So numerous were they, that we were not surprised to learn, that within the very small range of the Val de Bagnes, which is permanently inhabited, there is a population of 9000 souls. All the avenues to the church were crowded with well dressed, respectable looking men, the women being chiefly within the building. Our arrival and accoutrements excited some surprise, but we were allowed to pass unmolested by ill-bred curiosity, to one of the principal houses of the place, belonging to M. Gard, to whom M. Studer had been recommended on his former visit, and who, though a person of some consequence in the place, condescends, as is not unusual in similar circumstances in many countries, to make his house one of public entertainment, and the resort of the better class of peasantry, who, when the service was over, came and called for their *chopine* of wine, as they would have done in any common inn.

It was vain to think of proceeding any farther in a hurry. The demeanour of the people was intelligent, independent, and almost sarcastic. A guide was our first requisition ; and it was evident that though there would be no difficulty in procuring one who was acquainted with the pass into the Pays d'Aoste, his accompanying us would be considered rather as a favour, and must be upon his own terms. These, however, were in due time adjusted with the usual success and conciliation with which M. Studer always contrived to effect these negotiations, which he kindly undertook to superintend ; and after a considerable delay, which had not, however, the effect of enabling us to escape the hottest hours of a very warm day, we set forth under the guidance of Jean Pierre Feilay, who had been recommended by M. Gard, and who presented a fair specimen of the manly bearing and somewhat haughty independence which I have mentioned as characteristic of the inhabitants of this valley. After half an hour's walk from Chable, we reached Champsec, a small hamlet, in a great measure destroyed by the catastrophe of the

inundation of 1818. Here our guide lived ; and as he had some
domestic arrangements to complete, we lost the greater part of another
hour in waiting for him. At last all was complete, and we were fairly
in marching order. A little way beyond, we gained the northern side
of the Dranse ; and having passed the village of Lourtier, the last in
the valley, the path ascends rapidly. The river is discharged through
a sort of chasm, which shows evident marks of the devastating force
of the torrent on the occasion alluded to. The character of the
scenery becomes more grand, the walnut trees and irrigation disappear, and we are once more in the region of pines and savage rocks.
We remarked here a pretty illustration of the friction of glaciers as
distinguished from that of water. The sides of one of the ravines
through which the stream struggles is distinctly marked on its bold
limestone surface by the long grooves which have been considered as
peculiarly characteristic of the abrasion of glaciers. Though the descent
is very steep, and the wall of rock almost vertical, these chiselled and
polished grooves are worn out in a nearly horizontal, slightly declining
direction, and are *continuous* for many yards or fathoms. Superimposed upon these, on the very same surface, are the marks of wear
resulting from the action of floods, probably charged with great masses

Glacier and Water Marks on Limestone.

of debris. The water-marks are
rough and contused, quite in contrast with the smooth prolongation of the other. They also slope
downwards at an angle similar to
that of the river bed, whilst, as
has been said, the others are nearly
horizontal.

A succession of basins and rocky chasms diversifies the length of the
valley during several hours. I have seldom felt heat more oppressive
than during the first part of this walk, while toiling up the steeps above
Lourtier. Having, for several weeks previously, been almost constantly
on the ice and at a height of 6000 feet above the sea, the contrast of
temperature was, I suppose, more strongly felt. The chasms presented
wild cascades, containing the whole body of water in the Dranse ; but

the picturesque effect was certainly very much injured by the dingy and opaque appearance of the glacier stream, which rendered the sheets dull and lustreless, instead of sparkling and transparent. The valley above Chable is very confined, and almost untenanted; there are but a few châlets inhabited, during a small part of the summer, higher than Lourtier. Hence the Val de Bagnes, which is very long, acquires a wilder and more lonely appearance than many valleys more remote, and more difficult of access. Many cottages which once existed are now dismantled, and it was near one of these that we stopped to take our mid-day meal beside a brook; a little higher the defile became suddenly narrow, and presented a bold and picturesque outline. The Mont Pleureur stood before us on the left, from which descends the well-known Glacier of Gétroz. Still more on the left is the little frequented pass called the Col d'Orsera, leading to the valley of Hérémence, which had been traversed by M. Studer in 1841. The Dranse emerges from a dark defile, impassable on the left, and only to be traversed on the right by taking a high line above its level; from thence the water, swelled to its fullest in the month of August by the contributions of the various glaciers which we were soon to approach, emerged, sometimes in thundering cascades, sometimes pausing in still deep pools as it passes under a fine and romantic stone arched bridge, called Pont de Mauvoisin, by which we were to pass from the right bank of the river, which, since Champsec, we had continually followed, to its left bank, on which alone we could pass the defile. The bridge here, like almost every other in the valley, was carried away by the Debâcle of 1818, and the present lofty stone one has been since built, with a solidity which is rarely met with in such sequestered spots, where but a very few persons pass during the entire year. A few huts in front —the last built with any degree of solidity—concluded the picture.

The bridge passed, we slowly gained the elevation of rock on the other side. A carefully made path continues for some way farther, and traverses one of those steep inclines of shingle annually swept by avalanches, which require the track to be made afresh every year. This path continues on the left bank of the Dranse at a great height above it, affording at the same time a striking view of the Mont Pleureur, and the glacier which has been the principal cause of so much devastation.

I felt some disappointment in viewing the Glacier de Gétroz, of which I had heard so much, and of which the disastrous effects had been so great. I had expected to see one as vast and beautiful as the Glacier of La Brenva, for example, where, falling into the Allée Blanche, it forms a natural bridge above the torrent, or that of Miage, whose stupendous moraine has formed a lake, as the ice of Gétroz did. Instead of this, I found the defile narrow and confined, and though savage, scarcely picturesque. The proper Glacier of Gétroz is situated at a great height amidst the defiles of the Mont Pleureur, so that its extent cannot be appreciated, or its beauty admired, even from the elevation of the path opposite. The real source of the mischief is a secondary, and very uninteresting looking glacier, which, in its present diminished form, scarcely attracts attention in the depth of the valley, and resembles the masses of unmelted snow which so often choke elevated defiles during a great part of the summer. It is in reality composed of the fallen fragments of the true glacier, projected in the form of avalanches over a cliff of enormous height, where the true glacier terminates, whose mass, as it advances, is broken off, and falls headlong into the abyss. The *glacier remanié* which results is soiled, and imperfectly consolidated, and still forms a partial bar to the river Dranse. It must continue to do so, as long as the stream has no independent outlet, for the defile is so narrow, and the falling masses of the glacier so extensive, that the outlet must inevitably be choked in winter and spring when the Dranse (which owes its origin almost entirely to the glaciers still higher up the valley) has too feeble a current to keep its way clear.

The story of the debâcle of the Val de Bagnes in 1818, is too well known to require to be detailed here, and I have no new facts to add. It is sufficient to call to mind, that twice in the 16th century a similar mishap occurred, and indeed it is difficult to conceive why it should not have been much oftener repeated. The year 1818 had been, as we have seen, remarkable for the extension which most of the glaciers in Switzerland had experienced after a series of cold winters, and in this year the ice beneath the Glacier of Gétroz accumulated so much, as to have formed, by the stoppage of the Dranse, a lake no less than half a league long, 700 feet wide, and at one part, 200 feet deep. The impending danger was perceived,—the bursting of the lake with the return of spring was a certainty. M. Venetz, the intrepid engineer of the Valais, and the founder of the modern Geological Theory of Glaciers, proposed to avert it by cutting a canal through the ice, which should gradually drain the lake. Between the 10th of May and the 13th of June this was effected, and it was trusted that the channel would be sufficiently deepened to let the water gradually escape. But water already at 32 has only a feeble action in eroding ice, and the result was, that the cascade tumbling over the icy barrier worked back upon it so fast, that the gallery or canal, which had been originally 600 feet long, was destroyed, and fell away in fragments. Nor was this all, the cascade working on the soil beneath had loosened it so as to detach the remaining ice from the mountain, and thus precipitated the catastrophe. A deluge of 500 millions of cubic feet of water were let loose in the space of half an hour, to sweep through a tortuous valley full of defiles,—literally with the besom of destruction. A flood five times greater than that of the Rhine at Basle filled the bed of a mountain torrent. It was an awful, but a grand lesson for the geologist. The power of water was exerted on a scale such as Hutton and Playfair would have desired to see, could it have been exerted without the destruction of life and property. Bridges yielded; that at Chable dammed back the torrent upon the village, but happily gave way just as the houses seemed doomed to ruin. In this short space of its course, (from Gétroz to Chable,) the fall is no less than 2800 feet. Its acquired velocity was therefore enormous,—at the commencement of its course 33 feet in a second. Its power to *overthrow* buildings, and to *carry with it* trees, hay-stacks, barns, and gravel, cannot surprise us. But its transporting force upon blocks has probably been over-rated.[*] Enormous masses were certainly *moved*, especially in the

---

[*] On the Debâcle of Bagnes, see *Bibliothèque Universelle*, 1818 ; *Edin. Phil. Journal*, vol. i. ; LYELL's *Geology*, 1st edit., vol. i. ; CAPTAIN HALL's " *Patchwork*," vol. i.

neighbourhood of Martigny, as described by Captain Hall and Mr.
Lyell, who were both on the spot soon after the event. But there is
no kind of evidence that these granite masses were brought down from
the higher valleys by the torrent. On the contrary, I believe that
there is no question but that they lay (having been transported by
ancient glaciers. or in some other mode) within a very short distance
of their present positions, and that some of them were merely rolled
over a few times by the force of the current. I apprehend that the
Debâcle of Gétroz gives no countenance whatever to the opinion, that
blocks of 20 or 30 feet of linear dimensions can be transported to
any distance even by such stupendous currents.

When we passed the Glacier of Gétroz, there were workmen (for
whose use chiefly, no doubt, this road is kept in repair,) employed
in dividing the ice into blocks, by the ingenious process of Venetz, in
order to be carried off by the stream, and prevent future accumulations. The process consists in turning streamlets of water (not ice
cold) by means of wooden canals upon the ice, so as to saw it through
in the required direction, which is.effected with rapidity and certainty.
This operation is annually repeated, requiring the combined labour of
several men for many.weeks each summer. The expense is borne by
the Canton. There is but one way of permanently avoiding the risk
in future, namely, by constructing a tunnel, or cutting one through
the rock, by which the torrent may have a certain egress, independent
of the state of the glacier; but this has been considered as too expensive and difficult an operation under the circumstances.

Our way now lay up the bed of the former so formidable lake. The
bottom of the valley is flat and monotonous, the river wandering from
side to side, amidst rolled pebbles. Descending to its level, we re-
crossed to the eastern bank. Our walk from Chable had cost us nearly
four hours, and an hour and a half later we reached our humble rest-
ing-place for the night, the châlet of Torembec, 5300 feet above the
sea.

The accommodation offered in the upland and unfrequented châlets
is everywhere nearly the same, and may therefore be worth describing
for once. There are usually two buildings, quite distinct, the day and

the night apartment. The reader must not, however, suppose that these correspond in the remotest degree either in appearance or in furnishing to the correlative establishments of a drawing-room and a bed-room; the first contains neither tables nor chairs, the latter neither mattress nor pillow. The morning room is more properly a manufactory of cheese and butter than a place of ordinary accommodation. The fire is kept up for the purpose of heating the milk, which is done in copper cauldrons, whose size, and weight, and bright polish contrast strongly with the want of every ordinary convenience of life. A repetition of copper and other vessels for holding milk and raising cream occupy most of the spare room in the apartment; the floor is of earth and uneven, but, except in Piedmont, not usually dirty. The fire-place is a hole in the ground, the fuel is juniper, or scraps of larch wood where these can be had; and a sort of moveable wooden crane, from which the copper pot is hung, is one of the most artificial accommodations. There is no chimney, and therefore the fire is usually made near the door; nor are there windows of any description. For light, they use a little fat, burning with a wick in a small vessel, but often merely a bit of the more resinous pine wood, which they keep on purpose. There is no such thing as a table, unless the top of a chance barrel be admitted as the representative of one; nor are there any chairs, though the *one-legged* milking-stool, which affords an inconvenient repose to a weary traveller, is an indulgence which he probably owes solely to its indispensability in the great and overweening object in which all the uses and habits of a châlet centre,—the keeping and feeding of cows, and the procuring and manufacture of milk. Morning, noon, and night, the inhabitants think but of milk; it is their first, last, and only care; they eat exclusively preparations of it; their only companions are the cattle which yield it; money can procure for them *here* no luxuries; they count their wealth by cheeses.

The absolute want of culinary utensils is surprising and embarrassing. The only pot is sometimes that employed for heating milk, and of *copper;* at other times, there is also an iron one; but except certain wooden skimming-spoons, nearly square, and five or six inches wide in the mouth, there is often no other kind or description of dish, vessel, platter, spoon, or ladle. Where the civilization is a little greater (as at Torembec), there are a few *écueils*, or wooden bowls. Of course these deficiencies only created amusement to us, and the rice we had brought was boiled with milk and salt (which is kept for the cattle) in the only iron pot, and made a most substantial and not unpalatable mess for five hungry men, with a surprisingly small consumption of our stock. The evening meal being concluded, we betook ourselves to early rest. The sleeping apartment, I have said, is usually, as in this case, a separate hut, without window, fire, or chimney, built of loose stones, and with a door about three feet high, the floor being covered with grass more or less dry. On this we arranged ourselves in *parallel order*, covering ourselves with a sufficiency of the hay. It might have been hoped, that here we should have escaped the torments of a bad bed,—I mean the vermin ; but we had the inconveniences of a hay-loft without its inestimable advantage—cleanliness ; and in the course of the night I was forced to rise, and, stumbling over the bodies of four or five of my insensible companions, seek relief for a while in the open air, which was exceedingly mild.

We were astir by five. But it is impossible, generally speaking, to depart in a hurry from a châlet, any more than from a fashionable hotel. It was half-past six before we had breakfasted, and made up our packages : and having left our hosts satisfied by a moderate gratuity, our caravan was once more under way with the glaciers in our front. Before leaving the subject of châlets, I may observe that the character of the inhabitants is not undeserving of notice. I have always received, both in Switzerland and Savoy, a gentle, and kind, and disinterestedly hospitable reception in the châlets, on the very bounds of civilization, where a night's lodging, however rude, is an inestimable boon to a traveller. These simple people differ very much (it has struck me) from the other inhabitants of the same valleys—their own relatives, who, living in villages during the busy trafficking season of summer, have more worldly ways, more excitement, wider interests, and greater selfishness. The true *Pâtre* of the Alps is one of the simplest, and, perhaps, one of the most honest and trust-worthy of human beings. I have often met with touches of character amongst

them which have affected me, as I may elsewhere notice ; but, generally, there is an indescribable unity and monotony of idea which fills
the minds of these men, who live during all the finest and stirring part
of the year in the fastnesses of their sublimest mountains, seeing scarcely
any strange faces, and but few familiar ones, and these always the
same ; living on friendly terms with their dumb herds, so accustomed
to privation as to dream of no luxury, and utterly careless of the fate
of empires, or the change of dynasties. Instead of the busy curiosity
about a traveller's motives and objects, in undertaking strange journeys, which is more experienced in villages the more remote they be,
these simple shepherds never evince surprise, and scarcely seem to
have curiosity to gratify. Yet far are they from brutish or uncouth ;
they show a natural shyness of intermeddling with the concerns of
strangers, and a respect for their character testified by their unoffi-
cious care in providing and arranging what conveniences they can produce. Their hospitality is neither that of ostentation nor of necessity. They give readily what they have, and do not encumber you
with apologies for what they have not. Every traveller will see in
this description, strong opposition to the Swiss character as usually
displayed ; my remarks are confined to my experience in the higher
châlets of the Alps. Of course, I do not mean to state that exceptions
are not to be met with.

The same *menage* exists merely on a larger scale, where the *Alp* or
pasture ground is greater. In many an extensive range of cow-houses
is attached to the enclosure of the châlets. In some places, the cows
are brought in to be milked ; in others, this operation is more picturesquely performed by ranging the cows—(*Ranz des Vaches*—whence
the popular name of some Swiss airs)—on green sward terraces on a
hillside, where they may be seen to the number of some hundreds, tied
each to a little stake, whilst the shepherds busy themselves amongst
them with their milk-pails and one-legged stools. But to return to the
Val de Bagnes.

At half-past six, we left Torembec, and ascended the remaining
part of the valley, which opened itself a little higher (the now small
stream of the Dranse being again crossed to its western or left bank)

into a scene of greater majesty than it had yet presented. A corner was turned, the valley trending more to the south-east, and several glaciers hitherto concealed came into view. The recollection of the heat of yesterday made these a welcome sight, and I looked forward with pleasure to setting foot on ice again.

The first glacier visible on the right hand descended in 1821, as our guide Feilay informed us, so far into the valley as to approach the torrent. It has now retreated to a great height on the mountain side. Again, on the opposite or eastern bank a vast glacier descends from the lofty chain which separates the Val de Bagnes from that of Hérémence. It is called the Glacier de la Brēna, and is probably that marked "les 28" in Wörl's Map. It now terminates on the bank of debris, which it has carried down on the farther side of the torrent, but we were assured, that in 1822 it had extended so far as to cross the torrent, which made its way under it, and to rise to a great height on the western side. Indeed, this was matter of ocular evidence, for our path touched the extremity of the enormous frontal moraine which it had thrown up,—a mound of rocky fragments, from whose top we could clearly survey the vast area, of many acres in extent, which the glacier has uncovered during the last twenty years, strewed with fragments, and doomed to sterility. The material of the moraine is a true granite, the first we had met with in this valley, for below the rock is a kind of gneiss. According to our guide, the ice then presented a front seventy feet high. A little farther in advance, an extensive glacier, named Glacier de Durand, descended from the Mont Combin on our right, which it was impossible to avoid; we therefore prepared to cross it, which we did without difficulty. It descends quite into the valley, and crosses the stream as the Glacier de la Brēna had done, leaving a free passage beneath. The mere crossing of a valley by a glacier, if it be of any moderate breadth, is not of itself sufficient to produce a catastrophe like that of Gétroz. Here is one example; the Glacier of La Brenva in the Allée Blanche is another, and that of Allalein in the valley of Saas. It is probably the circumstance that the dam was formed by the *éboulement* of the Glacier of Gétroz, and not by the glacier itself, which occasions its particular danger. A channel once formed under a gla-

EYE SKETCH
OF THE
GLACIER OF CHERMONTANE
AND
COL DE FENÊTRES.

Mont
Avril

Col de Duran

Chalet de
Chermontane

Drance R.

Col de
Fenêtres

Chanrion

Lac

Grand Glacier de Chermontane

To Biona?

Ottemma

Avas

Trumma
de Bouc

Col
(leading to
Mevgenne)

Snowy Peak

Glacier
d'Arolla

Bramile
Veins

Mont Collon

Col de
Collon

Chalet

Evărayon

Val Biona

Chalet

Probable Passage to
the Val Tournanche

EYE SKETCH
OF THE PASS
OF THE
MONT COLLON.

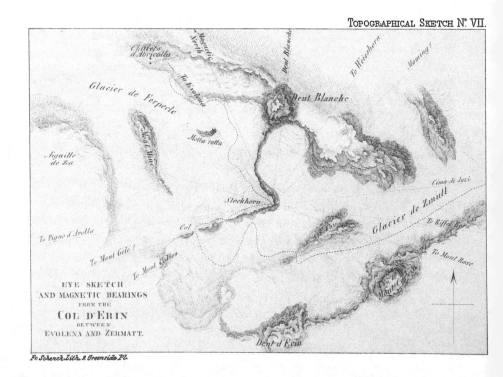

Chalets
d'Abricolla

Magnetic
North

Dent Blanche

To Weisshorn

Morning?

Glacier de Ferpecle

To Evolena

Dent Blanche

Aiguille
de Za

Mont Miné

Motta rotta

Cima di Jazi

Glacier de Zmutt

To Pigno d'Arolla

Stockhorn

Stockhorn

To Riffel Horn

Col

To Mont Gelé?

To Mont Collon

To Mont Rose

Mont C

EYE SKETCH
AND MAGNETIC BEARINGS
FROM THE
COL D'ERIN
BETWEEN
EVOLENA AND ZERMATT.

Dent d'Erin

cier is kept continually open, as the glacier advances gradually on-wards, but the falling in of ice may produce an abrupt stoppage.

The Glacier of Durand presents an even and clean terminal slope, of a convex form, with few fissures, and shows the system of veins which I have elsewhere described as proper to that form.

This glacier crossed, we arrived at the upper Châlets of Chermon-tane, at the foot of the glacier of the same name, which fills the entire head of the Val de Bagnes, and nearly touches the Glacier de Durand, (see the Map, page 1, and Topographical Sketch, No. V., which shows the pass of the Col de Fenêtres.) From these châlets (which are still on the western side of the valley, and at the foot of a hill called Mont Avril) there is a very fine view; the Glacier of Chermontane is a magnificent sea of ice, nearly or quite unexplored. It appears to have three great tributaries; one descending from behind the mountain called Ottemma, and where there is every appearance of there being a Col or pass: we thought that we clearly saw the summit level. In this direction, it is probable that a passage might be effected to the Glacier of Lenaret in the Val d'Hérémence, or to that of Arolla, at the head of the Vallée d'Erin; but the descent on the other side would be more difficult. The second branch passes between the Trumma de Bouc and the Mont Gelé, derived partly from a very lofty snow-capped peak, and partly from a short branch imme-diately behind the Mont Gelé, which can be of no great extent, since, from its direction, it must speedily reach the Val Pelline; and, indeed, we were informed that the shortest way to the village of Biona was in that direction, but our guide had never passed there. The third great arm of the glacier stretches up to the Col de Fenêtres, between the summit of Mont Gelé and Mont Avril, by which we were to pass. The Glacier of Chermontane terminates a little below the Châlet of that name. On its farther side is a pretty pasturage, called Chamrion, (*Champ Rond*) where there are two small lakes—one formed in a hol-low of the hill, the other between the slope of the hill and the ice of the glacier, somewhat like the lake Mörill on the Glacier of Aletsch.

I have already said that the upper part of the Val de Bagnes is little visited. I find no notice of it in the writings of De Saussure;

but Bourrit, in his lively work on the glaciers of Savoy, describes his having reached the Châlet of Chermontane, where he slept two nights, and visited the neighbouring glacier, of which he gives a somewhat pompous account, and a most exaggerated drawing of the lake ; but he did not attain any summit or Col : indeed, I have not met with a description of the Col des Fenêtres, from personal observation in any work. It was by this pass that Calvin fled in 1541 from persecution in Aosta, where he had been established for five years.* Though M. Bourrit speaks much of the discoveries which he made during his visit to Chermontane, they appear to amount merely to this : that he ascertained the existence of a great glacier, but neither its extent, its practicability, nor the connections of the ramified valleys which meet near its head. Formerly, it appears that this Col, like many others in the higher Alps, was easier passed than at present, and was even a common route of commerce. At that time, it is stated,† the Glacier of Durand did not extend so low as to require to be crossed, but was avoided. So small is the communication now, that there is not even a station of custom-house officers on the pass, though there is at Valpelline.

We did not stop at Chermontane, or even go to the châlets, but keeping on our way at a higher level, along the slope of Mont Avril, we gradually ascended towards the Col de Fenêtres, always on turf, and without any difficulty. The ascent was tedious, and we skirted the glacier without going upon it, for the greater part of the way. The Glacier de Fenêtres is but little inclined or crevassed ; in its higher part we traversed a portion of it without difficulty, so as to gain the Col more quickly. We reached the summit in four hours of easy walking from Torembec. For its height—which appears to be 9213 English feet, by M. Studer's observations—this must be considered as an easy pass, presenting, in good weather, not a shadow of danger.

The view towards Italy is wonderfully striking. The mountains beyond Aoste, and the glaciers of the Ruitor, are spread out in

---

* Compare Bourrit, II., 76, and Baruffi Viaggio in Piemonte ; Lettera, 27ᵐᵃ, p. 19.

† GODEFROY, *Notice sur les Glaciers*, p. 63.

the distance, and beneath we have the exceedingly deep valley of Ollomont, communicating with the Val Pelline, which is itself a tributary of the Val d'Aoste. It is enclosed by ridges of the most fantastic and savage grandeur, which descend from the mountains on either side of the Col on which we stood,—on the north-east, from the Mont Combin, rising to a height of 14,200 English feet ; on the south-east, from the Mont Gelé, which is 11,100 feet high, and almost too steep to bear snow, presenting a perfect ridge of pyramidal Aiguilles stretching towards Valpelline. The side of Mont Gelé towards the Col presents an adhering snowy coat so steep, that, seen in front, it appears almost vertical ; measured laterally with a clinometer, its angle was found to be 55° ; this appeared to be loose snow. Our course to Valpelline required us to skirt the foot of the peaky ridge just described : the descent was unusually rapid, and without particular difficulty. We passed a small lake partly bordered with snow, and soon after gained the pastures. Here we made a hearty meal by a brook, which exhausted a good part of our available provisions, and we thence dismissed our guide, who had plenty of time to recross the mountain by daylight. It was a considerable way before we reached any châlet, but when we did so, we caught a charming view of the bottom of the valley of Ollomont, which had hitherto been mostly concealed, covered with exquisite verdure, studded with houses, and traversed by lively streams, all seen as on a map, for our elevation was still 2000, if not 3000 feet above it. Beyond, the mountains near St. Bernard were apparent ;—below, the village of Vaux which we mistook for Ollomont. There we found copper works abandoned ; they appear to have been very extensive and complete ; the ore is a sulphuret, in the (metamorphic ?) gneiss of which the whole of this district is composed. There are several other villages, and Ollomont itself, composed of but a few scattered houses, distinguished by a church, is pleasingly situated. But here, as at Aoste, the enjoyment of natural beauty is rendered impossible by the loathsome deformity of the inhabitants ; we were really shocked to find that none of the villages through which we passed seemed to contain one reasonable human being ;—goîtres and cretinism appeared universal and inseparable. Repeatedly I tried to obtain an answer to

a simple question from the most rational looking of the inhabitants—
but in vain. This astonished and shocked us, for we were still at a
height of 4000 English feet above the sea, where these maladies commonly disappear; and we looked forward with despair to the prospect
of obtaining a guide for the difficult and unknown country which we
were next to traverse, from amongst such a population. But in this,
as in very many similar cases, first appearances are not to be interpreted to the letter. It was still the fête of Notre Dame du mi-Août,
and the effective population had mostly gone down to Valpelline, the
chief place of the district, and others perhaps were with their herds in
the mountains.

The scenery continued more and more engaging. In the course of
four hours walk we had passed from ice and eternal snow to the charms
of Italian scenery and climate, with more than Italian verdure. We
looked anxiously about for the village of Valpelline, which we expected
to have seen from a distance,—we feared that our maps had deceived
us, and that we had yet a considerable walk before us, when suddenly,
on turning a corner, we found ourselves in the valley of Valpelline ;
the church, with a spire of the Italian taste, and a few scattered houses,
mantled with vines and peeping out amidst walnut trees of exquisite
beauty, proclaimed the little capital of the district. In descending
we noticed large fragments of true syenitic granite, which appeared to
have their origin at no great distance, which we hoped that our next
day's walk would reveal ; in the meantime we entered the village.

CHAPTER XV.

FROM VALPELLINE TO EVOLENA BY THE COL DE COLLON.

ASCENT OF THE VAL PELLINE OR BIONA—GEOLOGY—SYENITES—CHALETS OF PRARAYON —HEAD OF THE VALLEY—ASCENT OF THE COL DE COLLON— REMAINS OF TRAVELLERS LOST IN A TOURMENTE—GLACIER D'AROLLA— ITS STRUCTURAL BANDS—MAGNIFICENT VIEW OF MONT COLLON—OPPOR- TUNE MEETING WITH PRALONG—HISTORY OF THE VICTIMS—ARRIVAL AT EVOLENA.

" C'est le domaine des glaces et des neiges, le palais de l'hiver, le royaume de la mort."
A. DUMAS.

THE village or hamlet of Valpelline offered little prospect of comfort- able accommodation, but we recollected a letter with which M. Biselx had provided us at Orsières, addressed to a proprietor and householder of the place, by whom we were received in a manner which I am sure that neither M. Studer nor myself will ever forget. The unexpected appearance of travellers by so unfrequented a pass, and accompanied only by strangers, (for it will be recollected that we had sent back our guide to Bagnes) produced a momentary hesitation. The wife of the gentleman to whom we were recommended had not returned from church, and an awkward pause took place at the door of the house, which was locked, whilst our arrival excited some curiosity amongst the loitering groups around. At length the lady came, and hearing our story and recommendation, instantly set about every arrangement which true hospitality could devise to ensure our comfort whilst we remained, and to speed our journey when we departed. The afternoon

S

was not far advanced, and we spent it in repose,—in a short stroll
through the beautiful meadows surrounding the village,—and in convers-
ing with our host and his sons, well educated and sensible boys, whilst
our excellent hostess busied herself in preparing supper and in arranging our apartment, which was the best the house afforded.   Meanwhile
we made inquiry, not without anxiety, as to the possibility of finding
a trusty and skilful guide who should conduct us across a glacier-pass
which we understood to connect the head of the valley of Valpelline,
which is in Piedmont, with the Vallée d'Erin in the Vallais. This had
always appeared the most doubtful step in our expedition.   Though
we had reason to believe that such a pass existed, we had no information of any traveller who had actually passed it, and we had been led
to think that though guides might be found on the Swiss side, it would
be much more difficult to procure them in Italy.   The specimen we
had seen of the natives of Ollomont increased our doubts ; but the very
circumstance of the fête, which had drawn so many to Valpelline, gave
us the greater choice of guides, and our host kindly aided us in the
selection, and by his authority and consequence in the place, procured
us a most satisfactory guarantee for the capacity and fidelity of any
one who should accompany us.   Amongst the visiters at Valpelline
that day, was a tall athletic and handsome man, below middle age, who
passed for being the strongest man of the whole valley, and whose usual
residence was some leagues higher up.   With him our arrangement
was soon made ; he promised to remain all night, and to accompany us
next day to the head of the valley of Biona (as the higher part of the
Val Pelline is called,) whence starting early the following morning, the
glaciers might be crossed to Erin.   He assured us that he was perfectly acquainted with the pass, which he called the *Col de Collon*.

The village of Valpelline is near the opening of the valley of the
same name, and only from two to three hours' walk from the Cité
d'Aoste.   It possesses the Italian character of scenery and products,
although 3040 English feet above the sea.   The morning of our
departure proved the prelude to a very hot day.   We were tempted to
rest longer than usual in our comfortable quarters, and as we had but a
short journey before us we were in no hurry to depart.   Madame A——

had anticipated all our wants. She had even prevented our servants from attempting to procure any of the necessaries which we wanted for our arduous journey, by insisting on providing them, much more effectually of course, from her own stores. The cordiality and genuine kindness of all her arrangements left us no room to offer any return but our truly heartfelt thanks for her generosity, and we quitted this worthy family with regret, being accompanied by one of the sons for a mile or two on our way.

The valley was always narrow, but at Oyace, a little way above Valpelline, it seems too close, and the village of that name is planted upon a rocky barrier which crosses the ravine, and which we found to be composed of true syenite, the same as M. Studer first noticed in boulders the day before, when descending upon Valpelline. There appear to be from point to point amongst these wild hills, outbreaks of syenitic rocks which have more or less metamorphosed the neighbouring sedimentary deposits, and have confounded all mineralogical characters in the result of this supervening action. Such at least was the opinion of my learned companion, whose long and close attention to the excessively intricate phenomena of Alpine geology entitles it to the greatest weight ; and to which any observations which I had an opportunity of making in his company induce me entirely to subscribe. It is well known that M. Sismonda, the intelligent geologist of Turin, has endeavoured to separate the rocks of this part of the Alps into primitive and metamorphic, the one of which he has coloured red, and the other blue. So far as we could observe, this separation seems indistinct and inconclusive ; and with the single exception of the true unstratified syenites,—such as those of the Glacier of La Brena in the Val de Bagnes, and that of Oyace,—the felspathose rocks seem to admit of no subdivision, but must be classed under the common denomination of Gneiss, whether primitive or metamorphic.

The boulders already mentioned, and others which occur from time to time in the valley, appear to be all derived from the neighbouring mountains ; and it is exceedingly remarkable, and quite in contrast to the appearances in the Val de Bagnes, that we found few or no striated and polished rocks, nor great masses of transported materials.

Between the villages of Oyace and Biona, we visited a vein of limestone, interstratified with the felspathose rocks in a direction parallel
to the length of the valley, and re-appearing at intervals up even
to its very highest part, where, as here, it is burnt for lime. Very
near this, copper is found in the same rock as at Ollomont.

The village of Biona is the last of any size in the valley,—the last,
I think, which has a church. The valley takes henceforth the same
name, Biona. We halted here, and made a hearty meal in the open
air upon fresh eggs and good Aostan wine. We then resumed our
march, as the day became cooler, and the scenery, at the same time,
still more picturesque and interesting. An excellent foot or mule
path leads all the way up the valley, a convenience which the traveller
owes to the Jesuits of Aoste, who have extensive property in the higher
pastures of Biona ; and it was at the châlet belonging to them that we
proposed passing the night. The village of Biona is 5315 feet above
the sea, by M. Studer's observation. Farther on, the larch trees
descend into the valley, and the river passes through some picturesque
defiles. The views looking back were very pleasing, and in front, at
the head of the valley, rose a lofty chain of mountains, (a mere appendage, however, to the great chain,) separating the valley of Biona
from the Val Tournanche ; over which we afterwards learned that a
passage may be effected, though not without difficulty.

At length we reached the Châlets of Prarayon, which belong to
the Jesuits of Aoste, and are marked by a lofty crucifix in front. They
are pleasingly situated in a green meadow near the head of the valley,
and about six hours walk from Valpelline. There was no one visible,
and it was some time before we obtained admission into the smaller
and humbler building, the larger one being locked up. Whilst supper
was preparing, I walked up alone to the head of the valley, which I
was anxious to explore, for our guide informed us, that our next day's
journey did not lie in that direction, but that we should have to return
upon our steps a little way, and then turn sharply to the northward.
It was an hour's walk to the commencement of the glacier, which fills
the top of the valley, and which descends directly from the great
chain. Having gained an eminence on the south-east side of the val-

ley which commanded the glacier, I saw that the ascent of it must be in some places very steep, though, I should think, not wholly impracticable. I recognised the limestone which we had found farther down the valley. Returning to the châlets, I found our evening meal prepared ; and I observed the temperature of boiling water to be 201°.58, whilst M. Studer's barometer stood at 608.3 millimetres. The height above the sea is 6588 feet. The general direction of the Valpelline is N. 60° E., (true ;) but for the upper two leagues N. 75° E., as far as the foot of the glacier, after which its course is N. 5° W.

We passed a comfortable night in a clean hay-loft, and slept longer than we intended, for we were not ready to start until 6 A.M. The morning was very favourable. Our guide, " l'homme fort de Biona," as he was called, or " l'habit rouge," the soubriquet which we had given him, from the curious practice of wearing scarlet cloth coats, which is common in the Pays d'Aoste—gave us at first no small concern. He was in low spirits last evening, and in no hurry to start to-day, and apparently not averse to draw unfavourable presages of the weather. We began to fear that he had undertaken more than he could perform, and that the way was perhaps known to him only by report. But our doubts gradually vanished. He took to the hill with that instinctive confidence which showed that he understood his business, and the farther we advanced, the more readily did he go on, and became more communicative. We afterwards found that he had been really unwell from the results of a drunken fit, from which he had not escaped when we first engaged him, and also that some doubt whether we should be able to follow him over the glacier and rocks, and a fear that he might be brought into trouble through our means, had probably oppressed him. We found him gentle, docile, robust, and trustworthy. During a part of this day's journey, he carried not only all our provisions, but no light share of the contents of Klaus's hotte or basket. His name was Biona, as well as that of his native place.

As we had been told the night before, we returned a little way upon our steps ; then, following a water-course used for irrigation, we turned sharply to the right. All our maps were here at fault. That of Wörl especially, the most detailed, presents no kind of resemblance

to the outlines even of the great chain, and the passage must have
been put down at random. It will be seen by the Topographical
Sketch, No. VI., which probably approximates to the real arrangement of the mountains, though in some degree conjectural, that the
pass is through the first lateral valley of the Val Biona below its
head. We there find a deep gorge, completely glacier-bound at its
upper end; but from the nature of the rocks, it admits of an easier
ascent than the glacier at the top of the Val Biona. We passed some
wretched shepherds' huts; and following an impetuous stream, we
came to the foot of a glacier descending on our left, which has blockaded
the valley with its prodigious moraine, and left a swampy flat above.
This passed, we kept to our right hand, having in front of us another
great glacier, which descends from the Col de Collon, and more to the
left a great and steep glacier, which appears to descend from the group
of mountains connected with the origin of the Glacier of Chermontane.
The direction of the valley we ascended was at first N. 20° W., (true,) and
when we came in sight of the glacier which we were to follow, it turned
sharply to N. 25° E. Pursuing a very steep and laborious ascent over
rocks, (without, however, any danger,) we reached the glacier, where
it was much more level than in its lower part, and obtained a distant
view of the Col. The ice was not much fissured, and we proceeded at
ease—only we came at length to where it was covered with perpetual
snow, and there we required to proceed with caution. We left upon
our right hand, the mass of mountains which separate this pass from
the head of Valpelline, and on our left new and hitherto unseen
chains began to display themselves, and rocks rising above the Col
or pass, which we were surprised to find marked by a very small iron
cross, shewing that it is well known to the country people, although
unfrequented by travellers. The only traveller whom I am aware
of as having passed here is M. Godefroy, the author of an Essay on
Glaciers,* already quoted. We now also learned the secret of our friend
" l'habit rouge," being so well acquainted with this obscure route,

---

* *Notice sur les Glaciers,* p. 65.

for he admitted that he had frequently passed it with bands of smugglers, who avail themselves of all the less frequented passes for introducing the articles of free commerce in Switzerland into Piedmont. We reached the Col in three hours from the châlet, which was sooner than we expected; and as it was only nine o'clock, and a beautiful morning, we sat for a long time on the rocks on the west side of the Col, and enjoyed the noble scenery. Although the height is 10,333 feet above the sea, (barometer, 528.1 millimetres,) it is so much surrounded by summits still more elevated as to command no very distant scenery. But before us, to the north, rose the majestic form of Mont Collon, round which swept the very extensive glacier which we had yet to traverse in its entire length during several hours; and to the eastward, beyond snow-fields of seemingly great extent, rose snowy peaks, which afterwards appeared to me to be the same as I saw from the Col of Ferpêcle, and over which it is just possible that a passage might be effected from the Val Biona to that of St. Nicolas, though, from the distance, it might probably be impossible to accomplish it without sleeping out on the glacier.

As we were far above the limits where water is found on the glacier, I used my portable furnace to melt snow for the use of the party, and afterwards to ascertain the temperature of boiling water, which I found to be 195°.15. We spent an hour of great enjoyment, for we now saw our way clearly, and all doubts were at an end of accomplishing a passage which, not to have performed, would have materially deranged our travelling plans; we then set forth in a cheerful mood to descend the long stretch of glacier which lay before us. There were few crevasses,—though whilst on the snow we walked with precaution and in a line, but without ropes :—we descended rapidly, whilst the majestic form of Mont Collon rose with increased grandeur before us. When we were fairly abreast of it our guide set up a wild and sonorous shout which the rarely wakened echoes of those stupendous precipices sent pealing back again in tones yet more fantastic. He added that this echo was well known to the smugglers, and that the reverberation of Mont Collon served to guide them in foggy weather, in a track which must be then singularly perilous, from the great breadth and

monotony of the glacier here, and the number of branches into which
it divides in its higher part, any one of which might easily be mistaken
for another.

Whilst we were amusing ourselves with the discordant shouts of the
party and responses of the mountain, our attention was suddenly led to
a very different matter. A dark object was descried on the snow to
our left, just under the precipices of Mont Collon. We were not yet
low enough to have entered on the ice, but were still on snow. This
proved to be the body of a man fully clothed, fallen with his head in
the direction in which we were going. From the appearance of the
body as it lay, it might have been presumed to be recent; but when it
was raised, the head and face were found to be in a state of frightful
decay, and covered with blood, evidently arising from an incipient
thaw, after having remained perhaps for a twelvemonth perfectly congealed. The clothes were quite entire and uninjured, and, being hard
frozen, still protected the corpse beneath. It was evident that an un-
happy peasant had been overtaken in a storm, probably of the previous
year, and had lain there covered with snow during the whole winter
and spring, and that we were now, in the month of August, the first
travellers who had passed this way and ascertained his fate. The hands
were gloved, and in the pockets, in the attitude of a person maintaining
the last glow of heat, and the body being extended on the snow, which
was pretty steep, it appeared that he had been hurrying towards the
valley when his strength was exhausted, and he lay simply as he fell.

The effect upon us all was electric; and had not the sun shone forth
in its full glory, and the very wilderness of eternal snow seemed glad-
dened under the serenity of such a summer's day as is rare at these
heights, we should certainly have felt a deeper thrill, arising from the
sense of personal danger. As it was, when we had recovered our first
surprise, and interchanged our expressions of sympathy for the poor
traveller, and gazed with awe on the disfigured relics of one who had
so lately been in the same plight with ourselves, we turned and surveyed, with a stronger sense of sublimity than before, the desolation
by which we were surrounded, and became still more sensible of our
isolation from human dwellings, human help, and human sympathy,—

our loneliness with nature, and as it were, the more immediate presence of God. Our guide and attendants felt it as deeply as we. At such moments all refinements of sentiment are forgotten, religion or superstition may tinge the reflections of one or another, but, at the bottom, all think and feel alike. We are men, and we stand in the chamber of death. Our friend of Biona, though he was the first to raise and handle the body, from which the others rather shrunk,—and though he examined the rigid clothes for the articles which they contained, and with our consent took out a knife and snuff-box from the pocket, and a little treasury of mixed Swiss and Piedmontese small coins, concealed in a waistband all entire and untouched, (by means of which we could identify the person and restore the money to his friends,)—though he performed all this with seeming indifference, we had no sooner left the spot than he declared that he would rather make a circuit home by the Great St. Bernard than return alone by this spot. Indeed it might well require resolution in a solitary man, with the chances of weather, to pass alone a Col like this, where, supposing him caught in a *tourmente*, it would require no vivid sensibility to raise the image of the last sufferer before him, and hasten the moment of despair, when the spirit yields to the pressure of hunger, fatigue, and bewilderment, and subsides insensibly into the sleep which knows no waking.

A very little farther on we found traces of another victim, probably of an earlier date;—some shreds of clothes, and fragments of a knapsack; but the body had disappeared. Still lower, the remains of the bones and skin of two Chamois, and near them the complete bones of a man. The latter were arranged in a very singular manner, nearly the whole skeleton being there in detached bones, laid in order along the ice,—the skull lowest, next the arms and ribs, and finally the bones of the pelvis, legs, and feet, disposed along the glacier, so that the distance between the head and feet might be five yards, a disposition certainly arising from some natural cause, not very easy to assign.

The glacier now enters a regular valley, and leaves the high slopes. It is bounded by Mont Collon on the left, sweeping for some miles round its base, and on the right by rugged cliffs, chiefly of gneiss, in which we

could distinctly see well characterized granitic veins, shooting in irregular zig-zags through the mass. The glacier on which we now were
is the Glacier of Arolla, that which occupies the head of the western
branch of the Vallée d'Erin. It is very long. Probably we might
have continued most easily all the way along the ice towards the
centre; but our guide advised us to follow the right bank along the
moraine, an excessively rough and fatiguing scramble, for a great distance, on angular moving blocks, without a trace of a path. This was
by far the most tedious and disagreeable part of our day's journey;
but M. Studer was rewarded by finding a mixture of gabbro or diallage
rock, in immediate connection with real granite and metamorphic
gneiss, to which he attached considerable importance.

The structure of the Glacier of Arolla is perfectly normal, presenting bands or veins nearly parallel, and vertical throughout a great
part of its length, which sweep round in the conoidal forms, proper, as
we have seen, to the lower termination or unsupported part of the
glacier. The lower extremity is very clean, little fissured, and has
from below a most commanding appearance, with the majestic summit
of Mont Collon towering up behind. The frontal bands are very distinct, and even at a distance of a mile or more, those very marked
ones which, in describing the Mer de Glace of Chamouni, I have
called " dirt bands," and which, perhaps, are the annual rings or
marks of yearly growth of the glacier, are beautifully developed, and
recur at intervals marked with almost mathematical precision.

The stream which descends the valley rises from under an arch of
ice at the foot of the glacier. The bottom of the valley is wide, gravelly, and waste. A number of desolate and stunted pine trees occupy
the western bank, and seem chilled by the near approach of the ice ;
many are dead, and some fallen. They serve to give a scale to the
majestic scenery behind. Their species is the *pinus cembra*, the hardiest
of their class which grow to any size in Switzerland, and they are consequently to be met with at great elevations. This pine has various names.
In the Patois of Savoy, and many other places, it is called " Arolla,"
whence the name of the valley and glacier. It is also called " Arve,"
and " Zirbelnusskiefer." It yields an edible fruit, and the wood is

Drawn from Nature by Professor Forbes. J.Tristan lith.

MONT COLLON AND THE GLACIER OF AROLLA; VALLÉE D'ERIN.

Day&Haghe Lith:rs to the Queen.

soft and well fitted for carving, for which it is preferred, especially in the Tyrol and eastern Alps. This wood of pines lies exactly between the foot of the glacier of Arolla and a small detached one descending from the mountain called Pigno d'Arolla, a summit on the western side of the great glacier.

I ought to have mentioned, that in quitting the northern foot of Mont Collon, during our descent, we left upon our left hand a great tributary glacier, steep and difficult of access, which separates the Mont Collon from the Pigno d'Arolla, and which may possibly communicate with the icy mountain of Chermontane, beyond the head of the valley of Hérémence. We staid some time to contemplate the wonderful majesty of the scene, of which I made a sketch, from which Plate VI. has been executed, and we then proceeded down the valley.

The Châlets of Arolla were a little way lower, across the torrent on our left, and the shepherd who kept them, perceiving the unusual sight of visitors, came down to meet us, and courteously invited us to rest ourselves, which, as the day was not too far advanced, and the way was now plain, we willingly did, and partook of his cheese and hard bread, with excellent butter. The châlets had even a finer view of the glacier than that which we had quitted, and thus looking in front of it, I saw very plainly the succession of structural bands disposed with the remarkable regularity already alluded to. One of our first inquiries was connected with the fate of the unfortunate men whose relics we had observed; and it appeared that our entertainer, Pralong by name, had himself been one of the party to which the most recently deceased of these men belonged. They had started in the end of October last year (1841) to cross the Col into Piedmont, in all twelve men; but being overtaken by a tremendous storm, they at length resolved to return; but too late for three of their number, who, worn out with fatigue, and benumbed with cold, were left behind,—the imperious calls of self-preservation requiring their abandonment. Our informant assured us, that he himself was the last to quit these unfortunate men in succession, when every effort to stimulate or assist them had been tried in vain. We understood that two of the bodies had already been recovered; the third was, no doubt, the one that we first

saw. The articles which Biona had taken from the body were afterwards recognised in Evolena, and the money (which did not amount to
more than three or four French francs) was faithfully paid over to the
Curé, and measures were taken to have the body brought down for
interment.

Our new acquaintance of Arolla gave us other information, which
interested me as much. Having complimented us on the successful
passage we had made, he asked if we were not desirous of attempting
the more arduous passage from Evolena to Zermatt, which, he
assured us, that he and his father had more than once performed, and
that they were indeed the only persons in the valley who had done so.
Now this passage had long piqued my curiosity, having a sort of
romantic interest, which attaches to what has been so seldom performed, as to render its possibility almost fabulous. It was certain
that it must carry the traveller amongst some of the highest and most
majestic peaks of this almost unknown district. Its elevation and
character I had already studied in 1841 from the side of Zermatt, and
had conceived the most lively curiosity to traverse these glaciers, and
to ascertain the relations of a group of mountains 13 and 14,000 feet
high, some of which are scarcely indicated on several of the latest
maps. My great doubt had been as to the possibility of finding any
guide in Evolena, and therefore, that the first man whom we met with
in the valley should be the very person who, I knew from Fröbel's
work, was reported to have some personal knowledge of this celebrated
pass, seemed a piece of good fortune not to be lightly thrown away.
After a short consultation with M. Studer, I found that the heavy
marching trim of the worthy Klaus, and his own wish to visit the
valley of Anniviers, would prevent him from undertaking this journey,
although we were both eventually bound for Zermatt ; therefore after a
few minutes' arrangement I determined my plan, engaged Pralong to
come down to Evolena next morning, and thence start with me in the
afternoon for the foot of the Ferpêcle Glacier, where we might sleep,
and attempt the passage the following day. Pralong desired nothing
better, and we soon started for Evolena. The walk to Haudères,
where the valley of Arolla joins that of Ferpêcle, the union of the two

forming the Vallée d'Erin, was very agreeable, and at times beautiful. At the hamlet of Chatorma we noticed striated and polished rocks, of which, as has been already said, we saw none in the Val Pelline. Below St. Barthelemi the way becomes steep, the torrent descends in rapids, and the banks are clothed with larch and pine wood; the ravine is altogether grand and picturesque. We then came to steep watered meadows, and at length, crossing first one stream and then another, we arrived at the hamlet of Haudères. Half an hour, which seemed a tedious while, over a fertile flat, divided into grass fields, and thickly studded with barns, brought us to the capital of the valley, the village of Evolena, which seemed to us the largest place we had seen for some time. A nearer approach shewed that the houses, which looked so imposing at a distance, were built of logs, and had dark and uninviting exteriors. But when we came to seek for accommodation, we found every anticipation we could possibly have made of discomfort and privation much exceeded.

CHAPTER XVI.

FROM EVOLENA IN THE VALLEY OF ERIN TO ZER-MATT IN THE VALLEY OF ST. NICOLAS, BY THE GLACIERS OF FERPECLE AND ZMUTT.

A NIGHT AT EVOLENA——WRETCHED ACCOMMODATION——DEPARTURE FOR ABRI-
COLLA——ASPECT OF THE GLACIER OF FERPECLE——A NIGHT IN THE CHALETS
——ASCENT OF THE GLACIER——THE MOTTA ROTTA——THE STOCKHORN——MAG-
NIFICENT VIEW OF MONTE ROSA AND MONT CERVIN——DANGEROUS DESCENT
——PRECIPICES——THE BERGSCHRUND——PRALONG RETURNS——THE GLACIER OF
ZMUTT——STRUCTURE OF THE MONT CERVIN——ARRIVAL AT ZERMATT.

WE knew too well what accommodation might be expected even in the *capital* of a remote Valaisan valley to anticipate any luxuries at Evolena. Indeed, M. Studer had already been there the previous year, and having lodged with the Curé, forewarned me that our accommodation would not be splendid. A change had, however, occurred in the establishment of the " Pfarrhaus," since 1841, by the introduction of the Curé's sister, who usually lived at Sion, a person of ungovernable temper and rude manners, who seemed to find pleasure in the arrival of strangers only as fresh subjects whereon to vent her spleen, and to show how heartily she despised the inhabitants of her brother's parish compared to the aristocratic burghers of the decayed town of Sion. Had this been all, and had our corporeal wants been reasonably attended to, we might have forgotten the ill-nature of expressions directed at random against ourselves and all mankind; but we experienced the greatest difficulty not only in procuring anything to eat, but even in being allowed to cook our own provisions. The Curé, a timid worldly man, gave us no comfort, and exercised no hospitality, evidently regarding our visit as an intrusion. Indeed, jaded by a fatiguing journey, without any prospect of beds, (for we had been told at once that we could not lodge in the *Curé*,) we wished ourselves a hundred times, in the course of the evening, at the deserted Châlets of Prarayon, where we had spent the former night; whilst the amiable family of A——— at Valpelline seemed, by contrast, to belong to another race of beings. The faithful Klaus, too, had been taken unwell during the latter part of the day; but there was no alternative but to sit round a table, attired as we were, for two hours, before a soup, prepared with our own rice, was presented to us. At a late hour in the evening we were told that one bed could be had in the village; we gladly left the Pfarrhaus, shaking the dust from our feet, and went to the destined lodging, where we found civil, and tolerably cleanly people, whose jargon, however, it was quite impossible to understand. There was actually but one spare bed in the whole village. We drew lots for the prize, which fell to me. It was clean, though neither soft nor even; but between two such journeys as I was undertaking, even to undress was a luxury, and I slept till late next morning, when I was awakened by M. Studer entering. Where he had slept never transpired. He had, however, spent a night of misery, and came to communicate his intention of departing immediately for the Val d'Anniviers, instead of passing the day in the neighbourhood of Evolena, as he had intended. I could not gainsay the propriety of his determination, although sorry to part. He left shortly afterwards, and we agreed to meet at Zermatt,—he going by Visp, I by Ferpêcle.

Before I had finished dressing, our worthy guide from Valpelline came to bid me adieu. During the latter part of our yesterday's walk we had become well acquainted, and his simplicity of character had touched us both. He had more than once expressed a wish to accompany us farther, as well as to avoid returning to his own country the same way. He urged nothing of the kind now, but quietly bid me good-bye and took the road to Haudères. When I saw him fairly gone, I could not but regret having parted with him so easily. I thought that he might be very useful in the more difficult journey which awaited me, my own servant being inexperienced, and the guide of Arolla, though he promised well, being quite unknown to me. I, therefore, ran after " l'habit rouge," and asked if he would accompany me to Zermatt, and return home by the Val Tournanche in Italy. To this he at once assented. There was no bargaining or hesitation, and he turned back with me.

In the forenoon, Pralong joined me according to promise. Having first dined, I started with my three men about two o'clock, with very fine weather, intending to sleep at the last châlets of Ferpêcle, and to cross the glacier the following morning. Before quitting Evolena I will say a few words respecting the Valleys of which it is the centre and capital.

The group of valleys of which we speak, and of which Erin is the chief, is situated between the Rhone and the great chain of Alps. Their openings into the valley of the Rhone are so small and unconspicuous that they are passed unnoticed by the traveller, rolling along in his private carriage, or that of the Simplon courier, almost without perceiving their existence ; yet opposite to three well known stages on that road, Sion, Sierre, and Tourtemagne, three several valleys proceed, the Val d'Erin, or d'Herens, (Eringer Thal,) the Val d'Anniviers (Einfischthal,) and the Vallée de Tourtemagne, (Turtmanthal.) Their magnitude and importance are in the order just stated. The Val d'Erin divides into two branches, the valley of Evolena and that of Hérémence, both of which terminate in great glaciers, to wit, the glaciers of Ferpêcle, Arolla, and Lenaret. The Val d'Anniviers divides into the Val de Torrent and Val de Zinal, with glaciers of the same names. The valley of Tourtemagne is uninhabited, except in summer, and terminates in a glacier at the foot of the Weisshorn.

These valleys have not only been hitherto unfrequented by *tourists*, but are almost unknown even to *travellers*, (to make a distinction commonly and not unjustly drawn in Switzerland.) De Saussure says nothing of them. Bourrit speaks of them so slightly that it may be doubted whether he ever was even so far as Evolena. Ebel mentions them only to acknowledge his want of information, and Simond is silent alike on their history and existence. Even at the time I am describing, although it was past the middle of August, the Curé informed us that we were the only strangers who had yet appeared that season at Evolena. A pleasant little work, by Fröbel, entitled, " Reise in die weniger bekannten Thäler auf der Nordseite der Penninischen Alpen,"* has given the first and only detailed account of them worth notice, and even his visit was one of but a very few days, and directed only to the more accessible points. His work is valuable from an improved map which it contains, (upon which the index map in this work is partly founded,) and which corrects many of the almost incredible errors of the best executed maps before that time, such as those of Weiss, Keller and Wörl. I should add, that a work published at Basel, also in 1840, by Engelhardt, under the title of " Natur schilderungen der höchsten Schweitzer Alpen," gives some account of these valleys, and confirms the unanimous testimony of travellers respecting the discomfort and incivility experienced at Evolena.

It seems to be admitted by all who have mentioned these valleys that their population is of a distinct race from their Swiss neighbours. Very different origins have been assigned to them,—that they came from the east, and were originally tribes of Huns and Alani, and that they settled here in the 5th century, is the most prevalent theory ; others pronounced them to be Saracenic, dating from the 9th century, whilst Fröbel inclines (chiefly upon etymological grounds, not perhaps very conclusive) to consider them a Celtic race. That they lived in a very independent manner, were heathens long after the conversion of their neighbours, became subject to the Bishop of Sion, and were christianized by his missionaries, is confidently stated. In modern times we know that they have shown a spirit of stubborn independence, and resisted, in their unapproachable fastnesses, the incursions of the French armies, at a time when the rest of the Vallais had submitted to the yoke of Bonaparte.

Their character does not appear to differ much from that of the Valaisans, or, indeed, of the Swiss generally. Their hospitality, according to Fröbel, is seldom disinterested, and an intense love of money predominates in all their transactions. A dollar which once finds its way to Erin is never changed, and never comes forth again. This feature, supposed generally to be an imported vice, conspicuous only on the great and frequented roads, is, therefore, not merely the result of English folly

---

* Berlin, 1840.

T

and extravagance; and my experience in other remote places confirms
the opinion. The character of the people is, farther, according to the
same writer, stiff and pedantic, not unfrequently producing a ludicrous
appearance of self-importance amidst an utter neglect of the common
comforts and almost decencies of life. Their food is not only coarse, but
scanty, and even unwholesome; their houses and apartments are amongst
the worst in the Alps : cleanliness is not amongst their virtues. Much
of this may be traced to laziness, which Fröbel says is the prevailing
vice : mules are abundant for country uses, and no man walks who can
possibly ride (even second) on a mule ; still less will any one carry a
common knapsack without complaint. Klaus's *hotte* was the wonder
of all who met us. Fröbel has, indeed, said so much about the impossibility of obtaining good guides in Evolena,* that I had despaired
of undertaking any considerable expedition, but Pralong seemed to be
rather an exception to the usual character, being active, civil, and far
from exacting ; he also displayed, on the whole, much personal courage
and resolution.

The language is barbarous, but I doubt whether it is more so, or
more decidedly national than in many other remote valleys of the
Alps. The word " fläthig" for *cleanly*, which Fröbel has mentioned
as distinctive, I have heard in the valley of Saas, where the population is, I believe, purely German. The name Evolena is said to
mean, in the native dialect, " tepid water," and may be derived
from a number of very beautiful springs, which rise from the fallen debris at the foot of the mountain slope immediately behind the town.
Borgne, means brook ; *biegno*, glacier, and *pigno*, mountain top, which
last Fröbel says is synonymous with the Spanish *penon*, the French
pignon, the mons *penninus* of the Romans, and the Gaelic *bein*.

These valleys, notwithstanding the seeming poverty of their inhabitants, annually export a great deal of produce. Evolena is eight hours'
distant from Sion. Its neighbourhood presents a very lively and fertile appearance, the valley being broad and well watered, covered with

---

* " Die Männer von Evolena sind schlechte Bergleute." " Zehn bis funfzehn Pfund
auf den Rücken zu haben ist einem Manne von Evolena eine unerträgliche Unbequem-
chkeit," p. 91.

pasture, and studded with barns and châlets up to a great height on both sides; for although the secondary ranges, those which divide Erin from Hérémence and Anniviers, are of considerable height, and of a fatiguing nature to climb, as those who have passed testify, they are fertile and grassy, affording excellent pasture. The cheerful appearance is indeed diminished when we approach and find what seem to be villages to be mere barns, or rather hay-lofts, without a single inmate, and when, in the inhabited places, we find so much want of comfort and cleanliness. But as I have said, the exports of dairy produce to the low country are large, and probably very greatly exceed the imports, although these must include most of what are commonly considered as the necessaries of life.

Besides the natural entrances to these valleys from the valley of the Rhone, which, as we have said, are narrow and unconspicuous, there are various passes to and from the higher parts of these valleys. In former times, the glaciers were, as we have also seen, undoubtedly much more accessible, and even the pass to Zermatt seems at one time, like the Col du Géant, to have been frequently used. From Hérémence, there is said to exist a passage to the Glacier of Chermontane, which may have been in the direction which we saw in crossing the Col de Fenêtre. There is also a long pass, but not over ice, into the Val de Bagnes, below Mont Pleureur, which M. Studer crossed in 1841. From Anniviers, it is very doubtful whether any glacier-pass exists; but from Tourtemagne, which is a valley inhabited only in summer, it is possible to cross the northern part of the Weisshorn into the valley of St. Nicolas above Stalden.

But to resume our journey. Having quitted Evolena at 2 P.M.,* I walked to Handères, where my guide, Jean Pralong, lived. This village is at the junction of the two *Borgnes*, three miles above Evolena. It was nearly deserted. Pralong took the key of his house from under a stone, and invited me into it. The entrance was rude and ill-furnished, the light and air coming in on all hands; but he conducted me up a

---

* The height of Evolena is 4532 English feet above the sea by M. Studer's observation compared with the barometers at Geneva and St. Bernard.

trap-stair to a very tolerable apartment, with clean-looking beds, which
we should have envied the night before. He offered me wine, and took
a supply himself for the journey, candles for our use at night, and
various other articles, including a rope to be used on the glacier. We
then started, and followed the east side of the eastern rivulet, that
descending from the Glacier of Ferpêcle. We followed narrow water-
courses to abridge our way, and during our ascent I was surprised to
notice the oriental plane-tree and the currant both growing wild. The
rocks exhibit traces of glacier friction, but neither here nor in the other
branch of the valley towards Arolla are the transported blocks numerous.

After two hours' walk from Evolena, we reached the châlets of Ferpêcle, the highest and only habitations of this part of the valley. Here
we proposed to get some hay to form our bed at night, which we conjectured might be a scarce commodity at the still higher station, where
we proposed sleeping. But this was not so easy a matter, for this
seeming village contained not a single inhabitant ; the greater part was
composed merely of hay-lofts, which, upon examination, proved to be
much better secured than at first sight seemed probable. But Pralong
was not daunted by the resistance of wooden bars and iron shackles,
and my geological hammer was unscrupulously applied to obtain an
entrance with the deliberate purpose of pillage. At length one door
was forced, and a good armful of dry clean hay was secured and carried
off, and all else replaced as before. We had now the lower end of the
Glacier of Ferpêcle immediately before us. The valley is very deep,
and the scene solitary and striking, but it is impossible to form here
any idea of the extent of the ice. Keeping always to the left, we began a smart ascent at first over the moraine of the glacier, which here
as elsewhere seems to have retreated of late years. At length we
gained a better path, traversing high pastures, and crossing the beds of
several vast torrents. Having now got considerably above the ice,
we advanced nearly on a level. We also saw rising beyond, groups of
jagged summits, which separate the Glaciers of Ferpêcle and Arolla, of
which the most conspicuous is a sharp pinnacle called Aiguille de Za.
These terminate towards the great chain in a range called the Dents
des Bouquetins. This led us to speak of those animals, and I asked

Pralong whether any were ever seen. He replied that they had long disappeared, and that the story went that long ago the government of the Vallais, desirous to preserve the race, declared the shooting of a Bouquetin to be a capital offence, from which time not one of these animals has been seen,—a practical proof, he probably meant to infer, of the impotence of extreme legislation. He also began spontaneously to talk about the glaciers, and the cause of their motion, and put several very pertinent questions. Amongst other things, he affirmed distinctly, that the glaciers advance indifferently in summer and in winter, and even that if the lower extremity be diminishing, it continues to do so—if advancing, to advance also—in winter as in summer.

As the evening fell we gradually approached—by a path which certainly seemed to lead to no human habitation, but to an endless wilderness of ice and rocks—the châlets of Abricolla, which we reached in an hour and a half from the châlets of Ferpêcle. The first symptoms of human art were two pyramids of stone, (*hommes de pierre* as they are generally called,) which directed us from a distance ; then two stone huts near together, and one or two others a little beyond. We soon found that there were inhabitants, and we were received with simplicity, and with that composure and seeming absence of curiosity which I have already mentioned as remarkable amongst the *Pâtres* of the higher Alps. A visit even from Evolena is a rarity, but most likely none of them had before seen or lodged a traveller and his guides, prepared to cross the glacier to Zermatt. Nevertheless, as their reception was far from repelling or suspicious, I was well satisfied with their tranquillity about my concerns and objects ; and preparing my arrangements for the evening, I left my guides, who all spoke different native tongues, to satisfy, as best they might, any latent curiosity of our hosts.

It was a charming evening, almost too mild to give quite a favourable prognostic for the weather. After sunset the moon, which was almost full, rose, and threw her light over a scene not to be surpassed. These chalets, placed on a broad grassy shelf of rich verdure, overhanging, at a height of several hundred feet, one of the noblest glaciers in the Alps, are not much less elevated than the Convent of the Great

St. Bernard,—a position sufficient in most cases to diminish the effect
of the higher summits, but which here only increases it, so stupendous
is the scale of nature at this spot. Rising abruptly from the glacier,
at no great distance on the left, is the grand summit of the Dent
Blanche,* which is called Hovenghorn as seen from Zermatt. Its
height is probably unmeasured, but is marked in Keller's map
13,000 French feet, which, I believe, is rather under the mark. To
the south the view is bounded by the ridge which I proposed passing,
from which the glacier descends in some places very steeply, and
with a striking effect, breaking over a rock called Mottarotta, which
divides its current for a short space. To the west of this a narrow
ridge of angular summits very abrupt and bare, divides the glacier
into two distinct branches. This is called the Mont Miné, and is
reputed to contain indications of ancient mines. I was surprised to
learn that sheep are usually conveyed across the glacier to graze upon
what seems a mass of broken rock. Between the Mont Miné and the
ridge formerly mentioned as separating the Glaciers of Arolla and
Ferpêcle, the western branch of the Ferpêcle glacier descends. This
ridge is far higher, and more commanding than the Mont Miné. It
has its origin at the Dents des Bouquetins, near the axis of the chain,
and it descends to the Aiguille de Za, and continues to its termination
above Handères, in the Dent de Visivi. From the considerable height
at which I stood, the glacier was seen (in its lower part at least) in plan,
and presented a view of the same description, but more extensive and
wild, than that of the Mer de Glace from the Montanvert. As now
seen by moonlight, its appearance was indescribably grand and peaceful, and I stood long in fixed admiration of the scene, the most striking of its kind which I have witnessed, unless, perhaps, I were to
except a moonlight walk over the great Glacier of Aletsch under very
similar circumstances. Amongst other things, I did not fail to remark
the wave-like bands, or "dirt beds," at regular intervals on the surface
of the glacier, in precise correspondence with what I had observed at
Chamouni from the Charmoz. Here they were, if possible, more strik-

---

* See the Topographical Sketch, No. VII., for this route.

ing, more numerous, and not less regular. Instead of 18 bands, 1 here counted 30, at intervals sensibly equal, and in forms like those figured on the map of the Mer de Glace. The moonlight was very favourable to this observation.

I soon after returned to the hut to supper. As might be expected, the cheer was not great, but cheerfully given. There could not be much less comfort than at Evolena; but it was at least freely offered. There was no temptation to prolong a stay within doors, unless to sleep. I retired early with my guides to the lodging prepared for us with the aid of the hay which we had brought. It was a small shed, about six feet square, and four high, attached to the principal hut, entered by a doorway through which one could creep with difficulty, and which was shut up with a piece of cloth. I was placed next the wall, and the others slept beside me. The shepherds themselves slept in a separate hut a little way removed. Before we went to rest, it was agreed that they should call us at 3 A.M., that we might be on foot before day, for all reports agreed, that whatever might be the difficulties of the journey, it was, at least, a very long one. In order to awaken us at the right time, they begged to have my watch with them for the night, a request which, in some other countries, might have been suspicious, (it was a valuable gold chronometer,) but which here I granted as readily as it was undoubtingly asked. As we lay down I was struck by the conduct of Pralong, who knelt down on the hay and said his prayers shortly, and without form or pretension of any kind; and we had not been long composed to rest, before we heard a solemn, and not unmusical voice proceeding from the neighbouring apartment. On inquiry of Pralong I found, that the practice of evening prayer is kept up amongst the assembled shepherds; a rare but touching solemnity amongst men of the common ranks,—for no women usually live in the higher châlets,—separated during so large a part of the year from the means of public worship.

I passed a sleepless, though far from an uncomfortable night. Pralong had spoken doubtingly of the weather in the evening, and I well knew that any thing like uncertainty in that respect could not be hazarded on such an expedition, for which I felt more and more disposed as I got better acquainted with the scenery of this interesting chain. Every change of direction of the moon's rays falling through the open walls and roof of our shelter I mistook for a cloud, and felt fresh anxiety lest the hour of rising should be overpast, as it had been at Prarayon. I was up before the rest, and whilst the stars were shining bright, the moon having set, I performed my hasty toilet. It was some time before breakfast could be got ready, and, as usual, an hour and a quarter elapsed before we were fairly under way, exactly at a quarter to 5.

It may not be out of place to mention here what was known respecting this pass, which has remained less celebrated than the Col du Géant, or the Strahleck, (both of which it exceeds in height,) because the valleys between which it communicates are, I believe, little known. I first heard of it from a guide at Zermatt, Peter Damatter, who told me, in 1841, that he had passed it, and that the town of Sion was visible from the top. He represented the distance as excessively great, so as with difficulty to be accomplished in a day. Venetz, the able engineer of the Vallais, (to which canton this country belongs,) wrote, in 1833, that this pass was so dangerous that he had never known but one man who had accomplished it;[*] whilst he mentions it as a proof of the great increase of the glaciers in modern times, that formerly it was in considerable use, and certainly, for the rare occasions that any one may be supposed to have business between Evolena and Zermatt, the circuit of three or four tedious days' journey by Sion and Visp is by no means cheering. Fröbel mentions, that some years before he wrote, several gentlemen of Sion effected the passage under perilous circumstances, having passed the entire day, from two in the morning until evening dusk, between the last châlets of Ferpêcle, and the first of Zmutt.[†]

---

* Mémoire sur de la Variation de Température dans les Alps de la Suisse, p. 7. I quote from a citation, not having the original by me.

† "Die Angaben über die Gangbarkeit dieses Passes sind sehr verschieden. Wie bei aller Gletscherpässen wird auch hier alles vom Jahrgange und von der Witterung abhangen. Der Herr Domherr Berchtold in Sitten welcher ich über denselben befragte, bemerkte mir, es sey in jedem Falle 'ein Aventure,' uber ihn zu gehen."—FRÖBEL, p. 73.

Making all customary allowance for exaggeration, I had good reason to take all precautions, and to start with the early dawn ; indeed we were scarcely off when Pralong intimated that he feared we were already somewhat too late.

It will be recollected that, besides Pralong, the guide of Evolena, I had the trusty Biona of Val Biona, and Tairraz of Chamouni, as my attendants. The provisions, and my personal effects, made a burden so light for each, that even an Eringer could not reasonably complain ; and taking leave of our hosts with thanks and remuneration, we hastened at a good pace to gain the glacier. But this was not the work of a moment. I have already said that the châlets of Abricolla stand on a shelf many hundred feet above the glacier ; and, what is always disagreeable, our first step to mounting was a steep and uncomfortable descent. We had not left the châlets ten minutes when we found a foaming torrent to be crossed. Now, a plunge up to the knees in a river even ice-cold is a trifle in ordinary travelling, and might be considered a refreshing commencement of a long day's walk ; but when that walk is to be of ten or twelve hours on a glacier, and over snows 11,000 or 12,000 feet high, such a freak might endanger life or limb. Accordingly, while Pralong and Biona spluttered through, I sought an easier passage higher up, which I at length found, and was followed by the wary Savoyard. Without difficulties worth mentioning we gained the surface of the ice, having lost, however, in level, a height of perhaps 1000 feet ; we then patiently and warily proceeded on our march,—

> To climb steep hills
> Requires slow pace at first : Anger is like
> A full hot horse, who, being allowed his way,
> Self-mettle tires him.*

But an unlooked for interruption occurred. My guides were all seized with sickness within a few minutes of one another. Their breakfast (boiled milk) had probably been prepared in a copper vessel, not cleaned overnight ; and though all hardy men, with robust

---

* Henry VIII.

stomachs, and accustomed to the universal milk diet of the Alps, they suffered distressingly from the poison. For myself, long experience had made me almost wholly avoid these messes, and every preparation of milk. I had drunk tea both night and morning, prepared in my portable boiler, and had filled my gourd with some of the same invaluable stomachic, which I now administered with effect to Tairraz and Biona, whilst Pralong declared that his casket, or *keg* as it would be called in Scotland, of red wine, was worth all the tea in the universe. Happily, I suffered no uneasiness, and the others, being probably accustomed to the disorder, made light of it, and gradually recovered; meanwhile we pursued our way. We were now (See the Topographical Sketch, No. VII.,) close under the rocks which bordered the glacier on our left, beneath the lofty peak of the Dent Blanche. Before us was the Motta Rotta, the rocky precipice already described as rising through the ice. At length the glacier became much crevassed, and we had a choice of difficulties, either to skirt the precipitous rocks on our left, or to make for the centre of the glacier on our right, with the chance of crevasses yet more impassable. Pralong, indeed, broached the notion of attempting the ascent of the glacier between the Motta Rotta and the Mont Miné, which, he said, would lead us more directly to the Col; but he did not know that such a passage had been attempted, and as, upon examination with the telescope, I perceived an enormous *Berg-schrund*, or well defined crevass, which separated the higher summits from the glacier steep, I preferred pursuing the direction in which he had already passed. We accordingly made for the rocks, and scrambled along and up them for a considerable way. We were preceded by a whole troop of Chamois, eleven in number, which we startled upon the ice, and which took immediately to the cliffs. At length it became difficult to say whether the rock or the glacier was the more formidable opponent, and we regained, with some difficulty, the surface of the latter, being now more than on a level with the châlets which we had left.

The sun was only now rising behind the ridge of the Dent Blanche, the ice was still hard frozen and slippery. The glacier was very steep and rugged, but the crevasses were exposed and the walking was more difficult than dangerous, although once I was only withheld by my companions from slipping into a chasm. But the snow line was soon gained, and the surface being still crisp, our footing was sure, and the bed of snow too thick to create any risk from crevasses. We were on the north or shady exposure, always the easiest to mount, and had a fatiguing climb up dazzling snow fields, about 30° of elevation abreast of the Motta Rotta, which was on our right. Pralong took the lead manfully, and was now quite recovered from his indisposition. The heights of the Motta Rotta gained, the Col might be said to be reached, for although snow fields of great extent separated us from it, they evidently presented no difficulty. It is, perhaps, only in this part of the Alps that such a prodigious extent of comparative table-lands of snow are to be found at such an elevation. New peaks began to rise before us, and especially the Mont Cervin, or Matterhorn, and the Dent d'Erin, whilst to the westward, the summits of Mont Collon, and the neighbouring chains peeped over the wilderness of snow and ice. The Col or pass, lay now, Pralong told me, considerably to the right, but seeing just before us a snowy summit, which alone concealed from us the view of Monte Rosa, and the great chain of Alps in that direction, I proposed, as we had gained this height at a very early hour, and with far less difficulty than I expected, to climb to the top of it to enjoy the view. Now, Pralong was not one of those teazing, pedantic guides who will never listen to any opinion, and who make it a point to thwart a proposition merely to show their consequence, the more so if it offer a chance of delay. I liked him for his confidence and good temper. He admitted that a traveller's opinion might be taken, at least as to the course which would please him best ; accordingly, we walked right over towards the precipice marked on the sketch as stretching from the Dent Blanche to the Stockhorn. As we approached it, I caught one of those glorious bursts of scenery of which all description must ever fail to realize the incommunicable grandeur, and one sight of which at once and instantly repays the traveller for days of toil and sleepless nights. Wandering on alone as near the verge of the snow-crowned precipice as I dared venture, (for there an unseen fissure in the compacted snow, some yards from the very ledge, might readily occasion the detachment of a mass, by the traveller's weight, into the abyss,) I gained the summit of the Stockhorn, of which I had considerably overrated the height from where I first proposed the deviation, and was seated on its top exactly at nine o'clock.

I wish I could convey an impression, however faint, of the view to the east. The morning was calm, the sky pure, and the sun bright; indeed, there was not a breath of wind, though I was here at a height of 11,760 feet above the sea, or 600 feet higher than the Col du Géant; and this stillness, combined with the reflected sun heat, made the air feel perfectly mild, although, to my surprise, I found the thermometer to be only 34°. The whole range of Monte Rosa, including that promatical summit,* scarcely inferior to it in height, called by some Montagne de Fée, and by others Mittaghorn, filled the eastern distance.

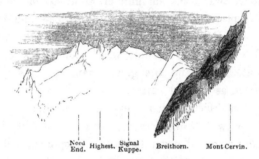

Nord End.   Highest.   Signal Kuppe.   Breithorn.   Mont Cervin.

View of Monte Rosa from the Col d'Erin.

From the great height at which I stood, there could be no doubt about which was the highest point. Although between 3000 and 4000 feet higher, the distance was so great as to bring the eye apparently almost on a level, and in no direction is the relation of these much contested summits better seen. The summit which I thus judged to be the highest, is exactly the "Höchste Spitze" of Von Welden, 15,158 English feet above the sea, of which more hereafter.

The whole lustre of the morning sun shone shadowless upon these

---

* What Von Welden has called "Berg X."

snowy heights, and upon the vast surface of the Glacier of Zmutt, of which only a portion can be included in the Topographical Sketch, and which lay completely, as in a map, at my feet, separated from me by stupendous precipices—" a vast vacuity." It is the cliff attempted to be shown in the sketch, of which the Stockhorn on which I was seated forms at once the salient angle and the highest point. A branch of the glacier, it will be observed, comes close to the foot of the Dent Blanche, and to the base of the precipice. The Dent Blanche, thus seen in its precipitous height from top to bottom, had a magnificent appearance, and from the height which I afterwards ascertained of the point on which I stood, I cannot doubt that its reported height (nearly 14,000 English feet) is not overrated. Beyond the Dent Blanche, appeared the elegant and commanding summit of the Weisshorn, whose height, recently determined by M. Berchtold of Sion, is 14,812 English feet, and which sinks into comparative insignificance the Gabelhorner and other rugged mountains, which separate the head of the Val d'Anniviers from the Glacier of Zmutt. But amongst the objects nearer at hand, even the Dent Blanche was not the finest. Right opposite, separated from me just by the breadth of the Glacier of Zmutt, were the Mont Cervin and the Dent d'Erin, the former of 14,766 feet,* the latter conjecturally 14,000 feet above the sea. Of the former, a representation as seen from Zermatt, will be found in Plate VII. : its unscaled and unscaleable pyramid is, beyond comparison, the most striking object in the Alps. The Dent d'Erin forms distinctly a part of the same range, united by a continuous and inaccessible precipice, and they are not isolated and unconnected masses, as represented in Fröbel's map. To the westward were seen the mountain groups of the head of the Val Pelline;—the Mont Collon, and the Pigno d'Arolla, the Dents des Bouquetins, and the seemingly interminable ice-fields over which (as I have said above)

---

* De Saussure. M. Berchtold's measurement is not sensibly different—namely, 13,839 French or 14,750 English feet, as stated by Engelhardt in the proceedings of the Swiss naturalists for 1841. The numbers given in Engelhardt's " Schilderungen" are many of them inaccurate.

a passage might possibly be effected to the Col de Collon, above which I thought that I perceived the Mont Gelé, near the Col de Fenêtres ; but in this I might easily be mistaken. It is probable that the Monts Velan and Combin might be seen in the same direction, but clouds rested on that part, and on that alone, of the horizon. I apprehend that Mont Blanc must be concealed by the mountains last named. To the north was the Glacier of Ferpêcle, which we had ascended, stretched out in all its length, flanked by its Aiguilles, and descending into the depth of the valley, in which we easily traced the village and church of Evolena, but Sion is certainly not visible.

Of all the views which I have seen in the higher Alps, none can compare with that from the Stockhorn of the Col d'Erin, (as I propose to call this pass, which has not yet received a name.) The unequalled view of Monte Rosa, and the centrical position with respect to three summits of the second (if not of the first) order, the Mont Cervin, Dent Blanche, and Dent d'Erin, which seem all so near as almost to be tangible, are sufficient to mark its character. The Weisshorn and the Cima di Jazi, as well as Mont Cervin, all border on 15,000 feet ; so that, counting all the peaks of Monte Rosa but as one, we see at once at least five distinct mountains higher than the Finsteraarhorn, long esteemed the highest in Switzerland proper. Compared to the Col du Géant, the view is here more vast and savage, and the individual objects finer and closer ; though the distant view of the chain of the Alps gives to the former a delightful and peculiar charm.

Before leaving this part of our description, I must say one word on the geography of this part of the chain. By Wörl's Map, or that of Keller until the edition of 1842, it would appear impossible that such a pass can exist as that which I am now describing. The chain of Alps (I write with Wörl's Map before me) is represented as turning from the Mont Cervin abruptly to the N.W.,—as including the Dent Blanche at the southern foot of which the Val Pelline is made to take its rise (!) and then, as bending back again towards the head of the Glacier of Arolla. Since the Dent Blanche is rightly placed between the Glacier of Ferpêcle and that of the Torrent-thal, it evidently would have been impossible to reach Zermatt from Evolena without crossing into Italy, and recrossing near the Mont Cervin. Now, without detailing other varieties of error, the reality is, that the main chain of Alps is here well defined, and nearly straight, extending from Mont Cervin through the Dent d'Erin to the nameless summits south of Mont Collon, and at the true head of Val Pelline or Biona. The whole north face of Mont Cervin and the Dent d'Erin is a united and inaccessible precipice, which falls into the Glacier of Zmutt, which extends far to the westward of both, not rising (as even Fröbel inaccurately represents it) immediately behind the Mont Cervin, but in the great ice-mass to the westward of the Dent d'Erin. Now, just where the Glacier of Zmutt takes its rise, is the commencement of a great lateral chain on so stupendous a scale, as to create little surprise that it has often been mistaken for the great chain. The Glacier of Ferpêcle descends from its north-western flank, where it forms the Col d'Erin and the Stockhorn, upon which we conceive ourselves stationed. It then expands itself into the mass of the Dent Blanche, which sends forth the ramification of the Dents d'Abricolla and Zatalane, which separate the valleys of Erin and Anniviers. From the Dent Blanche the chain takes an easterly direction, forming the summit called Moming in Erin, Triftenhorn at Zermatt, (where the Dent Blanche is called Hovenghorn,) which separates the valley of Zinal and that of Zmutt. This part of the chain seemed to me quite impassable. Then follow a range of peaks called Gabelhörner, which continue the chain in a north-easterly direction, parallel to the valley of St. Nicolas, until we reach the culminating point of the Weisshorn, a seemingly inaccessible peak of 14,812 English feet, which is often mistaken for Monte Rosa, especially from the Pass of Gemmi, whence it and the other parts of the chain just mentioned have been elaborately figured in von Welden's work as the *actual* Chain of Monte Rosa, and received specific names accordingly, although the real Monte Rosa is some thirty miles distant, and wholly concealed !

It will thus be distinctly understood that the passage of the Col d'Erin is not that of the great chain, but only of this ramification of it.

M. Studer having taken his barometer with him to Anniviers, I had only the sympiesometer and the boiling water apparatus to depend upon for the determination of the height. I consider the latter as the most certain, and as probably not erring more than 50 feet from the truth. It gave (by comparison with the barometer at Geneva) a height of 11,770 feet, the temperature of boiling water, being 192°.45, (or 191°.93 corrected,) and that of the air 34°. I melted snow, and caused the water to boil with great ease, even at this height, and thus supplied the party with plenty of water to drink, which otherwise it would have been impossible to procure.

Stretched upon the snow, we made a hearty meal; and the hour and a half which I spent here in observing my instrument, taking magnetic bearings of the principal objects, sketching the outline of Monte Rosa, and trying effectually to impress upon my memory a scene which I scarcely expect ever to see equalled or under circumstances so favourable, went quickly by, when Pralong modestly invited me to depart, as our task was far from accomplished; indeed, as it appeared, the most difficult part was to come.

Our object was now to descend upon the Glacier of Zmutt, of which, as I have endeavoured to explain, and to represent upon the Topographical Sketch, No. VII., the lower or more level part swept along the base of the Mont Cervin and Dent d'Erin, whilst a higher stage of it rose to the foot of the lofty precipice above which we stood. Now, whilst the *top* of this precipice sunk from the summit of the Stockhorn, westwards to the Col, and then rose a little, the glacier and the *foot* of the precipice rose rapidly and continuously to the westward, so that the top and bottom of the precipice became at length blended together, under a snowy sheet. To reach this point, however, would have been a long *détour*, and the glacier appeared dangerously crevassed. Having, therefore, descended from the Stockhorn to the Col, (which was not a great deal lower,) Pralong proposed to attempt descending the cliff, by which he recollected to have passed when he last crossed, and to have successfully reached the glacier below. We began cautiously to descend, for it was an absolute precipice : Pralong first, and I following, leaving the other guides to wait about the middle, until we should see whether or not a passage could be effected. The precipice was several hundred feet high. Some bad turns were passed, and I began to hope that no insurmountable difficulty would appear, when Pralong announced that the snow this year had melted so much more completely than on the former occasion, as to cut off all communication with the glacier, for there was a height of at least thirty vertical feet of rocky wall, which we could by no means circumvent. Thus, all was to do over again, and the cliff was reascended. We looked right and left for a more feasible spot, but descried none. Having regained the snows above, we cautiously skirted the precipice, until we should find a place favourable to the attempt.* At length, the rocks became mostly masked under steep snow slopes, and down one of these, Pralong, with no common courage, proposed to venture, and put himself at once in the place of danger. We were now separated by perhaps but 200 feet from the glacier beneath. The slope was chiefly of soft deep snow, lying at a high angle. There was no difficulty in securing our footing in it, but the danger was of producing an avalanche by our weight. This, it may be thought, was a small matter, if we were to alight on the glacier below; but such a surface of snow upon rock rarely connects with a glacier without a break, and we all knew very well that the formidable " Berg-

Passing the Bergschrund.

schrund," already mentioned, was open to receive the avalanche and its charge, if it should take place. We had no ladder, but a pretty long rope. Pralong was tied to it. We all held fast on the rope, having planted ourselves as well as we could on the slope of snow, and let him down by degrees, to ascertain the nature and breadth of the crevasse, of which the upper edge usually overhangs like the roof of a cave, dropping icicles. Were that covering to fail, he might be plunged, and drag us, into a chasm beneath. He, however, effected the passage with a coolness which I have never seen surpassed, and shouted

---

* Upon the rock from whence we finally descended I left a bottle containing the names of the party.

U

the intelligence that the chasm had been choked by previous *ava-lanches*, and that we might pass without danger. He then (having loosed himself from the rope) proceeded to explore the footing on the glacier, leaving me and the other two guides to extricate our-selves. I descended first by the rope, then Biona, and lastly Tair-raz, who, being unsupported, did not at all like the slide, the ter-mination of which it was quite impossible to see from above. We then followed Pralong, and proceeded with great precaution to sound our way down the upper Glacier of Zmutt, which is here suf-ficiently steep to be deeply fissured, and which is covered with per-petual snow, now soft with the heat of the morning sun. It was a dangerous passage, and required many wide circuits. But at length we reached in a slanting direction the second terrace or pre-cipice of rock which separates the upper and lower Glacier of Zmutt, and which terminates in the promontory marked Stockhi in the map. When we were fairly on the debris we stopped to repose, and to congra-tulate ourselves on the success of this difficult passage. Pralong then said that he wished to ask a favour of me. To my astonishment, this was that he might be allowed to return to Erin instead of descending the Glacier to Zermatt. He was afraid, he said, of change of weather, and did not wish to lose time by going round by Visp. Of course I readily granted his request, and paid him the full sum agreed upon. To return all alone (and it was now afternoon) over the track we had just accomplished was a piece of spirit which would scarcely have entered the imagination of any of the corps of guides of Chamouni. I almost hesitated at allowing him to expose himself, but he was resolved and confident, and having given him most of the provisions, and all the wine, we saw him depart.

We had still a long, though not a dangerous, stretch of glacier be-fore us. We had, in the first place, to descend the precipices behind the Stockhi to the lower level of the Glacier of Zmutt. Though steep, they were not dangerous like the last, and though the way was new to my companions as well as myself, we found no particular difficulty. We had now no alternative but to pursue the surface of the Glacier of Zmutt for several miles, which proved a fatiguing walk enough, the ice being intersected by crevasses, and in many places almost covered with vast boulders.   During this descent I had an opportunity of examining

Mont Cervin from the N. W.

closely the structure of the Mont Cervin on this side, which probably no mineralogist has had before. There seems no reason to doubt that it is entirely composed of metamorphic secondary rocks.   The lower part is of the system of *green slates*, which abound in this part of the Alps, and which here pass into serpentine and gabbro, as the moraines testify, the higher part of grey and *white slates*, remarkably contorted, and probably calcareous.   The middle strata of the Mont Cervin appear to form, by their prolongation, the Stockhi on which I stood, and the Col d'Erin and Stockhorn are composed of a repetition of the green slate, which contains so much felspar that it may be called gneiss.   The whole height of the Mont Cervin, down to the level of the glacier, is one continuous precipice, which must be between 7000 and 8000 feet.   The conformation of the Dent d'Erin is similar to it.

The gradual appearance of the moraines upon the Glacier of Zmutt was very striking.   I mean that they are slowly developed upon the surface of the ice, as I have described on the Mer de Glace of Chamouni.   They come from many quarters, and with a prodigious volume ; from the Dent d'Erin, the Mont Cervin, the Stockhi, the Stockhorn, and from other promontories divided by glaciers which fall from the range of the Dent Blanche and Triftenhorn, they accumulate at last upon so narrow a space of glacier as, from a distance, to appear to cover it entirely. The usual nearly longitudinal vertical structure was developed in the ice where we first descended upon it.   Both banks of the glacier were too precipitous to attempt to climb them, and for a long way we had to pick our steps as we best could on the ice and among the moraines.   At length we gained the right bank, not far above the first châlets of Zmutt, with which I was already acquainted by my visit of the previous year.   Immediately after, we entered the larch woods, and crossed the river where a very deep ravine is spanned by a most picturesque and insecure bridge, which passes to the village of Zmutt on the left bank of the stream.   I walked very leisurely, enjoying the fine evening, and half an hour after reached Zermatt, where I took up my quarters in the clean house of the village doctor, named Lauber, which serves as an inn.   I arrived at half-past five, P.M., or in somewhat less than thirteen hours, from Abricolla, including various halts.

CHAPTER XVII.

THE ENVIRONS OF ZERMATT.

In 1841 I visited Zermatt, in company with Mr. Heath, by the usual
route from Visp in the Rhone valley. It is about eight hours' walk;
one and a half hours to Stalden, where the valleys of Saas and St.
Nicolas separate, (see the general map;) two and a half to St. Nicolas; two to Randa, and two to Zermatt.* Between Visp and Stalden
the country is very pleasing, especially where the river is crossed at
Neubrück, whence there is a very fine view of a small portion of the
snowy range which separates the valleys of Saas and St. Nicolas.
Near Stalden are earth pillars, capped by boulders which have protected the soil beneath from the rain, which has washed all the neighbouring parts away, and left these standing, not unlike the marks left

---

* By some strange oversight it is represented, as a feasible excursion, in both editions of
Mr. Murray's valuable Hand-Book of Switzerland, to leave Visp in the morning and to
cross the Pass of Mont Cervin (Col de St. Théodule) into Piedmont the same day. Now,
the ascent of the St. Théodule from Zermatt is alone nearly five hours' heavy work,
(a rise of above 6000 feet,) so that were any luckless traveller to take this advice he
would find himself, *after thirteen hours' walking,* in the midst of a vast glacier 11,000 feet
above the sea.

by workmen to show the extent of an excavation. Similar columns
are likewise to be found near Usegne in the Val d'Erin, at St. Gervais, at Botzen in the Tyrol, and near Queyras in Dauphiné. The
boulders here seemed to be Gabbro or diallage rock. From Stalden
to St. Nicolas the valley is somewhat monotonous; but the Weisshorn
is a striking object. I did not trace here any decided marks of glacier action.

Between St. Nicolas and Randa several wild and bridgeless torrents
have to be crossed, which, in bad weather, must make this route nearly
impassable. I noticed particularly the mode in which a violent torrent accumulates boulders, forming a mound of blocks on either hand,
which serves, in some measure, to restrain its fury, whilst the level of
its bed is continually raised by the detritus which it accumulates;
and when, by extraordinary freshes, the barrier is broken, the country
on either side is, of course, deluged. I only speak now of the wildest and most powerful torrents descending at a great angle, and which
act sufficiently on blocks to roll them with the aid of gravity for a
great way, and chafe them into irregularly rounded masses, with a
noise which every one who has visited the Alps recalls as one of the
most striking of natural sounds, accompanied, as it always is, with
an impression of irresistible force. Now these rocky accumulations
have a very striking resemblance to the moraines of glaciers, and this
is a circumstance which it is well to be aware of, and which has not,
I think, been prominently stated. In *form*, these mounds resemble
moraines, the external, and even the internal slope, being in both cases
usually determined by the *angle of repose* of the blocks. The *materials*
of both are also alike;—angular blocks, more or less rounded by friction,
never quite smooth or polished, angular gravel, and sharp sand. In
the *disposition* of the materials,.I have not observed that regularity of
arrangement which is said to distinguish water-action from that of
glaciers. On the contrary, the deposit of these torrents seems to be
wholly devoid of layers of coarser or finer materials, and, as in true
moraines, the largest blocks often lie uppermost. I may mention the
great torrent descending from the Dent du Midi, which devastates the
country above St. Maurice, as another example of this.

The village of Randa lies amongst extensive meadows, and although on the opposite side of the valley from the Bies-Gletscher, descending from the Weisshorn, (which is now left behind on the right,) it has twice materially suffered from the lower part of that glacier giving way and filling up the whole bottom of the valley, in 1737 and 1819.* Above the village of Tesch the valley contracts, and a rocky barrier has to be surmounted. From thence a grand view of the Mont Cervin opens; and soon after, the village of Zermatt, charmingly situated in a green hollow, well flanked with wood, and enclosed by snowy summits, comes into view. It is at the rocky barrier just mentioned that I noticed the first clear traces in this valley of ancient glacier action in the polishing and striating of the surfaces,—a remarkably well defined result, which may be traced at intervals up to the very foot of the glacier. These striæ were distinctly found by M. Agassiz in 1839, under the glacier itself. This is one of those cases in which it seems impossible to deny this to be a conclusive proof of the ancient extension of the ice.

The village of Zermatt, (called Praborgne, in Piedmont,) is near the union of three glacier-bearing valleys,—the main valley, at the head of which is the great Glacier of Monte Rosa, called also the Glacier of Gorner, or Glacier of Zermatt; the valley immediately to the east, which contains the Glacier of Findelen, descending from the Saas-grat; and that on the right, or to the west, headed by the Glacier of Zmutt.

The river Visp takes its rise in these several valleys, and especially from the great Glacier of Monte Rosa, where it issues, as usual, from a cavern in the ice. I measured its temperature in 1841, at different points of its length, which I found to be,—

Under the glacier, . .	33°.3 F.
At Zermatt, . . .	35.5.
One hour below St. Nicolas, .	41.0.
Half an hour below Stalden, .	43.0.

The great Glacier of Monte Rosa terminates at present about three

---

* ENGELHARDT *Naturschilderungen,* p. 175. AGASSIZ, *Etudes,* p. 158.

miles above Zermatt. The lower part is too steep to be ascended, and it presents the phenomenon of conoidal bands, not only falling forwards until the frontal dip is nothing, but actually sloping outwards as in glaciers of the second order. It rests on serpentine and talc slates, and these rocks present exquisite proofs of glacier polish on the sides of the Riffelberg up to a very considerable height above the western bank of the glacier. Its breadth is here not great, and the surface is crevassed in a remarkable manner, as it rounds the promontory of the Riffel, like the rays of a fan which M. Agassiz has well represented in his Atlas, as well as the medial moraines on its surface, (which are numerous, and well defined,) and their origin at the rocky promontories separating the glacier streams which descend from the Breithorn, the Lyskamm, and Monte Rosa. It is difficult to ascend the rocks on the eastern bank of the glacier, but it is possible; and when the upper surface of the ice begins to be commanded, there are one or two ruined huts in which a shepherd seeks a temporary shelter, and which may serve as a landmark. Near this the serpentine rocks are beautifully excavated in nearly horizontal striæ, whilst below, in the immediate neighbourhood of the ice, I found not only grooves but scratches well marked on the serpentine and talc slates. These scratches visibly crossed one another in two series, *under a considerable angle*, and this must be recent work, because the weather soon wears this rock.

When the upper level of the glacier is viewed, either from the Riffel or from the path up to the Col of Mont Cervin, it presents a noble scene. It is a very vast ice river, whose surface, at the height from which it can be most conveniently observed, appears nearly even, though diversified by fissures and by structural bands like those on the Mer de Glace, which, so far as I can judge from the general view which I obtained without walking over it, are most distinct upon the southern half, and present complete loops bounded by the medial moraine, whilst the northern half (I mean of the breadth of the glacier) has probably a similar structure, although less distinct, and in one part, near the foot of Monte Rosa, is evidently much contorted. The tributary glaciers descending from the Breithorn have also a well-developed system of bands, quite normal.

Pl. VII.

Drawn from Nature by Professor Forbes., T Picken lith.

Day & Haghe, Lith.rs to the Queen

MONT CERVIN, FROM THE RIFFELBERG NEAR MONTE ROSA.

But the Riffelberg may be more easily ascended by its northern slope; there are, indeed, several paths, but it is a stiff walk of nearly three hours from Zermatt. The view corresponds to that of the Montanvert of Chamouni. Though much more vast, I doubt whether the impression of this glacier, and the chain beyond, is altogether so interesting as the other. Monte Rosa is indeed very high and very large, but it presents too many points and too many masses of nearly equal height; the view wants concentration to make a picture, and variety of form. I except, however, the Mont Cervin or Matterhorn, which is seen from hence, but in an opposite direction from Monte Rosa, and which I have already noticed as beyond comparison the most striking natural object I have seen,—an inaccessible obelisk of rock not a thousand feet lower than Mont Blanc! It is represented in Plate VII. The summits of Monte Rosa, distinctly seen from the Riffelberg, are the "Nord End," and "Höchste Spitze," of von Welden's map; then follow to the westward, the somewhat heavy looking range of the Lyskamm and Breithorn, terminating in the Petit Mont Cervin and Col de St. Théodule, a snowy chain of 11,000 feet, which is connected with the Mont Cervin. To the west is perfectly well seen the Glacier of Zmutt, the Col d'Erin, and the range of the Hovēnghorn, (Dent Blanche,) and Gabelhörner as far as the Weisshorn. To the north is the lateral, though very elevated and all but impassable, range which separates the Valley of Saas from that of St. Nicolas, which is called the Saas-grat, and of which the culminating point is (according to Berchtold of Sion,) no less than 14,574 feet above the sea. It is, I believe, variously called by different writers and guides, the Dom, Montagne de Fée, Mittaghorn, and perhaps by some erroneously, Cima di Jazi. Nearer at hand are the Strahlhörner, and close to the north foot of the Riffel is the Glacier of Findelen already mentioned, which unites to the eastward with the great Glacier of Monte Rosa, and which must be ascended in order to reach the Weissen Thor, a very remarkable pass leading to Macugnaga, which I shall mention later. Though the Glaciers of Findelen and Gorner have thus a common origin, the former has been retreating, (at its lower end,) the latter advancing for many years; this is a difficulty of which I know no plausible explanation. Peter Damatter, my guide, both in 1841 and 1842, asserts positively, that the Glacier of Gorner advances in winter, and *more* in winter than in summer; but by this is to be understood that the lower extremity advances faster into the valley; being, of course, protected from thawing influences, its advance would be more perceptible. Upon questioning him closely, he declared that he had seen the glacier press on the snow before it; and that, in January 1840 in particular, it had advanced towards a fixed mark no less than 50 *klafter* (fathoms) in three weeks; a result, however, which we must be allowed to doubt.

The top of the Riffelberg is a peak, or "horn" as it is called in German Switzerland, which long passed for inaccessible, as no guide at Zermatt had attained it. In 1841, I attempted it by the western side, and arrived within a few fathoms of the top, when I was stopped by a cleft and a precipice which was not to be ascended without incurring a needless risk. In 1842, however, some English students of Hofwyl, clambering about the rocks, found a circuitous path on the eastern side, by which the top may be gained without much difficulty; I, accordingly mounted it with Damatter, who had learned the way, and proceeded to take some bearings from the summit, which is a narrow rugged space. At first I thought Kater's compass pointed wrong; the sun, which was near setting, appeared due north! Then I took another compass and got the same result. It was clear that there was an enormous local attraction of the hill on the needle. We would charitably wish this to be considered as a possible explanation of some portion of the inconceivable errors of the more esteemed maps of this part of the Alps; errors which something like an oversight of 60°, as in the present case, would seem alone capable of accounting for. I will, however, preserve the bearings I took, including an azimuth of the sun, which may serve to correct the others, and which may possibly be of use. They are expressed in degrees, round the circle from N. by E.

Stockhorn, (Col d'Erin,)	.	.	5¼°
Dent Blanche, (Hovēnghorn,)	.	.	23
Triften-horn, (Moming,)	.	.	54
Weisshorn,	.	.	70
Mont Fée? (Saas-grat,)	.	.	139
Monte Rosa, (Höchste,)	.	.	204
Petit Mont Cervin,	.	.	289¼
Sun's Azimuth, 21st Aug. 1842, 4h. 34m,	.	349
Mont Cervin,	.	.	351

Now, the Riffelhorn bore 120½° from the Stockhorn. Supposing that observation correct, the Stockhorn ought to have borne 300½° from the Riffel; but it appeared to bear 5½°, (or 365½°,) consequently the local error was 65°! It appears, therefore, that the slaty beds of the Riffel are highly magnetic, probably from octohedral iron, which is found in large crystals on the neighbouring glacier of Findelen.

I take this occasion of adding a few remarks upon the relations of the rocks in this part of the Alps, which have been only incidentally mentioned. In doing so we must carefully distinguish statements of fact from theoretical statements. The former include the general distribution of rocks of certain mineralogical characters throughout the chain, as for instance granites, and the position and arrangement of the stratified rocks connected with them. The nomenclature of these rocks, and the limits of formations, may at present be considered as in some degree hypothetical.

The granite of the Alps appears at intervals along the chain as if it were continuous below, but breaking forth only here and there, and affecting various other rocks with which it is intermingled, constituting, as M. Studer has most prominently brought into notice, a series of distinct centres, rather than long lines or axes of elevation. At the same time, we undoubtedly find a linear arrangement amongst these granitic groups, and frequently indications of true granite where the rock does not occur in mass, as in the granite veins of the higher part of the

Val Biona, and of the valley of Arolla, and at Valorsine. It is not unfrequent that a secondary or parallel outburst of granite takes place, so that the chain appears to have two if not three axes. This is well-marked in the range of the Aiguilles rouges, near Chamouni, and something of the same kind will probably be found in the Val de Bagnes, the head of the Val d'Anniviers, and in the valley of Saas. M. Studer has indeed mentioned the Dent Blanche as a granitic centre. I do not recollect to have seen granite blocks on the Glacier of Ferpêcle, but I cannot be certain. The Stockhorn is certainly slaty. Now, though slaty rocks containing felspar are often in contact with Alpine granite, it is impossible to consider them as representing universally the gneiss formation of other countries. In the first place, the granite as often, or oftener, *overlies* the slate than the contrary. These slaty rocks may be distinguished by mineralogical character, but scarcely by any other. They are quartzose, or micaceous, or calcareous, or contain serpentine, and are in colour white, grey, black, or green, and these colours are amongst the most distinctive characters which they present. Thus, there are the black-slates of Fiz and the Bonhomme, near Chamouni, while the greater part of the mountains we have lately been describing are composed of regularly stratified alternations—felspar-slate, (gneiss,) quartz-slate passing into quartz-rock, talc-slate (*schistes verts*) passing into serpentine and diallage, and calcareous slates passing into dolomite, which last occurs in several repetitions in the section of strata formed by the valley of the Visp (or Viège.) The talc-slate also passes into pot-stone, which is worked near the town of Viège, (Visp,) and above Evolena. The Col of Erin is a felspathose slate or gneiss ; the Stockhi is a white quartzose slate, probably containing lime, and this appears to constitute the whole of the higher part of the Mont Cervin, whose unapproachable precipices will for ever prevent the geologist from a nearer survey. I have already said, that the middle strata are contorted, and probably calcareous, and that the lower part, together with the Hirli—an accessible promontory at the foot of the pyramidical part on the side of Zermatt—is composed of the green slates passing into serpentine and Gabbro. The mineralogical descriptions of De Saussure of this part of the Alps are intelligible and exact, and, with the exception of his attempt to classify the rocks amongst the regular primary deposits, may be considered to be nearly as precise as any that could now be given. For the sake of condensation, I may add, that the Riffel, and nearly all the chain of Monte Rosa, are composed of similar beds, which generally rise towards the eastern points, on the north side to the south-east, on the south side to the north-east. De Saussure says distinctly, that the beds of the Mont Cervin rise to the north-east at an angle of 45°; my impression was, that they are less inclined or nearly horizontal—but De Saussure is no doubt correct.* His opinion of the arrangement and materials of the beds composing it I find to coincide accurately with my own observations on the spot.

The highest part of Monte Rosa, judging from specimens brought from the last accessible point by M. Zumstein, is mica-slate. The whole system of Monte Rosa, as already said, rises to the east, and the first regularly crystalline rocks we meet with are near the Pizzo Bianco, above Macugnaga, which will be mentioned farther on. With respect to the age of these various rocks, few geologists are as yet disposed to decide with much confidence. I have already observed,† that the division between true primitive gneiss and mica-slate, and rocks of the same mineral character, which may be traced continuously into beds containing lias fossils, seems to be an arbitrary distinction, and one upon which no two observers could exactly agree. The age of the felspathic and micaceous slaty rocks, may be considered as open to discussion. The others— namely, the grey and green slates which I have described, are included by M. Studer under the general name of *Flysch*, a widely spread formation in Switzerland, but whose superposition is too irregular and uncertain, and the series of formations too imperfect to afford any clue to its age, whilst the one or two fossils which have been found in it seem to point to an age newer than the lias, and older than the medium chalk formation. What an overturn of all ancient ideas in geology to find a pinnacle of 15,000 feet high, sharp as a pyramid, and with

---

* *Voyages*, § 2243.　　　　　　　† Page 275.

erpendicular precipices of thousands of feet on every hand, to be a representative of the older chalk formation ! and what a difficulty to conceive the nature of a convulsion (even with unlimited power) which could
produce a configuration like the Mont Cervin rising from the Glacier of
Zmutt.*

Some pretty minerals are collected near Zermatt, principally from
the moraines of the Glacier of Findelen.   The most remarkable is one
of the talc family, a silicate of magnesia called *Pennine*, which occurs
well crystallized in talc-slate.   It is blackish green by reflected light,
and by transmitted light it is dichroitic, being of a brown orange in
one direction, and of a bright green in another.   On the Riffel, I found
a large vein of an imperfectly characterized mineral, which M. Studer
considers to be a variety of kyanite.   A considerable variety of garnets,
particularly the black kind, are found at Findelen, as well as octohedral iron.

---

* I am happy to learn from M. Studer, that he has at length completed a first approximation to a geological map of the Swiss Alps, the result of twenty years' observation.

CHAPTER XVIII.

FROM ZERMATT TO GRESSONAY BY THE COL OF MONT CERVIN.

DETAINMENT AT ZERMATT—ASCENT OF THE PASS OF MONT CERVIN—THE COL
—FORTIFICATIONS—THE DESCENT—HIGHLY ELECTRIC STATE OF THE
ATMOSPHERE—CUSTOM-HOUSE OFFICERS—BREUIL—VAL TOURNANCHE—
CHAMOIX—COL DE PORTOLA—VAL D'AYAS—BRUSSONE—COL DE RANZOLA
—ARRIVAL AT GRESSONAY—M. ZUMSTEIN.

> Never till now,
> Did I go through a tempest dropping fire.
>   *   *   *   *   *   *
> A common slave (you know him well by sight)
> Held up his left hand, which did flame and burn
> Like twenty torches join'd ; and yet his hand,
> Not sensible of fire, remain'd unscorch'd.
> *Julius Cæsar, Act I., Scene 3.*

In 1841 I had been prevented from crossing the celebrated pass of Mont Cervin along with my friend, Mr. Heath, and in 1842 another accident threatened again to make Zermatt the limit of my journey. A trifling injury to my foot, received on the Mer de Glace at Chamouni, and which had not appeared to get worse during the severe walking which I had since performed, assumed a more serious appearance during a day or two of comparative repose which I passed at Zermatt, waiting for M. Studer's arrival from Visp. I became a close prisoner, for nearly a week, at the little inn at Zermatt, where I was fortunate in finding much comfort and attention from the worthy

Madame Lauber. The weather had altered for the worse, which diminished my regret at the detention, and I had the advantage of M. Studer's company. My friend was, however, resolved to lose no more time than the weather rendered necessary in resuming his journey, and as my foot was now convalescent, I consented to accompany him, on a morning of somewhat doubtful promise, when we were called by the faithful Klaus, whilst the stars were still shining bright through the wild drift of cloud. The impatience of the guides on such occasions is not the least of the evils of detention. My Savoyard, who spoke not a word of German, pure or impure, (and nothing else was understood at Zermatt,) suffered the horrors of *ennui* to an extent which might be thought to belong exclusively to the loungers of our great cities ; and but for a small speculation in the minerals of Findelen, which he fully counted upon disposing of with a profit of 200 per cent. at Chamouni, he very probably would have insisted on walking off to enjoy the daily fund of summer's gossip of his native valley.

We set forth about half-past 4 A.M., and having crossed the torrent of Zmutt, wound slowly up the steep pastures which skirt the western edge of the Glacier of Gorner. We gradually attained a considerable elevation above its surface, before crossing another torrent, which descends from the Boden-Gletscher, which we left upon our right. In the preceding year I had ascended thus far, and crossed the Boden Gletscher to the foot of the Hirli, where there is a gloomy *tarn* called the Schwartzen See, beyond which is a fine view of the northern precipice of the Mont Cervin, and of the Glacier of Zmutt. Those who do not propose to pass the Col of St. Théodule may thus make a very interesting excursion, and return by the châlets of Zmutt. Now, however, we kept right onwards, and a little after seven we reached the edge of the glacier which we had to traverse. Its surface is tolerably level, it is very extensive and desolate, not being included between bold walls as in the lower glaciers, but occupying a sort of vast table land, at an average height of nearly 10,000 English feet above the sea. We had an opportunity of appreciating its desolation, for we were repeatedly enveloped in the rolling mists which swept over the Col, and which appeared to boil up tumultuously from the side of Italy, which we were approaching, and to be repelled on the Swiss side by an uncertain north wind. This wind secured us, however, a fine view of Monte Rosa, and of the chain of the Weisshorn; but I learned nothing new of the topography of either from this point, nor does the panorama admit of comparison with that from the Col d'Erin. Even the Matterhorn, (Mont Cervin,) which, however, we saw imperfectly, loses its apparent height, since here it rises only from the ridge, already at a height of 11,000 feet. Having walked already for a long time over snowy flats, we entered a kind of defile as we approached the Col. The mists closed round us, and a stranger might very easily have entirely lost his way, for the defile presented many accessible points; our guide, Damatter, however, took the matter very coolly, and brought us safely to the Col. The weather was damp and raw, and we had no view. We had been five hours and a quarter constantly ascending from Zermatt. We hastened to observe our instruments. The temperature of the air was 35° Fahrenheit, and the barometer stood at 511.53 millimetres, 3½ millimetres above what I had observed it at the Col du Géant. The height above the sea comes out 10,938 English feet, by a comparison with the barometer, both at Geneva and St. Bernard.

The Col du Mont Cervin, or St. Théodule, consists of felspar slate, or gneiss, and exhibits well preserved traces of a rude fortification, called "Fort du Saint Théodule." De Saussure says, that it was erected two or three centuries ago by the inhabitants of Aoste, to prevent an invasion of the Valaisans. "Ce sont," he adds, "vraisemblablement les ouvrages de fortification les plus élevés de notre planète. Mais pourquoi faut il que les hommes n'aient érigé dans ces hautes regions un ouvrage aussi durable que pour y laisser un monument de leur haine et de leurs passions destructives." Certainly there is nothing more jarring to the impressions of stern grandeur and vast solitude than the not unfrequent occurrence of military works in many parts of the Alps,

"High heaven itself our impious rage assails;"

x

the pass of the Col de la Seigne, at the head of the Allée Blanche, and more than one of the very savage Cols near Monte Viso, bear witness to this strange anomaly.

We were disappointed of the fine view which we ought to have seen towards Aoste. Fortunately the clouds cleared so far as to let us see our way across the remaining part of the glacier on the Italian side, which is much steeper than the other, and consequently traversed by extensive rents, which being covered knee-deep with snow freshly fallen during the last few days of bad weather, were, in some places, not a little dangerous. The pass of the Mont Cervin appeared to me, on the whole, a more considerable undertaking than I had expected. Knowing that it is frequently traversed in favourable seasons by horses and mules, I expected to have found the glacier both shorter and easier. This season, indeed, no beast of burden had crossed, and it appeared almost inconceivable how they ever could ; but such is certainly the fact, and we saw more than one trace of animals which had perished in the passage. Another circumstance which led me to expect an easier passage than we found it, was the ludicrous outfit of our friend Peter Damatter, the guide of Zermatt, who, instead of bringing a good ice pole and cord, as he ought to have done, being aware of the fresh snow, had provided himself merely with an umbrella. He was glad to borrow a stick from one of the party to sound his way on the Italian side, although we alleged, that he used it with little dexterity ; but the snow was literally knee-deep, and we encountered several wide crevasses, into one of which Tairraz had almost fallen, although he was the last of the party who had trodden in the guide's footsteps. Had he unfortunately done so, we should have had difficulty in extricating him for want of a rope. When he recovered his footing, he looked as pale as a sheet, but proceeded quietly.

At length we were free of the glacier, and recovered a track by no means obvious, which leads to the Châlets of Breuil, leaving upon our left hand the longer and more difficult route by the Cimes Blanches, conducting to St. Giacomo d'Ayas. The atmosphere was very turbid, the ground was covered with half melted snow, and some hail began to fall. We were, perhaps, 1500 feet below the Col, or still above 9000 above the sea, when I noticed a curious sound, which seemed to proceed from the Alpine pole with which I was walking. I asked the guide next me whether he heard it, and what he thought it was. The members of that fraternity are very hard pushed indeed, when they have not an answer ready for any emergency. He therefore replied with great coolness, that the rustling of the stick no doubt proceeded from a worm eating the wood in the interior! This answer did not appear to me satisfactory, and I therefore applied the *experimentum crucis* of reversing the stick, so that the point was now uppermost. The worm was already at the other end! I next held my hand above my head, and my fingers yielded a fizzing sound. There could be but one explanation—we were so near a thunder cloud as to be highly electrified by induction. I soon perceived that all the angular stones were hissing round us like points near a powerful electrical machine. I told my companions of our situation, and begged Damatter to lower his umbrella, which he had now resumed, and hoisted against the hail shower, and whose gay brass point was likely to become the *paratonnerre* of the party. The words were scarcely out of my mouth when a clap of thunder, unaccompanied by lightning, justified my precaution.

At length we got below the level of the clouds, and the first shelter we reached was the wretched retreat of two Sardinian *douaniers*, who had lighted a fire under a portion of the remaining arch of what had once been a pretty solid edifice, probably a cowhouse; stones being plentiful, and wood the reverse, this mode of roofing had been adopted. They received us with civility, and allowed us to dine by their fire; and as we had been on foot for eight hours, we were entitled to some repose. The absolute discomfort in which this class of men live is greater than in almost any other profession. Hard diet, constant exposure, sleepless nights, combined with personal risk, and still more galling unpopularity, great fatigue, and perpetual surveillance, are the ordinary accidents of their life. Liable to suspicion when they quit the wildest and most inaccessible parts of the chain where a smuggler may by possibility pass, posted for hours together on a glacier 9000 feet above the sea, and, like animals of prey, taking repose during daylight in some deserted hovel in their wet clothes—one cannot but conclude the smuggler's life to be luxury compared to the protracted sufferings of their detectors. On many frontiers the *Douaniers* are a slovenly and self-indulgent race; but on others I know that this is no exaggerated picture of their lives, even in the finest season of the year.

We descended to Breuil, a group of châlets pleasingly situated at the first green level in the Valley, where we arrived at two o'clock, having staid an hour with the *Douaniers*. The scenery of the head of the Val Tournanche, in which we now found ourselves, is very striking. The Mont Cervin, which, owing to the clouds, we saw imperfectly, is the most conspicuous object, and next to that the excessively rugged range which stretches away from the Dent d'Erin as a centre, and forms the boundary between the Val Biona and Val Tournanche; the Dent d'Erin itself rose in terrible majesty. We noticed what appeared to be an ancient moraine descending from between the Mont Cervin and the Col we had passed, exactly where, according to the general belief, the former passage existed,—namely, close under the Mont Cervin itself.

We performed the road from Breuil to the village of Val Tournanche leisurely, having the afternoon before us. In one place the valley becomes contracted, and the torrent dashes through a picturesque ravine, which exhibits distinct traces of glacier action as well as the friction due to water. Transported blocks are not numerous. Below this occur the first permanent habitations, and Val Tournanche itself, which would be a pretty village any where, seems a paradise to one descending from such savage scenery as we had left. The valley, though narrow, and partly bordered by precipices, has yet an undulating grassy bottom, with well watered meadows. The heights are clothed with pines, and the cottages peep out through walnut trees, as well as the spire of the village church, which has an Italian character. There is no inn, but we were received and hospitably entertained by the Receveur des Douanes, with whose subordinates we had dined.

Next morning we proceeded to cross the lateral chain which separates the Val Tournanche from the valley of Ayas or Val Challant, to which, as has been already said, there is a direct passage from the Col of St. Théodule. This is the first of a series of ridges which require to be passed in succession by one who would make the circuit of Monte Rosa. These lateral passes, though none of them difficult, are generally steep and fatiguing, and render this expedition a far more serious one than the circuit of Mont Blanc. There are usually several passes of these ridges: in the present case, having accidentally met with the Curé of a parish which lay in the way of the pass to Ayas, we left Val Tournanche, in company with him, at seven o'clock, A.M., and descended the valley a little way farther; we then took a foot path to the left, and soon found ourselves in a wood, which covers the precipices of that part of the valley. Our Curé was a stout walker, and a useful guide, for our footpath (which was a *short cut*) soon split into numberless tracks, and as we gradually got amongst the rocks, we were glad that we were not left to waste time by discovering a way for ourselves. We ascended gradually higher and higher, and all the while, as we walked parallel to the course of the valley, the torrent was working itself deeper and deeper, so that from each fresh crag we found a greater interval between us and it, until at last, turning a rock, we stood above a precipice at least 2000 feet high, to which here and there a clinging pine seemed to give more steepness, by offering a scale for measuring the abyss. This point gained, we rejoiced in the beauty of the morning, and of the herbage spangled with drops from the early mists; and as we turned round we saw behind us the Mont Cervin rising in unclouded grandeur. We then passed from rock and wood to an open Alpine pasturage, which seemed cut off by these precipices from the world beneath, and here was the home of our Curé, a little village, appropriately named Chamoix, one and a half hours distant from Val Tournanche.

From thence, a gentle though pretty long ascent took us to the Col de Portala, composed of limestone, and very precipitous on the eastern side, where it immediately overlooks the village of Ayas. The height of the Col, by M. Studer's observation, is 7995 feet, and that of Chamoix 6004 feet above the sea. The descent presented no difficulty, and from Ayas, two hours' pleasant walk, took us to Brussone. In the course of it we crossed a singular tract of country. It was evidently the site of a lake which had been formed by the damming of the waters by a tremendous landslip which had taken place from a mountain on the right.

At first we thought it a moraine, but we saw evidently that it was but a current of debris which had descended from the neighbouring hill, disengaged, like that of Goldau, probably by the force of water. The scale of it is immense.

We arrived at Brussone soon after three, and thought of going on to Gressonay, which would have been quite practicable. The beauty of the spot, however, tempted us to remain, notwithstanding the indifferent accommodation which the Lion d'Or offered; but, after all, we might have been much worse lodged. The village is beautifully situated on a frequented mule road from Chatillon to Gressonay, by the Cols of Ion and Ranzola; the lower part of the valley also communicates with the valley of the Doire, near Fort Bard. It appears, however, to be shut in by the highly picturesque mass of Mont Nery, a dark mountain, with snow lying in its higher ravines, and which, from its general character, is probably in geological relation with the mountains of Champorcher on the other side of the great valley.

The next morning was beautiful. It was Sunday, and as we slowly ascended the heights above Brussone, we met numbers of peasants descending to church, who greeted us in French patois. After 2½ hours, we reached the Col de Ranzola, which was higher than we expected, being 7136 feet above the sea, whilst the level of Brussone is 4431. It is a narrow opening in the ridge, from whence we suddenly obtain a view of the deep and narrow Val de Lys, and, soon after, of the village of Gressonay. As usual, (owing to the general western dip of the strata,) the east side of the Col is steep. Right opposite, we observed the Col of Val Dobbia leading from the Val de Lys to the Val Sesia. We ought to have enjoyed a view of Monte Rosa; but though the weather was fine, the mountain remained veiled in clouds. A rapid descent brought us, in about an hour, to the village of St. Jean de Gressonay, the principal place in the valley, where we at once perceived, by the appearance of the people in their Sunday costume, as well as their language, that we were amongst a new race. In fact the Val de Lys, and part of the neighbouring ones, are inhabited by a colony of Germans. We arrived at comfortable quarters, *chez* Luscos, in good time for the mid-day meal, and disposed ourselves to remain there for the rest of the day. Every thing betokened German neatness and order, and in a very short time, what appeared to us a sumptuous meal was set on the table, at which sat our host, himself the representative of one branch of the family of Luscos, one of the most dignified of the valley, and whose stately portraits, mingled with those of the reigning sovereigns of their time, graced the walls of the old baronial-looking hall with huge stone-arched fire-place, and numberless windows, in which we sat. The women, as usual, wore the more characteristic costume, and especially the caps of gold tissue, so common in some parts of Germany. The familiar language talked at table was German, though probably all the natives present could talk more or less French and Piedmontese. We were received with courtesy, and entertained less as guests at an inn than as at a private house, and we found that the charges bore a proportion to the favour thus conferred on us. Nevertheless, in such situations, a traveller is generally willing to purchase unusual comforts at a higher rate. An intelligent old man sat next M. Studer, and after a little conversation, he turned out to be M. Zumstein, the well-known ascender of Monte Rosa, whose acquaintance we had been prepared to make. He entered readily into conversation. When, after dinner, I handed him a letter which had been sent to me by the ever friendly care of the Chevalier Plana of Turin, he at once offered to devote himself to our service during our stay in the Val de Lys, and to accompany us to the glacier.

In pursuance of this plan, we proceeded next morning to Naversch, where M. Zumstein lives, forty minutes' walk above St. Jean, proposing to visit the glacier at the head of the valley, and to cross the Col d'Ollen to the Val Sesia next day. But the weather was too unfavourable. The clouds which had hung over Monte Rosa for two days now descended into the valley, and by the time we reached M. Zumstein's house, it rained heavily. We therefore paid him a long visit, and obtained some particulars respecting his journeys. His barometer, compared with M. Studer's, gave (29th August, 11 A.M.)—

	Fr. inch.	Lines.	Millimetres.
M. Zumstein's,	23	7.8 =	640.5
M. Studer's,			640.4
Temperature of boiling water, uncorrected,)			204° 20 F.

In the afternoon, M. Studer and I walked down the Val de Lys about five miles below St. Jean. We quitted the green slates and serpentine, which are the prevailing rocks of Gressonay, and found hornblende-slate with granite veins; the hornblende contains garnets, which are very characteristic of the mountains of Cogne, with which probably the Mont Nery is geologically connected. We also observed well-characterised *roches montonnées*, where the valley contracts into a narrow ravine, and its level suddenly falls. In general, blocks transported from any considerable distance are rare in this valley. We returned to St. Jean to dinner, and M. Zumstein spent the evening with us. M. Studer resolved not to wait any longer for fine weather, and to cross at once to the Val Sesia. I was unwilling, however, to omit examining the Glacier of Lys, and as our routes were to separate, at all events, very shortly, he keeping the southern side of the Alps, and I returning by Monte Moro into Switzerland, we determined, though with regret, to part here. The day was finer than the preceding ones, though clouds still lowered. M. Studer crossed the Col de Val Dobbia, whilst I re-ascended the valley to Naversch to join M. Zumstein who good-naturedly accompanied me to the glacier, though the day was far from fine.

CHAPTER XIX.

GRESSONAY—MONTE ROSA.

THE GERMAN VALLEYS OF MONTE ROSA—PECULIAR RACE OF QUESTIONABLE
ORIGIN—THEIR MANNERS AND DIALECT—TOPOGRAPHY OF MONTE ROSA—
ATTEMPTS TO ASCEND IT BY VINCENT AND ZUMSTEIN—THE HIGHEST POINT
STILL UNATTAINED—AN EXCURSION TO THE GLACIER OF LYS—ITS RETREAT
—ITS STRUCTURE—RETURN TO STAFFEL.

THE Valleys of Gressonay, Sesia, and Anzasca, all in the Sardinian
dominions, and to the south of the great chain of Alps, are inhabited,
in their higher parts only, by a race of men whose physiognomy, dress,
and language, alike bespeak a German origin.

Were the heads of these valleys in immediate communication with
those of German-Switzerland by easy passes, this would occasion little
surprise, accustomed as we every day are to see national limits trans-
gress natural or geographical boundaries, and the peculiarities of con-
terminous races to be softened by an imperceptible gradation. But in
the Piedmontese valleys of Monte Rosa, the case is quite otherwise: the
chain of Alps is their prison, not their portal; for from two out of
three of them, no human foot has certainly ever passed directly from
Italy to Switzerland, or the reverse. The German colony must, therefore, have been introduced through the Italian territory, and their choice,
or their necessities, have driven them to the mountain fastnesses, which,
perhaps, reminded them of those of their native land.

De Saussure has, as usual, nearly exhausted what it is of importance to say respecting the possible origin of these mountaineers. He
has classed the existence of the German colony as one of the nine pecu-

liarities of the district;[*] he has stated, in a few sentences, what may be conjectured as to their origin, and in a few more he has adroitly sketched their character.

De Saussure supposes that they were Valaisans who crossed the Monte Moro, (a pass from the Val Anzasca into Switzerland,) at a remote period, in order to occupy the higher valleys of Monte Rosa, whose rough surface and rude climate had repelled the more delicate Italians. He describes the people as simple, timid, and even rude, but honest; their greatest fault, a want of hospitality, which he found embarrassing at a time when inns were even rarer than at present.

It may be affirmed that the manners of the German settlers have improved since the time of De Saussure, which leads us to believe that their fault arose from their ignorance and isolation. I met everywhere with respectful, and even touching attention. Any traveller speaking the German language, is certain to be well received; and it is interesting to observe the tenacity with which these descendants of an unknown stock cling to the usages and the speech which form the only evidence of their birth, for history and tradition are both silent on the subject. Though most of the inhabitants—at least the men— speak several languages, acquired during their earlier years of expatriation, they invariably prefer speaking German, which many of them do with fluency, and without accent; far better, in short, than most persons of a similar class in German-Switzerland. The expatriation to which I have alluded, arises from their practice of going forth from their valleys at an early age to push their fortunes in wealthier lands, and especially in southern Germany. But, almost invariably, they at last return to marry,[†] and to settle in comfort at home. Hence, ease and independence is still more marked here than in Switzerland. Some of the earlier writers, as Scheuchzer, distinguish Gressonay as the " Merchant-Valley," [‡] *par excellence;* and at one

---

[*] *Voyages,* § 2243.

[†] They have an expressive proverb to this effect :—

"Weiber und Steine muss man lassen wo sie wachsen."

[‡] Krämer-Thal.

time the race of pedlars in southern Gremany were termed " Gressonayer" collectively.

Their habits are cleanly and active, and their houses, built in the true German taste, would alone, and at once, distinguish them from their Italian neighbours. I spent a Sunday at Gressonay, as already mentioned, which gave me an opportunity of seeing the holiday costume of the women, which resembles some of the gayest in Switzerland, especially the abundance of gold and silver lace, and the metallic helmet-caps. In religion they are strictly Roman Catholics : their churches are adorned with frescos in the Italian taste.

Since De Saussure called particular attention to the German settlers of Monte Rosa, several German authors have written respecting them. Of these the chief are Hirzel-Escher, Von Welden, and Schott. Of these works now before me, the last is the most elaborate as respects the question of population ; * but it is tedious from its detail, and disagreeable to read, from an affectation of singularity in the spelling and printing of the German language.

Schott has given specimens of the Patois of each of the various *communes* of the German valleys—namely, Issime and Gressonay in the Val de Lys—Alagna, the highest village, which alone is German, in the Val Sesia—Rima in the Val Sermeuta—Rimella in the Val Mastalone—and Macugnaga in the Val Anzasca. That of Gressonay appears to be the least impure German ; and indeed it is there alone that the striking externals of the German race are to be found in perfection : nearest to it in this respect is the valley of Anzasca. In every case, the Patois is a corrupt mixture of Roman and Teutonic roots, of which the author has given an elaborate vocabulary. It is curious to observe, that in the proper names of these valleys, the family names have preserved pretty generally their German character, as Ackermann, Beck, Schwartz, Zimmermann, Zumstein, whilst the Christian names are chiefly Italian.

---

* Die Deutschen Colonie in Piemont, ihr Land, ihre Mundart, und Herkunft. Von Albert Schott. Stuttgardt, 1842.

The second of the works above named, that of Von Welden, is interesting from the topographical details which it gives of the complicated environs of Monte Rosa, which, till then, were very imperfectly understood,—and not less so from the details of successive attempts to reach its highest summit, made by M. Zumstein, (a native of these valleys,) and described in his own words.

The vexed question of the comparative height of Mont Blanc and Monte Rosa, was undecided before the survey of Von Welden, which was published in 1824.* It required an elaborate operation to determine its absolute height, on account of the complication of peaks of nearly equal elevation which form its summit, all of which cannot be seen from perhaps any point external to them, and which must nevertheless be separately and minutely observed, in order to ascertain which is really the highest. Thus De Saussure, as appears evidently from his own view, (*Voyages*, Tom. IV., Pl. V.,) measured not the highest peak, but only the third in height, now called the Zumsteinspitze. He made it 2430 toises, or 15,540 English feet above the sea.† This was within 200 feet of the height of Mont Blanc; but later and more precise observations all agree in making even the highest point considerably lower. Von Welden finds it to be 14,222 French, or 15,158 English feet, which agrees nearly with the mean of the results of Carlini, Oriani, and Corabœuf.‡

Monte Rosa is a union of several mountain chains, rather than one summit. The map, page 1, though on a small scale, will give an idea of their arrangement. From it, or from any map based upon Von Welden's, it will be seen, that a vast inaccessible ridge stretches nearly east and west, commencing at the Col du Mont Cervin, between Zer-

---

* Der Monte Rosa, eine topographische und naturhistorische Skizze. Wien, 1824.
† *Voyages*, § 2135.
‡   Carlini,   2348 toises.
    Oriani,   2388 ——
    Von Welden, 2370 ——
    Corabœuf,   2379 ——
         Brugière, Orographie de l' Europe, p. 208.

matt and Breuil, and terminating in the Cime de la Pisse, to the east of Monte Rosa. This chain includes the Petit Mont Cervin, the Breithorn, and the Lyskamm. Another vast ridge, though a shorter one, meets this nearly at right angles, stretching from Monte Rosa northwards, towards the Cima di Jazi. It also crosses the chain to the south, so as to form the ridge of the Col d'Ollen between the Val de Lys and Val Sesia. The union or *knot* formed by these two chains is the locality of the elevated summits properly called Monte Rosa. Of course four cavities or angles are left when the transverse chain meets the longitudinal one. The one of these to the north-eastward, which is the most precipitous, and which, indeed, has been compared by De Saussure to a crater, forms the head of the Val Anzasca, and embosoms the Glacier of Macugnaga; the north-western one, vaster, but less precipitous, gives birth to the great Glacier of Gorner, or of Zermatt; the south-western angle contains the Glacier of Lys, which descends from the Lyskamm * into the valley of Gressonay; the fourth, or south-eastern cavity, is occupied by the head of the Val Sesia, and has also extensive, though less prominent glaciers.

Thus Monte Rosa is in ground plan like a four-rayed star or cross. All the highest summits are ranged along the northern and southern rays, especially the former. The point of union of the rays is not the most elevated, though, in some respects, it is the most generally commanding top. It is the most conspicuous from the Italian side of the Alps; it has been called by Von Welden "Signal Kuppe." It is the fourth in point of height. The three higher lie all immediately north from it; the first in order is the "Zumsteinspitze," the highest which has been ascended, which is a snowy blunt summit, mistaken by De Saussure for the highest. Next follows the highest; a sharp rocky obelisk, well seen from the Col d'Erin, (see page 300,) and from Monte Moro, (see next chapter.) It is connected with the Zumsteinspitze by a longitudinal very sharp icy ridge like a house roof, which, on the eastern side, descends with appalling rapidity to an abyss which is scarcely

---

* *Kamm*, a comb-shaped or jagged ridge of mountains.

equalled for depth and steepness in the Alps. Beyond the highest, or " Höchste Spitze," is the second highest, called by Von Welden " Nord-End," which, like the last, has never been scaled. The difference of height of these four summits is trifling, amounting to only 34 toises, or little more than 200 feet, from the highest to the lowest. Three other summits of somewhat less height form the southern arm of the Cross, namely, the " Parrotspitze," " Ludwigshöhe," and " Vincentpyramide," the last of which, and also the lowest, was the first ascended of the group.

Having now endeavoured to give a distinct geographical idea of the position of this group of mountains, (which I have seen and sketched in almost every direction from whence they are visible,) I will add a very few words respecting the attempts which have been made to ascend it, which have excited far less interest than those upon Mont Blanc; and such is the confusion prevalent on the subject, that some guides of Chamouni maintain, that they have ascended the summit of Monte Rosa from the Col of Mont Cervin, which is a good deal more ridiculous than if they proposed to scale Mont Blanc by ascending the Glacier of Argentière.

The explorers of Monte Rosa, in its wilder recesses, were M. M. Vincent and Zumstein, the former the earlier, the later the more persevering and successful. I can only mention briefly the results of their journeys, which may be found contained in an interesting series of papers by M. Zumstein in Von Welden's work.

The first ascent of the lowest summit was by M. Vincent alone, in August 1819, whence his name was justly given to it. Then he and M. Zumstein together repeated the ascent, with more favourable weather. The chief difficulties experienced were from a huge ice cleft, or *Bergschrund*, and from the labour of cutting 600 steps with a hatchet on a steep ice slope. The ascent on this, as on all other occasions, was made from the side of Gressonay, near the Col d'Ollen, where gold mines are worked above the limits of perpetual snow, and where, therefore, a shelter, however rude, could be obtained, at a height of 10,800 feet, certainly the highest temporary habitation in Europe.

The second journey, that of 1820, was performed by Zumtein alone, with the purpose of making for the summits farther to the north, and also the highest. He was accompanied by a surveyor, with a theodolite, who was commissioned by the Turin Academy to make observations for the improvement of the maps of Monte Rosa; but the Italian surveyor being unused to such excursions, the labour and expense of the journey were unavailing, although it clearly appears from the narrative, that had Zumstein himself been able to make the observations, he would have had ample time and opportunity for doing so,—one proof amongst many of the necessity (which De Saussure saw and acted on) of the director and chief of such an expedition being not only an experienced mountaineer, but himself capable of undertaking all the experiments and observations which he desires to be made. Under such circumstances, the zeal and sense of responsibility of the traveller and discoverer himself, are alone equal to the task of making observations of any value, or rather, not positively mischievous by their inaccuracy. The most perfect land-measurer, the most experienced laboratory assistant, are alike thrown out when they are expected to make their contacts, verify their zero points, record degrees, minutes, and seconds, with as much deliberation balanced on a dizzy pinnacle or exposed to a pinching frosty gale, as in their ordinary localities, and with the usual appliances.

M. Zumstein left the peak which he had before ascended, and several others, on his right hand, following the elevated snow valley which separates the high range of Monte Rosa from the Lyskamm. It appears that these vast snow-fields may be traversed without danger, unless from the chance of being overtaken by night or bad weather at so great a distance from shelter. The valley of Zermatt is visible from them; and we find that some peasants of Gressonay, who reached this point as early as the time of De Saussure, brought back startling reports of an unknown pastoral valley discovered by them amidst the wilds of Monte Rosa;* the fact being merely that they saw the woods

* Compare De Saussure, *Voyages*, § 2156, and Zumstein, in Von Welden, page 124.

and meadows of Switzerland, backed by the icy chains of the Col
d'Erin, Dent Blanche, and Weisshorn.

So distant are the higher summits of Monte Rosa from the gold-
miner's hut whence the party had started in the morning, that the day
was spent before the loaded guides and the timid surveyor could be got
forward to the foot of the higher peaks. Here Zumstein had the courage to determine upon passing the night in a cleft of the ice at the
height of 13,128 French, or 13,992 English feet above the sea,—undoubtedly the greatest height at which any one has passed a night in
Europe.

The next morning, the summit bearing the name of Zumstein was
attained without much difficulty. Here, too, the opportunity of
making observations was lost, for whilst waiting for the ever-tardy
engineer, the horizon became clouded. The party perceived, however,
that they were not, as they expected, upon the highest point, which
was 750 yards farther north, and 200 feet higher. It appeared to
them to be inaccessible in this direction. The barometer stood at $16\frac{1}{8}$
French inches. None of the party experienced the exhaustion and
other symptoms so often felt on Mont Blanc. They returned to the
huts after having been forty hours on the snow. Twice afterwards,
M. Zumstein repeated his visits to this peak, but without succeeding
in making farther progress.

I shall conclude this chapter with some account of an excursion in
the valley of Gressonay, where we stopped at the close of the last chapter, in the friendly company of M. Zumstein, the mention of whose
name naturally suggested this digression.

The Valley of Gressonay, or Lys-Thal, is more contracted and
mountainous than I had expected to find it, and this is characteristic
of several of the valleys which diverge from Monte Rosa, which seem
to be mere cracks or rents without diverging branches of any extent.

The sides are steep without being precipitously grand. Near St. Jean the valley is flat and fertile : at Castel, half an hour's walk above, it rises suddenly amongst rocks to a higher level. The distant view of Monte Rosa, which ought to be the centre of interest, was indeed wanting, for it remained impenetrably covered with clouds. Nevertheless, with M. Zumstein for my guide, I left Naversch, forty minutes' walk above St. Jean, for the Glacier of Lys. At the hamlet of La Trinité, which is situated in the midst of a little plain, one hour from St. Jean, a small valley branches to the right, which affords the easiest road to the Col d'Ollen leading to the Val Sesia on the east. We continued a due northerly course, passing several cottages, which, though small were clean and cheerful. In the lower part of the valley are many houses of considerable pretension, and at least three storeys high, which are all built of wood, and inhabited by the wealthier natives, who have returned with fortunes acquired in foreign countries, to pass the remainder of their days at home. Amongst these is Baron Peccoz, who acquired his nobility from the King of Bavaria, and who, having made money in trade in Germany, passes the greater part of the year at the very head of the valley of Lys, where he can indulge what is, with him, an insatiable passion for Chamois hunting. His substantial dwelling is the very last permanent habitation in the valley, at a spot called *Am Bett*, and within half an hour's walk of the glacier. He entertained the sons of the King of Sardinia, and their suite, on a visit, which they made some years since to Monte Rosa. Having an introduction to him, through M. Plana's kindness, I might have availed myself of his hospitality, but he was absent upon his favourite sport, and M. Zumstein was good enough to secure for me humbler, but most comfortable, quarters for the night, in the cottage of a worthy peasant of the valley. At a place called Staffel, the serpentine unites with the chlorite slate, and higher up is replaced by red gneiss. At a spot called Cour de Lys, are some traces of glacier action, namely, polished rocks, which, it has been observed, are rather rare in this valley. At Castel there are some blocks which appear to have been transported; but this evidence is doubtful where the geology is so monotonous.

Y

At length we reached the glacier, at a distance of not more than two and a half hours' walk from St. Jean. It has retreated continually since 1820, and has left a vast enclosure—sharply defined by its moraine—a perfect waste, having (as I judged) not less than a square mile of area. Within this area is a kind of rocky precipice, above which the glacier has now retired: it is composed of gneiss, including quartz veins, and though these have never before been uncovered by the ice in the memory of man, M. Zumstein assures me that he has found marks of blast holes where metallic veins had been sought for, probably gold, which is still worked in the neighbourhood.

We ascended on our right the eastern moraine of the glacier, I mean its *ancient* moraine, which extends yet far beyond that of 1820, and with some labour and fatigue we gained the level of a kind of *plateau*, which intervenes between the crevassed ice descending from Monte Rosa and the final slope of the glacier, at its lower end. Here the view ought to have been very grand, but we were now completely in the clouds, with a drizzling rain. I wished to cross the glacier, in order to examine its structure, and a rise in the mist favoured us. The glacier stream is here composed of two great ice flows, derived from the two sides of a promontory, called *die Nase*, or the Nose, and the eastern one is itself the result of two others, so that three streams of ice appear distinct where we crossed the glacier, with the usual belted structure, vertical near the sides, and under the medial moraine, and presenting a three-fold convexity in its front, as I have observed in other very wide glaciers, where the individual structure is not immediately lost. The bands were very well developed. I pointed them out to M. Zumstein, who candidly admitted, that much as he had been amongst glaciers, he had never noticed them before.

The moraines of the Glacier de Lys are composed exclusively of gneiss and syenite, without a trace of green slate or serpentine, so abundant below.

Having crossed the glacier, we took refuge for a while from the weather in one of the rude cabins, constructed by the shepherds, amongst the blocks of the ancient moraine. We then descended the west side; and I observed, in the moraine of 1820, several bands or heaps of stones, arranged transversely to the glacier, and parallel, like the ridges of a ploughed field. I am uncertain whether or not these were deposited in the *last* crevasses of the glacier before it disappeared.

We returned somewhat wet to the village of Staffel, and slept in the clean beds which had been provided for us. The guide whom I had desired to follow me from St. Jean to cross the Col d'Ollen next day, and to bring provisions, did not appear, and indeed the guides of this country seem to be not altogether sure. I ate cheerfully, however, the rye bread of the house, baked at Christmas 1841, and cut, with a hatchet, into morsels like sugar, of a size which could be put into the mouth at once. I found it not unpalatable, and even preferred it to fresh bread of the same kind.

CHAPTER XX.

TOUR OF MONTE ROSA CONCLUDED—FROM GRES-SONAY TO VISP, BY MACUGNAGA AND MONTE MORO.

PASSAGE OF THE COL D'OLLEN—ALAGNA—RIVA—THE COL TURLO—VAL QUARAZZA—ITS POLISHED ROCKS—MACUGNAGA—THE PEOPLE AND THEIR HABITATIONS—EXCURSION TO THE GLACIER OF MACUGNAGA—THE WEISS THOR, A REMARKABLE PASS—STRUCTURE OF THE GLACIER—GLACIERS OF THE SECOND ORDER—GEOLOGY OF THE CHAIN OF MONTE ROSA—PEDRIOLO—GIGANTIC FRAGMENTS OF ROCK—RETURN TO MACUGNAGA—PASS OF MONTE MORO—VIEW OF MONTE ROSA—DESCENT TO SAAS—GLACIERS OF SCHWARTZ-BERG AND ALLALEIN—GABBRO—SAAS—STALDEN—PEASANTS' THEATRI-CALS—VISP.

NEXT morning, after taking a cordial leave of M. Zumstein, I started soon after dawn from the hospitable roof of my entertainer at Staffel, with dull but fair weather, to cross the Col d'Ollen to Alagna, in the Val Sesia. A cheerful well-mannered peasant, named Joseph Skinoball, replaced my faithless guide as far as the Col, whence he turned back. During the ascent we left upon the left hand the gold mines of Indren, and the spot named " Die hohe Licht," so often referred to in Zum-stein's ascents of Monte Rosa. The Col d'Ollen might be reached either from La Trinité or from Staffel, or direct from the Glacier of Lys. In fine weather it would not be too long a day's work to go from St. Jean to the glacier, and thence to Alagna or Riva, to sleep. The Col is wild, and composed of jagged rock mingled with snow. I ascended in two hours and a quarter from Staffel. Water boiled at 195°.70, by the thermometer, whence I find the height to be 9758 feet above the sea. Keller makes it 1000 feet less.

From a little way beyond the Col there is a fine view eastwards, including part of the Lago Maggiore and the hills beyond. The descent to Alagna is very steep and long, (as it lies much lower than Gressonay,) but, at the same time, interesting. The Val Sesia is here very narrow, and is included between two serrated chains of mountains, of which the Zuber, on the western, and the Taglia Ferro and Monte Turlo on the eastern side, are conspicuous. The lower part of the descent to Alagna is through beautiful wood and green pastures. Alagna itself has a pretty church, in the Italian taste, and is most agreeably situated. I called on the Curé, who had ascended the Signal Kuppe, (one of the summits of Monte Rosa,) a month before. Alagna is a very poor place. A much more barbarous German is spoken than at Gressonay, and it is so completely on the German boundary, that at Riva, only half an hour's walk farther down the valley, Piedmontese is exclusively spoken, so that I was assured that a great part of the whole inhabitants of these two *communes*, especially the women, are incapable of understanding one another. There being no inn at Alagna, I descended the valley to Riva to sleep, although I should have to retrace my steps. I had, indeed, intended walking farther down the Val. Sesia, which is more pleasing than the Val de Lys,—for I arrived at Riva before noon; but a violent thunder storm, which lasted all afternoon and part of the night, prevented me. The result, however, was happy. It put an end to the recent uncomfortable weather, and the wind having changed, some of the finest days of the season succeeded, commencing with the 1st September. It is a singular, and not unimportant fact, which every native of these valleys whom I consulted agreed in stating, that the N.W. and N.N.W. wind brings fine weather, and that the E. wind, which in Switzerland (and even at Courmayeur) is dry, is here the wet wind. M. Plana mentioned the same as being true at Turin.

The following morning I was up before daylight, and left Riva at a quarter past five. The weather was beautifully clear, and the summits of Monte Rosa showed finely, with the morning sun above the deep wooded valley. Riva is situated at the foot of the Col de Val Dobbia, and is, therefore, nearly opposite to St. Jean de Gressonay. The church contains some paintings of a rude kind. I had soon retraced

my steps to Alagna, and there was introduced by the Curé to a shepherd of Biella, who was going to cross the Turlo pass, and who offered
to show the way. He was a merry fellow of the true Italian cast, with
a broad brimmed hat, and spoke only the Piedmontese jargon. He
had spent the night over the wine-skin, and pathetically lamented
the fatigues of the ascent, for which, indeed, he was not in very good
training, and before we reached the top he declared himself to be "prope
della morte." About three quarters of an hour's walk above Alagna,
we passed an extensive establishment connected with a gold mine, the
property of the Sardinian government ; but, like most of the others in
this neighbourhood, it has fallen completely into decay. The only
gold mines which I believe are now worked to any extent are those of
Pestarena, in the Val Anzasca. We crossed the stream soon after,
and commenced the ascent of the Turlo. At a little height, Monte
Rosa had a grand appearance, of which I made a sketch, Plate VIII.,
the chief summits visible being, (as I judged by the map,) Vincent's
Pyramid, Ludwig's Höhe, and the Signal Kuppe. A steep zigzag
leads to the higher châlets seated in an extensive hollow in the hill.
From hence, a seemingly endless ascent over smoothish rocks mixed
with turf, leads to the Col, which remains in view the whole way.
Monte Rosa is hid, and there is no variety of view. All travellers
consider this, and justly, as one of the most tedious passes in the
Alps, although it presents no kind of difficulty. The last part of the
ascent is over fallen masses of rock. I observed a group of chamois
to the right. The summit is marked by a cross. Here I found the
temperature of boiling water to be by my thermometer 196°.68 ; that
of the air being 36° at 11 A.M., from which I conclude the height to
be 9,141 English feet, instead of 8,400 as marked by Keller.

The view from the Col Turlo is a wild one. The ridge is itself jagged
and pinnacled in fantastic forms ; on the eastern side, the ground falls
(as usual) much more steeply, and the bottom of the Val Quarazza seems
at an immeasurable depth, separated by an extensive snow field. Monte
Rosa is still concealed by the mass of the Pizzo Bianco, which rises on
the left. A very steep descent, first over snow, and then over fallen
rocks, brought us, not without fatigue, down a height of several thousand feet. When we had reached the level of the highest sheep pas-

Pl. VIII.

Drawn from Nature by Professor Forbes _ T.Picken.

Day & Haghe Lith^{rs} to the Queen.

MONTE ROSA, FROM ALAGNA .

tures, my guide took his leave ; he gratefully accepted the trifle which I gave him for his safe conduct, and then he started off with the half-cheerful, half-plaintive exclamation,—" We shall meet no more but in Paradise ;" and so we separated.

Not long after, I reached the châlets of Plana, which, like most of those in the neighbourhood, are inhabited by Piedmontese, and not by the German settlers, and consequently are very filthy. I rested awhile on the rocks between the châlets and the river, which were very beautifully rounded and striated, I have no doubt by glacier action. The forms were smooth, undulating ones, and the polish fine ; the rock is a gneiss, approaching nearly to granite. I may mention that, in the Val Sesia,—that is, in the very small space of it which I traversed,—I observed no glacier marks on the rocks. In the higher part above Alagna, I noticed a very beautiful Syenite in blocks ; I also observed quartz-rock *in situ*, near the goldworks. Near the Col Turlo there occurs a beautiful Mica slate, with crystals of schorl, (which mineral I also found on the Glacier de Lys,) succeeded by a granitoid gneiss with large felspar crystals. The Val Quarazza, which is a tributary of the Val Anzasca, contains in its lower part granitoid blocks, probably transported by glaciers. I crossed the torrent a little below the châlets of Plana; the valley there becomes picturesque and wooded, and a series of cascades occur near the junction of the valleys. Turning to the left, by the village of Isella, I reached Macugnaga about 4 P.M., having travelled very quietly. This valley is very pleasing in its appearance, the houses are dispersed over its surface rather than grouped in villages, but Macugnaga is the last Commune. The people are agreeable, talking German ; the houses neat, and the hay-harvest gave a lively appearance to the scene. For a while I could not get access to the inn, until the landlord, a decrepid, hunchbacked, and blind man, though still below middle age, made his appearance from labouring in the hayfield, and by his pleasing manner, and his attention, soon gained my interest, and made me well satisfied with what his house afforded, which, indeed, was more than average comfort, considering the remoteness of the spot. There was a visiter's book, and I do not think that a dozen travellers of all countries had entered their names since the previous year. The landlord's name is Verra, and his wife is an obliging person.

On the 2d September I rose at five, intending to cross the Monte Moro into the Vallais. The weather was superb, and Monte Rosa clear. Whilst I dressed I began to regret my purpose; and when I descended to breakfast, and got a view of the head of the Valley of Macugnaga, in all its magnificence, I called to mind that I had seldom, if ever, regretted a day's delay in the midst of fine scenery, and had often cursed the infectious haste of travellers. Therefore, although I had lost two days at Gressonay, I called my Savoyard, and desired him to prepare for a trip to the neighbouring glacier. We were soon on foot, with an enchanting morning, the sun was not yet risen on the valley, which had a freshness very symptomatic of fine weather, and which I had not enjoyed for some time; the northwest wind had established itself. A little above the village stands the church of Macugnaga, and beside it a noble and thriving lime-tree, forming an excellent foreground to the vast scenery behind, which is, beyond all comparison, the finest view of Monte Rosa itself. From thence I passed to the village of Pecetto, with its church, which is the last in the valley, and both here and at Macugnaga, I was struck with the unusual taste displayed in ornamental gardens at the cottage doors, and with the great beauty and luxuriance of some of the choice flowers, especially carnations. The inhabitants I met, and who greeted me in German, were chiefly females and old men. All the young men leave the valley to seek their fortune in France, or elsewhere, as merchants. The inhabitants of the Val Sesia are, in like manner, chiefly *colporteurs* or hawkers. This circumstance explains a curious remark of De Saussure, who, wishing to have a heavy case of minerals transported to Vanzone from Macugnaga, inquired for a man who could carry them. He was answered that no man in the valley was equal to the task, but that a *woman* could easily do it, if it was the same to him. And it is certain, he adds, that two women can carry a mule's burden.*

---

* *Voyages,* § 2244.

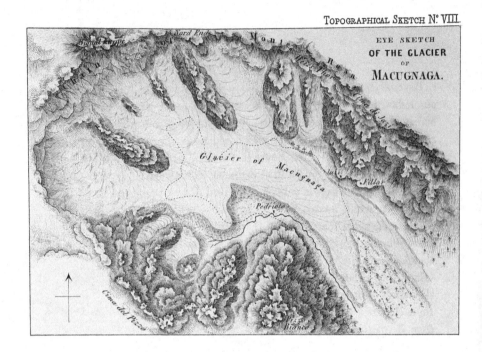

EYE SKETCH
OF THE GLACIER
OF
MACUGNAGA.

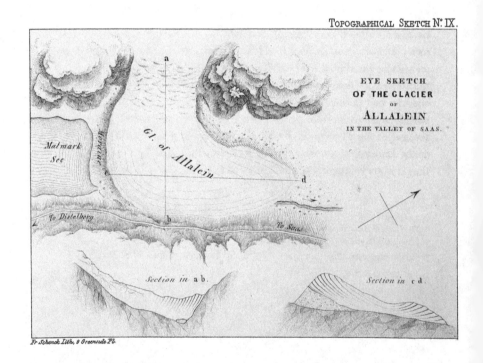

EYE SKETCH
OF THE GLACIER
OF
ALLALEIN
IN THE VALLEY OF SAAS.

Fr Schenck Lith, 9 Greenside Pl.

Beyond Pecetto a charming path lay through fields and woods, without habitations, but interspersed with barns; and the great glacier which occupies the head of the valley appears conspicuously. I ascended a steep wooded slope, which separates the lower end of the glacier into two, of which, however, by far the larger is on the right hand, the other being only a little overflow. This slope is very high and steep; the upper part is entirely composed of the ancient moraines of the glacier, which have a singular figure like artificial mounds, (see the Topographical Sketch, No. VIII.,) and embrace a charming well-watered pasture ground. From its upper part I crossed the main branch of the glacier on the right to the Châlets de Jazi at the foot of the mountain of that name.* From thence the view of the precipitous amphitheatre of Monte Rosa and the Saasgrat is very fine. Nearly above these châlets I knew must be the celebrated pass of the Weiss Thor from Zermatt to Macugnaga. The Piedmontese shepherd who occupied the châlet could give me no information respecting it, and the range appears on this side so absolutely precipitous, that I could hardly convince myself that any track could be found accessible to human foot. It is certain, however, that occasionally precipices are more practicable than they appear at a distance, and generally less vertical; and after a very careful examination, I detected a passage of the rocks, and only one, which it seemed possible to pursue. This pass is mentioned by almost every writer on Monte Rosa. De Saussure says that it is very dangerous,. but does not state that he conversed with any one who had performed it.† In Hirzel‡ and Von Welden,§ I find no particular addition from personal knowledge. Engelhardt¶ relates the account of the passage of the Weiss Thor by his guide at Zermatt, no doubt Damatter, who has repeatedly assured me that he once passed it, that it is very dangerous, much more so than the Col d'Erin. Schott**

---

* The Cima di Jazi appears to correspond with the Strahlhörner, when seen from Zermatt.
† *Voyages,* § 2145.   ‡ *Reise,* p. 32.   § *Monte Rosa,* p. 38.
¶ *Schilderungen,* p. 195.   ** *Die Deutschen in Piémont,* p. 61.

states that this pass was formerly more used than at present, and almost exclusively for the purpose of pilgrimage from the Vallais to the Monte Sacro at Varese, and this corresponds accurately with what I learned from the host Verra at Macugnaga. It is pretty certain that it has been crossed but once in the memory of men now living, and then by a pretty numerous company. I believe that no one in the Val Anzasca has passed it.

I continued along the western moraine of the glacier for some way above the châlets, and crossed the foot of the first tributary glacier descending from the Monte Rosa, or rather that part of it next the Weiss Thor called the *Nord End*. It has the usual scallop shell structure of steep glaciers. I then crossed to the centre of the glacier to examine its structure, and ascended the axis of it up to the limit of perpetual snow, (or névé,) having sent my companion to await my return on the eastern moraine.

The general structure of the Macugnaga glacier is quite normal, in single waves, as shewn in the sketch of a Ground Plan, No. VIII. Higher up, the glacier descends steeply on a twisted inclined plane, occasioned by the barrier which it has itself raised to its advance on the eastern side, by a stupendous moraine several hundred feet high, composed of huge blocks. The structure of the ice is beautifully developed as it sweeps round this spiral inclined plane, and is quite conformable to the cause which I have elsewhere assigned to it. Above this the glacier becomes more level. Its surface is thickly covered with snow, and this snow is evidently, in many cases, the result of avalanches which fall from the steeps of Monte Rosa upon the glacier, which De Saussure has stated to be one source whence glaciers derive their sustenance,—a fact which has been rather strangely denied. The snow, or névé, is usually disposed in bands or layers horizontally deposited, which most likely owe their origin to successive avalanches, or successive snow falls. I wish distinctly to state, that I attribute this stratification to nothing like the cause of the veined structure of glacier ice. I got some excellent sections of the glacier and névé together; the former underneath, presenting the usual vertical bands; the latter superimposed in true horizontal

Pl. IX.

Drawn from Nature by Professor Forbes. J. Picken lith.

GLACIER OF THE SECOND ORDER, NEAR MACUGNAGA.

Day & Haghe lith.rs to the Queen

Superposition of the Névé on the Glacier of Macugnaga.

strata. On the surface of the ice I found the remains of a gravel cone of vast extent, (see page 26.) I mention this as a glacier phenomenon of rather unfrequent occurrence.

From the higher plateau, at the summit of a stupendous precipice, several thousand feet in height, to which the snow clings difficultly, is seen the principal range of summits of Monte Rosa; first, on the left, the *Signal Kuppe*, then the *Zumstein-spitze*, marked in De Saussure's view from Macugnaga as the highest. From this to the *Nord End*, a very considerable distance, there runs a sharp snowy ridge, which is broken at several points by projecting rocks; the first is a trifling pinnacle, but the second is a tremendous rocky tooth, the *Höchste Spitze*, or highest summit, which appears to join on to the snowy ridge before mentioned in such a way as to leave great doubt whether, even supposing the foot of it to be attainable, it could be ascended. East from the Signal Kuppe is a secondary ridge, connecting Monte Rosa with the Cime de la Pisse of Von Welden, and which, at the same time, separates the valleys of Anzasca and Alagna. From this several secondary glaciers descend, and have a short course, with great moraines. I sketched them, (see Plate IX.,) as illustrating well the clamshell structure, and this form of glacier. From the Cime de la Pisse the ridge turns N. E., and joins the Pizzo Bianco, ascended by De Saussure. I had an opportunity of examining undoubted specimens of rock, which had descended with the glaciers from different parts of the chain. From the highest ridge, (*Zumstein-spitze* to *Nord End,*) the rocks are a fine grained gneiss, and a beautiful silvery mica slate. This latter rock was shewn to M. Studer and myself at Gressonay by M. Zumstein, as the highest attainable one. From the Signal Kuppe and Montagne de la Pisse there descends a gneiss, with large felspar crystals, such as I observed on the Col de Turlo. In general there is little chlorite, and no trace of serpentine or green slate, on this side of Monte Rosa.

I descended the steep moraine before alluded to, and at length perceived the smoke of a fire, which Victor had lighted below for his amusement. Nothing gave me so great an idea of the vast magnitude

of the scene by which I was surrounded, as the difficulty of distinguishing a human figure, and the apparent insignificance of the blocks of
stone, or, to speak more properly, fragments of mountains with which
the ground at the foot of the moraine was strewed. These masses which,
as seen from a distance, lay in indistinguishable heaps, were, I am confident, the largest detached blocks of stone which I have ever seen in any
position,—I mean, which had rolled, or been carried altogether from
their native bed. That beneath which Victor had prepared our dinner
was, I suppose, 500 feet in circumference, and it was 120 feet high.
Since its deposition from lying on an irregular bed, it had broken
through the middle, and left a serrated gap in the upper part.   It
was surrounded by several others scarcely less gigantic. These blocks
are described by De Saussure, § 2144, and by Engelhardt, who dis-
cusses whether they were brought down by the glacier, and form part
of the moraine. I incline rather to believe, that they fell from the
slope of the Pizzo Bianco. The scene was beautiful, and interesting,
and intensely solitary. These masses rest upon an alluvial well-watered
flat between the edge of the glacier and the natural side of the valley.
It is protected from the glacier by the vast barrier of debris already
alluded to, which checks its progress, and, in fact, forms the little valley in question, which is covered with the most vivid green, and which
forms the pasture or *Alp* of Pedriolo, the name of a few huts farther
down, and already deserted for the season. With these stones as a
foreground,—which, recalling past times and physical power, might be
termed the Druidical monuments of nature,—the extent of glacier behind, and the chain of Monte Rosa in the distance, all seems harmon-
ized to one scale of immensity, and the eye is satisfied.

I returned to Macugnaga by the track which leads over the rocks
at a great height above the glacier, from the Alp of Pedriolo, and
having passed two groups of wretched hovels by the way, I descended
a steep and intricate path, which brought me back to that which I had
left between Pecetto and the glacier.

The Glacier of Macugnaga, (called also *Anza Gletscher*, or the eastern Glacier of Monte Rosa,) appears to be as large as it has been for a
long time : it has not shrunk like the Glacier de Lys.

The following morning, at half-past five, I was on my way to the Monte Moro, the easiest passage of the great chain of Alps between the great St. Bernard and the Simplon, but yet impracticable for horses or mules. Still it appears formerly to have been passed by beasts of burden, for there is a carefully constructed pavement visible at various parts of the ascent, especially towards the top, which has been noticed by De Saussure* and other writers, and which it is impossible to mistake. It is on record, that this pass was in frequent use in the fifteenth and sixteenth centuries, and the road was maintained at the joint expense of the inhabitants of Saas and Antrona.† Although the absolute height of the Moro is greater than that of the Turlo pass, it is incomparably less fatiguing, being both shorter and more interesting. Indeed, I could not refrain from turning round continually to admire the magnificence of the view of Monte Rosa, which, though the

| Signal. | Zum-stein. | Höch-ste. | Nord End. | Weiss Thor. |

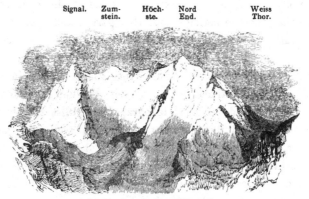

Monte Rosa from the Pass of Monte Moro.

point of view never altered, seemed to rise to a greater height in proportion as I ascended. In four hours I gained the top, and having melted some snow, I observed the boiling point, which was 196°.30 by my thermometer, having been 205°.35 at Macugnaga the evening before. The temperature of the air was 41°. Compared with the barometer at

---

* *Voyages*, § 2145. De Saussure seems never to have crossed the Monte Moro himself.
† Schott, page 63 ; Engelhardt and Venetz.

Geneva and St. Bernard, the height of Macugnaga above the sea appears to be 4,369 feet, and of Monte Moro 9,641 feet.

The descent to Saas is singularly easy and pleasant. There is a steep bed of snow crossed at first, but afterwards a gentle fall leads the whole way down to Visp in the Vallais. On the right hand is the great chain of Alps stretching away towards the Simplon. On the left is the redoubtable *Saasgrat*, a lofty chain of inaccessible snowy peaks, separating the valley of Saas from that of Zermatt or St. Nicolas, and from which a series of glaciers descend into the former. There is said to be a passage from the one valley to the other from the top of the Findelen Glacier to the north of the Cima di Jazi or Strahlhorn; which must enter the Saaser-thal near Distelberg, the highest group of châlets. Damatter assured me, at Zermatt, that there is no other practicable pass across the Saasgrat.

I must say a word here respecting the maps of this country, which are worse than those of perhaps any other part of the Alps, and are all nearly equally bad, though with a great diversity of errors, which, showing that the artists have copied neither nature nor one another, leaves us to consider them as pure fabrications. Thus, in the map of apparently most authority of any—Von Welden's—attached to a work professedly geodetical and topographical, whilst the Italian side of the mountain and its valleys are neatly and well laid down, the northern or Swiss side is a mass of pure invention, in which the most obvious features are no where to be found, and villages and glaciers, lakes and mountains, are jumbled into inextricable confusion. Take the easily accessible neighbourhood of Zermatt;—the great Glacier of Gorner is to be recognised only by its name, (Zermatt Gletscher,) and *debouches* on a lake which has no existence ; the Riffel and the Glacier of Zmutt are no where ! Nor is the valley of Saas better. The Matmark See, a lake below Distelberg, is supplanted by an imaginary glacier, composed of tributaries from all sides, and across which the path of the Moro is carried. A very pretty and detailed map of the Simplon pass and its neighbourhood, published by authority, replaces the great Glacier of Macugnaga by a great lake ! Wörl, in his map, has copied Von Welden's errors. Even the new government map of

Sardinia, of which a sheet has lately appeared, has perpetuated blunders even worse than Von Welden's, in exquisite engraving. Lakes are created, villages are displaced, and others which have no existence inserted where glaciers should be! The Italian side is, however, admirably executed, even though not quite precise in the details of roads and villages. On the whole, the most careful map of the Swiss part of the chain is that in Engelhardt's work; * but the author has unfortunately adopted a complex and impracticable system of projection, partly picturesque, partly geometrical, which greatly diminishes its value. I cannot help thinking, also, that in this as in other maps, the breadth of the Saasgrat is underrated at its upper part. It is a very pretty, though certainly not an easy topographical problem, to unravel the complication of this chain, of which the mountains are so inaccessible, so varying in their forms, and each called by several different names.—But to resume the descent to Saas.

Four glaciers are passed by the way. The first is of small size, on the right hand, and near the pass. It is steep, but even, and exquisitely ribboned in the usual manner. The second glacier is on the left, descending from the summit called on the Sardinian map Monte Moro. It chiefly struck me, from the small stream of *pure* water which flowed from under it, as was also the case in the last glacier.

The third glacier is below the châlets of Distelberg on the left. It is called Schwartzberg. It is very remarkable, from its shrunk and wasted appearance. The limits of a moraine of recent date stretch quite to the eastern side of the valley, (which is here wide,) where it has left one enormous block of green-slate, a cube of about sixty feet, slightly rounded on the edges.† As far as I could learn from some peasants who were passing, this block was deposited about twenty years ago. The glacier has now retreated quite to the other side of the valley.

The fourth glacier, called Allalein, is the most remarkable of any.

---

* Schilderungen der Höchsten Alpen, 1840.

† This is mentioned by De Charpentier (p. 41) under the name of the Blaustein; he describes it as deposited in 1818, and as having 244,000 cubic feet of contents.

It completely crosses the valley (which is here rather narrower) with its moraine, which, damming up the river, forms a lake called the Matmark See. The moraine supplies the well-known blocks of gabbro, containing Smaragdite, which are recognized so extensively over the plains of Switzerland, and which have no native locality in the Alps but here. They are brought down by the glacier from the inaccessible heights of the Saasgrat, which near this place rises to about 15,000 feet, so that the rock may probably never be found *in situ*. These masses are usually much rounded by attrition, notwithstanding their excessive hardness. The structure of the Glacier of Allalein is well developed, and quite regular. It resembles generally the Glacier of La Brenva in the Allée Blanche, and as in that case the river passes under it. It also resembles the Glacier of the Rhone in the way in which it pours into the valley, and its consequent structure, which is represented both in Plan and Section in the Topographical Sketch, No. IX. The veined structure is especially developed *in front*—*i. e.*, against the opposing side of the valley, where the pressure is greater than laterally, and consequently the ice, seeking the direction of least resistance, is gradually swayed down the valley, and takes the particular form shown in the map, which, together with the sections, will give a clear idea of its whole structure. The direction of the crevasses is generally radial, or perpendicular to the structural bands. I walked over a part of the glacier, but it is not easy to advance far. The front of it is, as I have said, pushed by the general mass against the eastern wall of the valley. The rock, which is here soft, is disintegrated and clayey, and it was interesting to see that the glacier had left vertical markings or striæ upon the clay which had lately been uncovered by its melting, exactly as it would have done on rock, and in the very same direction as I observed them in similar circumstances against fixed rock at La Brenva.

Below Allalein the road falls more rapidly, and a very wild gorge is entered, which continues for a mile or two. The little village of Almagell is the first reached. Here a path on the right leads into the Val Antrona. In half an hour longer I was at Saas, where I received a hearty welcome from Moritz Zurbrücken, the worthy host in whose house I spent a night last year. The journey had been a short and interesting one, and its fatigues were soon forgotten over a roast leg of chamois and a bottle of good wine.

The neighbourhood of Saas presents one very interesting excursion, which I made in 1841, to the Valley of Fée, which is a small branch of the Saas-thal, descending from the mountains to the west. The easiest ascent is by a foot-path, exactly opposite to the village of Saas, and which is distinguished by a series of station chapels at intervals. The valley of Fée, like most of those in this neighbourhood, joins the principal valley at a higher level; and when that level is gained, the view is very striking. The entire head of the valley is bounded by a vast glacier, descending from the three lofty mountains, marked in Engelhardt's map, Schwartzhorn, Féehorn, and Stuffen, or Dom. The village of Fée, which is inhabited all the year, lies in a beautiful green hollow, amidst meadows and trees, which seem to touch the regions of ice. Indeed, a few years ago, the glacier descended so as to threaten the destruction of the higher châlets and trees, and completely to obstruct the passages to an *alp* or pasture between two branches of the glacier which then closed round it. About 1834, the glacier began to retreat, and has continued to do so since, so that it is now at a very considerable distance from the châlets, which it had almost touched. But what interested me most in the valley of Fée were the admirable traces of former glacier action throughout its length. *Roches . moutonnées* of gneiss occur in the whole of the lower part of the valley, scooped out by horizontal grooves, perfectly continuous for some yards or fathoms, and which it is impossible to contend for a moment that water, however charged with stones, is capable of producing. Some of these grooves are like elaborate chiselling, and, on the whole, it would be difficult to find a better specimen of the phenomenon in question. It is remarkable, that in the valley of Saas, *above* the entrance of the valley of Fée, I perceived no such traces, which, however, appear at several points between Saas and Stalden. The rock of the higher valley, which is slaty and often friable, is certainly not favourable to the preservation of such surfaces. By continuing from Fée, along the western side of the Valley of Saas, a beautiful walk may be followed through the wood, nearly as far as

z

Almagell. The annual *fête* of the valley is held at Fée, on the 8th
September.

From Saas to Stalden, there is a great variety of scenery ; and in this
respect the Saas valley is much more interesting than the neighbouring
one of St. Nicolas. There is a series of green flats of small extent,
separated by gorges of greater or less depth ; one of these, in particular, about an hour's walk above Stalden, is extremely fine. The river
rushes through a very deep, narrow chasm, overhung with magnificent
larch trees, amongst the finest which I have seen in the Alps, and the
head of the valley is closed by a snowy peak, perhaps the Monte
Moro. It is also crossed by a little foot-bridge, upon which the traveller
may stand to view the scene, if he wish to increase its sublimity by no
visionary sense of danger in his own position ; for the bridge is so
weak that a heavy man might break it, and beneath is a furious torrent at a depth of perhaps 200 feet. The view *down* the valley is fine,
as well as up ; the Bietsch-horn, a very elegant mountain north of the
Rhone, stands in the opening. Where the valley of Saas is most contracted, the gneiss rocks, which form mural precipices, are striated
horizontally to a great height—probably 800 feet. Glaciers peep
through the ravines on the western side, but none of them reach the
valley.

Stalden is beautifully situated, as already mentioned, at the junction of the Valleys of Saas and St. Nicolas. I had an opportunity
of witnessing here a remarkable scene on my last visit. A comedy
was to be acted by peasants dressed in costume, who were to perform
on a stage erected in the open air. There were not less than forty
actors, the female parts being performed by men, and the costumes
were elaborately and ingeniously devised—in some cases not without
propriety and taste. I was able to remain long enough to see only
the opening of the piece named *Rosa von Tannenburg*, which was pre-
luded by a procession of the actors, amongst the most conspicuous of
whom were three devils attired in tight suits of black, with horns and
tails, the senior wearing goat's horns and the subordinates those of the
Chamois. The entertainment was under the immediate patronage, and
even. direction of the clergy. The morning mass at Saas was said that

day at four, instead of five o'clock, in order to allow the pastor and his flock to reach Stalden in good time, and one of the *vicaires* of Stalden (who correspond to our curates) seemed to be the master of ceremonies, for he was frequently seen in earnest conversation with the junior devil with the Chamois horns. I must add, that the scene was one of the most romantic which can be conceived. Behind the village was a truly natural theatre, with a green meadow for the pit, whilst a range of low cliffs, with a concave front festooned with ivy and brushwood, represented the boxes and gallery, and an audience of not less than two thousand persons, almost entirely peasants, with their gay costume, filled the allotted spaces. The sky was intensely blue, and the summits of the Weisshorn and other snowy Alps completed the picture.

I was obliged to withdraw sooner than I wished, in order to reach Visp in time for the Diligence which was to take me to Sion. Thus closed one of the most interesting journeys which I have had the good fortune to make. Since leaving Orsières three weeks before, I had not even crossed a road which admitted of the passage of a wheeled carriage.

CHAPTER XXI.

AN ATTEMPT TO EXPLAIN THE LEADING PHENO-MENA OF GLACIERS.

THE DILATATION THEORY CONSIDERED, AND COMPARED WITH OBSERVATION—THE GRAVITATION THEORY EXAMINED—THE AUTHOR'S THEORY PROPOSED—GLACIERS REALLY PLASTIC—CONDITIONS OF FLUID MOTION—COMPARED WITH THOSE OF A GLACIER—EFFECT OF VISCOSITY—THE VEINED STRUCTURE OF THE ICE A CONSEQUENCE OF THE VISCOUS THEORY—ILLUSTRATED BY EXPERIMENTS—COMPARISON OF A GLACIER TO A RIVER—CONCLUSION.

" Rien ne me paraît plus clairement démontré que le mouvement progressif des glaciers vers le bas de la vallée, et rien en même temps ne me semble plus difficile à concevoir que la manière dont s'exécute ce mouvement si lent, si inégal, qui s'exécute sur des pentes différentes, sur un sol garni d'aspérités, et dans des canaux dont la largeur varie à chaque instant. C'est là, selon moi, le phénomène le moins explicable des glaciers. Marche-t-il ensemble comme un bloc de marbre sur un plan incliné ? Avance-t-il par parties brisées comme les cailloux qui se suivent dans les couloirs des Montagnes ? S'affaisse-t-il sur lui-même pour couler le long des pentes, comme le ferait une lave à la fois ductile et liquide ? Les parties qui se détachent vers les pentes rapides suffisent-elles à imprimer du mouvement à celles qui reposent sur une surface horizontale ? Je l'ignore. Peut-être encore pourrait-on dire que dans les grands froids l'eau qui remplit les nombreuses crevasses transversales du glacier venant à se congeler, prend son accroissement de volume ordinaire, pousse les parois qui la contiennent, et produit ainsi un mouvement vers le bas du canal d'écoulement."

RENDU, Théorie des Glaciers, p. 93.

In the second chapter of this volume I stated the usually received opinions as to the cause of the formation and maintenance of glaciers. We found that authors are pretty well agreed in considering that the snow which falls on the summits of the Alps becomes converted into ice by successive thaws and congelations, but that the details of the process are by no means so well understood, and that the immediate cause of the descent of these frozen masses towards the valleys has been very differently explained.

The chief theories we reduced to two; the theory of DILATATION, and that of GRAVITATION. On the former the ice is supposed to be pressed onwards by an internal swelling of its parts, occasioned by *rapid* alternations of freezing and thawing of its parts, or rather by the continual formation of minute crevices, into which water, derived from the warmth of the sun, and the action of the air on the surface, is introduced, and where it is frozen by the cold of the glacier, whose bulk it thus increases. On the theory that gravity or weight is the sole cause of glacier motion, the ice, lying on an inclined plane of rock, is supposed to slide over it, by its natural tendency to descend, aided by the action of the earth's warmth, which, on the hypothesis of De Saussure, prevents it from being frozen to the bottom.

It may be proper now to enquire shortly what light has been thrown upon these two theories by the observations detailed in a former part of this volume.

Of the facts which have been established in Chapters VII. and VIII., with respect to the motion and structure of the ice of glaciers, two seem at least to be not opposed to the theory of DILATATION. I mean the *more rapid movement of the glacier at its centre*, (p. 146) and the *infiltration of its mass* by water permeating the capillary fissures, (page 174.) The former fact having been unknown to the supporters of the dilatation theory, has not been adduced by them in its favour; which it is, indeed, only thus much, that a body having a certain consistence and variability of form, when subjected to *any* pressure, whether internal or external, will yield soonest in those parts which are least retarded by friction. This fact, however, has no direct bearing on the *cause* of the pressure.

The latter fact would be entirely favourable to the theory of De Charpentier and Agassiz, could it be carried out in its consequences, in the manner which they suppose. But it is not enough that there be capillary fissures and crevices, and that these be filled with water,—*that* does not help the matter at all,—it must also be shown that that water undergoes conversion into ice, so as to dilate it at the time, and to the extent, required for the motion. I conceive that the observations which I have made, show such a cause of motion to be inconsistent with the phenomena; and this inconsistency is two-fold, first, from the direct evidence that, though the ice is permeated by water, yet the water freezes rarely, and to an insignificant extent; and, secondly, from the motion of the glacier in its different parts, and at different times, being at variance with what must have held true upon the theory in question.

1. The water included in a glacier is rarely in a freezing condition. I need not now repeat the arguments which have been adduced (page 36) to show that upon every principle of the doctrine of heat, especially the doctrine of latent heat, it is impossible that the transient cold of the night should in any circumstances produce more than a superficial and most imperfect congelation,—that to suppose any thing else would be to suppose in a glacier an indefinite supply of cold,* contrary to first principles, and to direct observations with the thermometer on the temperature of the ice, which has been found by M. Agassiz himself to be constantly, and at all depths, within a fraction of a degree of 32°. But besides this, the most direct observation shows that the nocturnal congelation, which is so visible at the surface, drying up the streamlets of water, and glazing the ice with a slippery crust, extends to but the most trifling depth into the mass of the glacier. This is so evident, upon consideration, that when fairly placed before him, M. de Charpentier has been obliged to abandon the idea that the diurnal variations of temperature produce any effect. In truth, there is positive evidence that no internal congelation takes place during the summer season, when the motion is most rapid, and when, therefore, the cause of motion must be most energetic. Of this I will give one striking example.

Towards the end of September 1842, when, it has been already mentioned, a premature winter had covered the Mer de Glace with snow,

---

* This argument has been well put by M. Elie de Beaumont, with his accustomed clearness.

and lowered the temperature of the air to 20° Fahrenheit, I had occasion to make an expedition over nearly its whole extent, in the direction of the Glacier de Léchaud, in order to observe the marks which had been placed in that direction, and to determine the motion of the higher parts of the ice. The excursion promised to be far from agreeable. The sky was lowering when I started from the Montanvert, and it soon began to snow, and continued to do so with little intermission during the day. The Mer de Glace had been covered with snow *for a week ;* at the Montanvert, to a depth of six inches, but in its higher parts of not less than a foot and a half. I was not sorry, however, to have an opportunity of ascertaining the conditions of the ice, under circumstances so critical for the theory of dilatation, for now, if at any time, the freezing and expansive effects of cold ought to be visible, the ice having been completely saturated by the preceding wet weather, and, it might be supposed, effectually cooled by five days of frost. As the walk promised to be laborious, if not difficult, owing to the thick coating of snow, I took with me David Couttet of the Montanvert, and Auguste Balmat, as usual, with the instruments and provisions. We started in a lowering morning at half-past six, and in less than an hour it began to snow, with a drifting wind, though fortunately without cold. To most persons, the journey would have been an alarming one, but we were all three so intimately acquainted with the surface of the ice, and the direction of the moraines, that we had no fear of losing ourselves. It required, however, all Auguste's intimate knowledge of the glacier to keep us clear of dangerous crevasses and holes ; for the snow was often kneedeep, and the glacier and moraines alike filled with innumerable pitfalls. We crossed the moraines, as usual, near the Moulins, and visited the stations B 1 and C. We then kept nearly under the ice fall of the Glacier du Taléfre, and reached with precaution the higher glacier of Léchaud, on our way to station E, where I anxiously wished to make an observation of the progress of the glacier. But now the bad weather increased so much, that we were glad to get behind a great stone and eat our breakfast, waiting for a favourable change. The wind blew in strong gusts from the Grande Jorasse, tossing the snow about so as to render all objects at a distance undistinguishable, thus

threatening to make our expedition ineffectual, for the rock called the
Capucin du Tacul, which was my index for the bearings on the glacier
from Station E, was hopelessly invisible. After some delay, the storm
abated, and the Pierre de Béranger, whose azimuth I had fortunately
taken as a check, showed itself. We therefore advanced up the glacier,
but again the storm thickened, and as we got to the foot of the rock
on which Station E was fixed, David Couttet (who had hitherto been the
chief encourager of the expedition) said quietly, " Nous allons faire une
bêtise," and proposed to return, for we were half blinded by the snow.
I begged, however, that we might at least stop and take shelter as
before. We did so, and profiting by a few minutes' pause in the drift,
I fitted up my theodolite, and took an observation of the motion of the
glacier since my last visit with due care and deliberation. We then
returned nearly as we had come, fortunately without accident, and
reached the Montanvert after nine hours' absence. What struck me
most in this expedition was, that even at the highest station, which is
7900 feet above the sea, and in this severe weather, the ice, far from
being frozen to a great depth, appeared charged with water as usual,
except at the surface. The stick which marked the point of the glacier
observed, and which I expected to find firmly frozen into its place, was
standing in water in its hole in the ice, and of course quite loose.
The *surface* of the glacier generally was dry,—there was not a rill of
water in the Moulins, or elsewhere: yet the congelation had scarcely
penetrated at all. Couttet and Balmat were all the time afraid of
treading into a watery hole, and thus getting their feet frozen, an accident which I thought very unlikely to happen; but they both did get
their feet wet in the course of the day. Hence there can be no doubt,
that, as Couttet very distinctly expressed it, the snowy covering kept
the glacier warm, just as it does the ground, and that the cold penetrates extremely slowly even when winter arrives. I may add, that near
the Tacul I found no difficulty in obtaining a draught of water by
breaking the crust of ice formed on a pool in the glacier under a stone.
It was on this excursion that I observed the blue colour of snow, mentioned on page 71, which was most distinctly perceptible by transmitted light, whenever the snow was pierced by a stick to a depth of six

inches or more. It was at one part of the glacier that this was most evident, which I attributed to the particular degree of aggregation which it had there, neither very dry nor very moist.

From the incidents just related, I think it seems to be demonstrated beyond a doubt, that, at least, any *transient* impression of cold is quite incapable of converting the infiltrated water into ice at any depth in the glacier.

2. At the same time that the preceding observations were made the rate of motion of the glacier was carefully observed; for I concluded, as a matter of certainty, that, if the dilatation theory were true, a sudden frost succeeding wet weather must inevitably cause the glacier to advance far more rapidly than in summer, or, indeed, at any other season ; for there could never possibly be more water to be frozen, nor could cold ever act with more energy than at the time in question. What the facts were, we have already seen in the seventh chapter, where it appears, both from the tables and figures, (pp. 139—141,) that the progress of the glacier was retarded during the cold weather which prevailed from the 20th to the 25th September, and that it re-advanced when the thaw had taken place some days later.

3. The motion of the glaciers during winter, established in the same chapter, (page 151,) is directly contrary to the conclusions invariably drawn by the glacier theorists from their supposed immobility ; since they consider, that while the glacier is completely frozen, and has no alternations of congelation and thaw, there can be no dilatation.

4. The experiments mentioned in page 132 shew, that the motion of a glacier during the day and night is sensibly uniform, which is contrary to the same view.

5. The rate of motion of the glacier at different parts of its length has been shewn (page 144) to be by no means such as the expansion of an elongated body, supported at one end, and pushed along its bed, would occasion.

6. The advocates of the theory of dilatation have rightly maintained, as a consequence of the theory, that the ice will expand in *all* directions, and consequently upwards, that being the direction in which the resistance is least of all. They thence conclude, that whilst the ice

wastes by melting at the surface, the surface will be raised by the inflation of the interior mass by the expansion of freezing water, and that its absolute level will thus be maintained, or will even rise, notwithstanding the daily waste. They profess to have made experiments which confirm this view; but I have already stated, (page 155,) that my own are entirely at variance with it, the absolute level of the ice lowering with great rapidity during the season of most rapid motion; a conclusion which is entirely confirmed by the observations of MM. Martins and Bravais, lately published.*

On these, amongst other grounds deduced from direct observation, I consider the dilatation theory maintained by Scheuchzer, De Charpentier, and Agassiz, as untenable.

In the next place, let us consider the sliding theory of Gruner and De Saussure, of which a sufficient account has been given in Chap. III.

As I understand the GRAVITATION theory, it supposes the mass of the glacier to be a *rigid* one, sliding over its trough or bed in the manner of solid bodies, assisted, it may be, by the melting of the ice in contact with the soil, which possesses a proper heat of its own, and which lubricates in some degree the slope, as grease or soap does when interposed between a sliding body and an inclined plane. It is only in so far as the theory is considered as applicable to a rigid body, that I have objections to state to it.

1. In the case of the greater number of extensive glaciers, there are notable contractions and enlargements of the channel or bed down which they are urged. Let any one glance at the Mer de Glace, and see two extensive glaciers meeting at the Tacul, forming a vast basin or pool, from which the only outlet has a less breadth than the narrowest of the tributaries; the idea of *sliding*, in the common legitimate sense of the word, is wholly out of the question.

2. We have already seen that the ice does not move as a solid body, —that it does not slide down with uniformity in different parts of its section,—that the sides, which might be imagined to be most com-

---

* *Annales des Sciences Géologiques, par Rivière,* 1842.

pletely detached from their rocky walls during summer, move slowest, and are, as it were, dragged down by the central parts. All this is consistent with motion due to weight or gravitation; but not with the sliding of a rigid mass over its bed.

3. The inclination of the bed is seldom such as to render the overcoming of such obstacles as the elbows and prominences, contractions and irregularities of the bed of glaciers, even conceivable, being, on an average of the entire Mer de Glace, only 9°, a slope practicable for loaded carts; but the greater part of the surface inclines less than 5°, which is below the steepest slope on the great highway of the Simplon, an artillery road.

4. It has been convincingly proved in Chapter VII., that the motion of the glacier varies not only from one season to another, but that it has definite (though continuous) changes of motion, simultaneous throughout the whole, or a great part of its extent, and therefore due to some general external change. This change has been shewn to be principally or solely the effect of the temperature of the air, and the condition of wetness or dryness of the ice. In order to reconcile this to the sliding theory, it should be shewn, that the disengagement of the glacier from its bed, depends on the kind of weather which affects its surface and temperature. In no part of the summer is the glacier actually frozen to its lateral walls; the difference, then, must be due to the action of the earth's heat in gradually melting away the irregularities of the inferior surface of the ice, in contact with the rocky bed on which it reposes. I have already said, that I consider such an influence of the proper heat of the earth to be distinctly included in De Saussure's theory, as it has been stated by himself, and understood by his successors.* It was, however, suggested to me very distinctly by M. Studer last summer, as not inconsistent with a motion by gravity without acceleration; and I admit the ingenuity of the thought, which,

---

* Any one who carefully reads De Saussure's § 535 in connection with § 533, will be convinced, that he gives all due weight (we should be inclined to say *more* than due weight) to the effects of subterranean heat in *detaching* the ice from its bed, *lubricating* it on its bed, and even *elevating* it over obstacles by the hydrostatic pressure of confined water.

as it will be seen in the sequel, I am disposed to allow, may be one *way* of glacier motion, though not exactly the *cause* of it. The same thought was afterwards suggested to me by Sir John Herschel, and more lately I have seen it stated, that Mr. Hopkins of Cambridge, the author of an ingenious pamphlet on the theory of glacier motion, has illustrated it by experiment.* But this is an effect which must remain nearly the same at all seasons, being due to the constant flow of heat from the interior.

5. The flow of heat from the interior is so *very* trifling that it may be doubted whether it is adequate to produce the particular effect of wearing off the prominences of the descending ice, or of moulding it to the form of the channel. In order to do so to any effectual extent, it would be necessary that prominences of many feet or yards in extent should be melted away in a moderately short space of time. Now, what is the fact? M. de Beaumont has estimated,† by the theory of Fourier, from the observations of Arago on the earth's temperature, that the quantity of central heat which reaches the surface of the earth, is capable of melting 6½ millimetres of ice, or exactly *a quarter of an English inch in the space of a year*. Now, even admitting, (as I think we may,) that if the surface of the earth were covered with ice, the flow of heat would be somewhat greater, still it must be admitted to be capable of disposing of portions of ice insignificant compared to the inequalities which oppose its downward progress.

6. This small quantity of heat is not always applied (as Professor Bischoff‡ and M. Elie de Beaumont have justly remarked) to *melt* the ice of glaciers. Below 32° it will simply tend to raise the temperature of the ice in contact with the soil, and powerfully adhering to it. The almost *pendant* glaciers of the second order, which are seen only at great heights, those, for instance, on the precipices of the Mont Mallet, (see page 79,) must remain permanently frozen to the rock.

---

* Since writing the above I have been indebted to Mr. Hopkins for a farther statement of his views, which I will farther allude to in Appendx, No. IV.

† *Annales des Sciences Géologiques, par Rivière*, 1842.

‡ *Wärmelehre*, p. 101, &c.

Nevertheless they do actually descend over it, for they continually break off in fresh avalanches. This is a fact which neither the theory of dilatation nor that of gravity, as commonly stated, is capable of explaining.

After the detailed though scattered deductions which have been made in the course of this work, from observations on the Movement and Structure of Glaciers, as to the cause of these phenomena, little remains to be done but to gather together the fragments of a theory for which I have endeavoured gradually to prepare the reader, and by stating it in a somewhat more connected and precise form, whilst I shall no doubt make its incompleteness more apparent, I may also hope that the candid reader will find a general consistency in the whole, which, if it does not command his unhesitating assent to the theory proposed, may induce him to consider it as not unworthy of being farther entertained.

My theory of Glacier Motion then is this :—A GLACIER IS AN IMPERFECT FLUID, OR A VISCOUS BODY, WHICH IS URGED DOWN SLOPES OF A CERTAIN INCLINATION BY THE MUTUAL PRESSURE OF ITS PARTS.

The sort of consistency to which we refer may be illustrated by that of moderately thick mortar, or of the contents of a tar-barrel, poured into a sloping channel. Either of these substances, without actually assuming a level surface, will *tend* to do so. They will descend with different degrees of velocity, depending on the *pressure* to which they are respectively subjected,—the *friction* occasioned by the nature of the channel or surface over which they move,—and the *viscosity*, or mutual adhesiveness, of the particles of the semifluid, which prevents each from taking its own course, but subjects all to a mutual constraint. To determine completely the motion of such a semifluid is a most arduous, or rather, in our present state of knowledge, an impracticable investigation. Instead, therefore, of aiming at a cumbrous mathematical precision, where the first data required for calculation are themselves unknown with any kind of numerical exactness, I shall endeavour to keep generally in view such plain mechanical principles as are, for the most part, sufficient to enable us to judge of the comparability of the facts of Glacier motion with the conditions of viscous or semi-fluid substances.

That Glaciers are semifluids is not an absurdity.

The quantity of viscidity, or imperfect mobility in the particles of fluids, may have every conceivable variation; the extremes are perfect fluidity on the one hand, and perfect rigidity on the other. A good example is seen in the process of consolidation of common plaster of Paris, which, from a consistency not thicker than that of milk, gradually assumes the solid state, through every possible intermediate gradation. Even water is not quite mobile; it does not run through capillary tubes; and a certain inclination or *fall* is necessary to make it *flow*. This may be roughly taken as an index of the quality of viscidity in a body. Water will run freely on a slope of 6 inches in a mile, or a fall of 1-10,000 part,* another fluid might require a fall of 1 in 1000; whilst many bodies may be heaped up to an angle of several degrees before their parts begin to slide over one another.

Thus, a substance apparently solid may, under great pressure, begin to yield; yet that yielding, or sliding of the parts over one another may be quite imperceptible upon the small scale, or under any but enormous pressure. A column of the body itself is the source of the pressure of which we have now to speak.

Even if the ice of glaciers were admitted to be of a nature perfectly inflexible, so far as we can make any attempt to bend it by artificial force, it would not at all follow that such ice is rigid when it is acted on by a column of its own material several hundred feet in height. Pure fluid pressure, or what is commonly called hydrostatical pressure, depends not at all for its energy upon the *slope* of the fluid, but merely upon the *difference of level* of the two connected parts or ends of the mass under consideration. If the body be only semifluid, this will no longer be the case; at least the pressure communicated from one por-

---

* According to Dubuat, (Hydraulique, tom. i., p. 64. Edit. 1816,) at a slope a great deal lower; but its exact value does not now concern us.

tion (say of a sloping canal) to the other, will not be the *whole* pressure of a vertical column of the material, equal in height to the difference of level of the parts of the fluid considered; the consistency or mutual support of the parts, opposes a certain resistance to the pressure, and prevents its indefinite transmission. It must be recollected, that, in the case of glaciers, the pressing columns are enormous, the origin and termination of many of the largest having not less than 4,000 feet of difference of level; were they, therefore, perfectly fluid, or suddenly converted into water, the lower end would begin to move with the enormous velocity of 506 feet a second, or would move over 44 *millions of feet* in 24 hours. Now, the velocity of the Mer de Glace is only about 2 feet in that time, a difference so enormous that the fluidity of a glacier compared to water will not appear so preposterous as it might at first sight do, considering the small degree of transmitted pressure required to be effectual.

Again, it has been attempted to show, (page 174,) that a glacier is not coherent ice, but is a granular compound of ice and water, possessing, under certain circumstances, especially when much saturated with moisture, a rude flexibility sensible even to the hand.

Farther, it has been shown that the glacier *does* fall together and chokes its own crevasses with its plastic substance, (p. 173.)

When a glacier passes from a narrow gorge into a wide valley it spreads itself, in accommodation to its new circumstances, as a viscous substance would do, and when embayed between rocks, it finds its outlet through a narrower channel than that by which it entered. This remarkable feature of Glacier Motion, already several times adverted to, had not been brought prominently forward, until stated by M. Rendu, now Bishop of Annecy, who has described it very clearly in these words,—" Il y à une foule de faits qui sembleraient faire croire que la substance des glaciers jouit d'une espèce de ductilité qui lui permet de se modéler sur la localité qu'elle occupe, de s'amincir, de se rétrecir, et de s'étendre, comme le ferait une pâte molle. Cependant, quand on agit sur un morceau de glace, qu'on le frappe, on lui trouve une rigidité qui est en opposition directe avec les apparences dont nous venons de parler.   Peut-être que les expériences faites sur de plus grandes masses donneraient d'autres résultats." *

Now, it is by observations on the glacier itself that we can best make experiments on *great masses* of ice, as here suggested.

The Motion of a Glacier resembles that of a Viscous Fluid.

All experimental philosophers are agreed as to the facts, that a fluid like water, heavy and slightly *viscid*, moves down an inclined plane or canal, with a velocity which varies according to the slope, and which varies also from point to point of the section of the stream.   The part of the stream which moves fastest is the surface, and especially the *central part* of the surface.   The velocity of motion diminishes *on* the surface from the centre to the sides, and *from* the surface towards the bottom.

The cause of these variations is admitted to be the *friction* of the sides and bottom of the canal or bed, which retards the fluid particles immediately in contact with them, and the adhesion of these particles to their neighbours, that is, their *viscosity*, communicates this retardation by certain gradations, which are not correctly known, to the interior mass of the fluid.   Hence,—

I. The centre and top of the stream move faster than the sides and bottom, especially if *the friction of the fluid particles over one another be less than their friction against the sides of the canal.*   If this be not the case—if the friction of the contained mass against the containing or supporting walls be *less* than the friction which exists amongst its own particles, the mass will *slide out of its bed*, and will so far act as a

---

* Theorie des Glaciers de la Savoie, p. 84.   Whilst I am anxious to show how far the sagacious views of M. Rendu coincide with, as they also preceded my own, it is fair to mention, that all my experiments were made, and indeed by far the greater part of the present volume was written, before I succeeded in obtaining access to M. Rendu's work, in the 10th volume of the *Memoirs of the Academy of Chambéry*, which I owe at length to the kindness of the right reverend author.

solid body. If it have a certain mobility amongst its own particles, it will, whilst sliding over its bed, alter, at the same time, the relative position of its own particles;—it will move partly as a solid, partly as a fluid. We may then fairly call it a *semifluid* or a *semisolid*.

II. From this it also evidently appears, that the greater the viscosity of the fluid, the farther will the lateral and fundamental retardations be communicated towards its centre, and the general velocity of the stream will be more nearly regulated by the limit of the mobility of its parts.

III. In every case, the greatest variation of velocity of such a stream will take place near the sides and bottom, whilst the higher and the central parts will move most nearly together.* The position of any particle moving with the mean velocity of the entire stream, has not, I believe, been determined; but Dubuat has practically found this singularly simple result, that the velocity of the top and bottom of a stream being known, the mean velocity of the entire stream is the arithmetical mean of these two velocities.

IV. The difference of the velocities of a stream at the top and bottom, depends upon the actual velocity of the stream, and increases as that velocity increases. The rate of increase appears to be as the square root of the velocity, and is independent of the depth.†

V. The velocity of the water in a stream increases with its declivity. If the bed of a river be highly inclined, the water flows rapidly; and again, if the embaying of a river by a strait accumulates the water above, there its declivity will be diminished.

VI. If any circumstance causes the viscosity or consistence of a fluid to vary, all these phenomena will vary proportionally. Thus, warm water is less viscid than cold, and a vessel will be sooner

---

* A slight consideration will show, that this might naturally be anticipated, yet some eminent writers have supposed the velocity to increase *uniformly* from the bottom to the surface of a stream. The doctrine of the text is fully confirmed by direct experiments upon the river Rhine, quoted in Mr. Rennie's *Report on Hydraulics*, Part II., *British Association Report*, 1834, p. 467.

† Dubuat, Art. 37, 49, 65.

emptied through a narrow aperture the higher the temperature of the liquid.*

Now, in all these respects, we have an exact analogy with the *facts* of motion of a glacier, as observed on the Mer de Glace.

First, We have seen that the centre of the glacier moves faster than the sides, (p. 146.) We have not indeed extended the proof to the top and bottom of the ice-stream, for it seems difficult to make this experiment in a satisfactory manner. In the case of a glacier 600 feet deep, the upper hundred feet will move nearly uniformly, on the principles already mentioned ; hence, crevasses, formed from year to year, will not incline sensibly forwards on this account, especially as the action of trickling water is to maintain the verticality of the sides. I conceive that this is a perfectly sufficient answer to an objection which, at one time, I myself urged against the hypothesis of the surface of the glacier moving most rapidly. Of the fact, I entertain no doubt, though I see much difficulty in obtaining a satisfactory proof of it.

I have no doubt that glaciers slide over their beds, as well as that the particles of ice rub over one another, and change their mutual positions ; but I maintain, that the former motion is caused by the latter, and that the motion impressed by gravity upon the superficial and central parts of a glacier (especially near its lower end,) pull the lateral and inferior parts along with them. One proof, if I mistake not, of such an action is, that a deep current of water will flow under a smaller declivity than a shallow one of the same fluid.† And this consideration derives no slight confirmation in its application to glaciers, from a circumstance mentioned by M. Elie de Beaumont, which is so true that one wonders it has not been more insisted on,—namely, that a glacier, where it descends into a valley, is like a body pulled asunder or stretched, and not like a body forced on by superior pressure alone.

---

* DUBUAT, Art. 3.

† It is well known that the mean hydraulic depth, or the ratio of the section of a stream to the perimeter of contact with its bed, is the most important element (together with the declivity) in determining its velocity, or the effectual moving force which acts upon it. Now, in the case of common friction, that of a solid body, neither the absolute nor the relative depth of the sliding body can have any influence in determining its motion.

Secondly, We have already seen (page 367) how enormous would be the velocity of a glacier if suddenly converted into a fluid, and how prodigious a force is absorbed, as it were, by the consistency or solidity of the ice. The moderate, though marked difference, found (page 146) between the lateral and central velocity of a glacier is in conformity to the second principle stated above, that the retardation due to friction will be more completely distributed over the whole section in proportion as the matter is less yielding.

Thirdly, The chief variation of velocity is, we have seen (page 146) near the sides.

Fourthly, We have found in page 147 a most remarkable confirmation of Dubuat's principle, that the amount of lateral retardation depends upon the actual velocity of the stream under experiment; whether we consider different points of the glacier, or the same point at different times.*

Fifthly, The glacier, we have seen, like a stream, has its still pools and its rapids. Where it is embayed by rocks, it accumulates—*its declivity diminishes, and its velocity at the same time ;*—when it passes down a steep, or issues by a narrow outlet, its velocity increases (page 145.)

The *central* velocities of the lower, middle, and higher regions of the Mer de Glace are (page 144)—

1.398   .574   .925

And if we divide the length of the glacier into three parts, we shall find (page 117) something like these numbers for its declivity†—

15°   4½°   8°

Lastly, When the semifluid ice inclines to solidity during a frost, its motion is checked ; if its fluidity is increased by a thaw, the motion is instantly accelerated, (p. 148.) Its motion is greater in summer than in winter, because the fluidity is more complete at the former than at the latter time. The motion does not cease in winter, (page 151,)

---

* Une chose étonnante, c'est que ni la grandeur du lit, ni celle de la pente n'influent en rien sur le rapport des différentes vitesses dont nous parlons, *tant que les vitesses moyennes restent les mêmes* où celle de la surface est constante."   *   *   *

DUBUAT, Art. 65.

† These numbers do not express the actual slopes at the points where the velocities were measured, but the slope of the inferior, middle, and superior regions of the glacier.

because the winter's cold penetrates the ice as it does the ground, only
to a limited extent, (pages 232, 360.) It is greater in hot weather
than in cold, because the sun's heat affords water to saturate the crevices: but the proportion of velocity does not follow the proportion of
heat, (pages 141, 150,) because any cause, such as the melting of a
coating of snow by a sudden thaw, as in the end of September 1842,
produces the same effect as great heat would do. Also, whatever cause
accelerates the movement of the centre of the ice increases the difference of central and lateral motion, (page 147.)

The Veined Structure of the Ice is a consequence of the Viscous Theory.

We have now to complete what was partly said in Chapter VIII.,
where we endeavoured to illustrate the phenomena of the veined or
ribboned structure of the ice, and to explain its cause.

This structure we have seen to consist in the recurrence of alternations of blue and white, or compact and aërated ice in a glacier, resembling the veins in chalcedony, the parts being thin and delicately sub-
divided.

We have seen (pages 177, 181) that this structure has all the appearance of being due to the formation of fissures in the aërated ice or consolidated névé, which fissures having been filled with water drained from
the glacier, and frozen during winter, have produced the compact blue
bands.

We have farther seen (pages 160, &c.) that this ribboned structure
follows a very peculiar course in the interior of the ice, of which the
general type is the appearance of a succession of oval waves on the
surface, passing into hyperbolas with the greater axis directed along the
glacier. That this structure is also developed throughout the thickness
of a glacier, as well as from the centre to the side, and that the structural surfaces are twisted round in such a manner that the *frontal dip*,
as we have called it, of the veins, as exhibited on a vertical plane cutting the axis of a glacier, occurs at
a small angle at its lower extremity, and increases rapidly as we
advance towards the origin of the
glacier, as shown in figure 1.

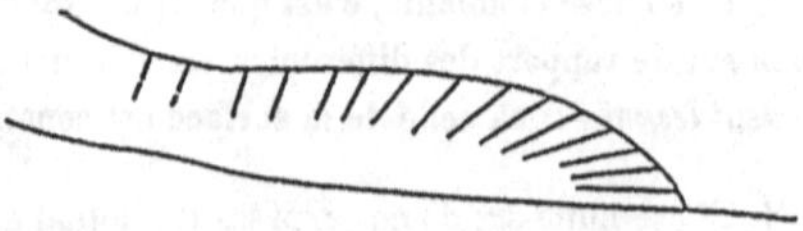

Fig. 1.

We have also considered glaciers generally as of three kinds, which, having a common structure, yet exhibit it in different forms or modifications. These three glacier forms may be termed the *canal shaped*, the *oval*, and the glaciers of the *second order*. Picturesque views of these are given in Plates II., IV., and IX., representing the *Mer de Glace*, the Glacier of *La Brenva*, and the secondary glaciers near *Macugnaga*. The annexed figures (one of which has been already used) show by views with ideal sections of such glaciers, the manner in which the structural surfaces traverse the mass of the ice. Figure 2. shows the conoidal structure

FIG. 2. *Showing the Structure of an oval Glacier by ideal Sections.*

of a glacier of the oval kind. Figure 3. shows this drawn out, as it were, into a canal-shaped glacier. On the right hand, in the upper part of the figure, a small glacier of the second order is shown, where it appears that its structure consists of a series of superimposed shells, nearly parallel to the soil, which might easily be confounded with the annual layers of the névé.

FIG. 3. *Showing the Structure of a Canal-shaped Glacier by ideal Sections.*

All these structures I explain on the common principle of the differ-

ence of velocity of the higher and lower, as well as of the central and
lateral parts of the ice; for wherever the parts of a stream, whether
liquid or semisolid, move with different velocities, there must be a force
applied to separate them from one another, as I have fully explained at
page 176.

But hear Dubuat, an eminent hydrostatical writer. Speaking of
ordinary rivers, he says, " La viscosité de l'eau ou l'adhérence que ses
particules ont entre elles, occasionne une résistance très-petite, mais
finie, qui s'oppose sans cesse à leur séparation : or, il ne peut y avoir
de mouvement uniforme dans l'eau, sans que ses filets ne prennent différentes vîtesses, selon qu'ils sont plus ou moins proches de la paroi,
qui retarde et rend uniforme le mouvement de toute la masse. Cette
inégalité de vîtesses ne peut avoir lieu sans une séparation mutuelle
des parties contiguës. La viscosité, ou, si l'on veut, la force avec
laquelle ces parties s'attirent, s'oppose à cette séparation ; il faut donc
qu'il y ait constamment une partie de la force accélératrice destinée à
vaincre cette résistance ; et lorsque la force accélératrice est assez petite
pour lui être seulement égale, le mouvement doit cesser, quoique la
pente soit finie. S'il existait un fluide dont les parties n'eussent
aucune adhérence entre elles, la plus petite pente possible suffirait pour
lui imprimer un mouvement ; mais les différents liquides connus éprou-
vant plus ou moins l'effet de la viscosité, la pente à laquelle ils commencent à couler est d'autant plus grande que l'adhérence de leurs
parties les éloigne moins de la nature des solides."*

From this we might expect that we
should have a separation of the ice particles, a rupture or fissuring of the substance of the glacier every where parallel
to the resisting walls or bottom, producing a cross section, as in the annexed
figure.

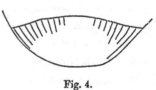

Fig. 4.

This is what actually takes place, and especially in glaciers of the
second order, where the retardation being almost entirely due to the

---

* DUBUAT, I. 58.

action of the bottom, the successive layers slip over one another with increasing velocity towards the surface.

But the question arises, how does this action produce the *frontal dip* of Fig. 1? why does not the canal-shaped glacier present a series of trough-shaped layers, as in Fig. 4, whose bottom remains parallel to the bottom or bed of the glacier. The reason appears to be this. The fluid is urged on (in the centre of the glacier especially) by its own *weight*. It is falling down an inclined plane by the force of gravity. It is, however, prevented from moving by the intense friction of the whole of the lower part of the glacier upon its bed. If the glacier be solid, there can be no motion, unless there be sufficient force to overcome this friction; and this we have seen to be one great (and we think insurmountable) difficulty, both of the hypothesis of De Saussure and that of De Charpentier. But the semifluid has another mode of progression,—the pressure *may* not overcome the friction of the bed, or else the fluid pressure at the lower end *may* be dragging the whole glacier over its bed, *that* is immaterial; but any particle in a fluid or semifluid mass, urged by a force from above, does not necessarily move in the direction in which the force impels it, it moves *diagonally*; forwards, in consequence of the impulse; upwards, in consequence of the resistance directly in front. Hence a series of surfaces of separation shaped (to use familiar illustrations) somewhat like the mouth of a coal-scuttle, or of a sugar-scoop, will rise towards the surface, varied in curvature by the law of velocity of the different layers of the glacier. Near the head or origin of the glacier, where the resistance in front is *enormous*, the tendency of the *separation planes*, which are those of apparent cleavage, will be very highly inclined. As the lower end of the glacier is approached, the resistance continually diminishes, the line of least resistance becomes more and more nearly horizontal; and finally, when the lower end of the glacier is reached, the planes fall away altogether, and the upper layers roll over the lower ones, now wholly unsupported. Such we have seen to be the actual phenomena of the Mer de Glace.

Imagine a long narrow trough or canal stopped at both ends, and filled to a considerable depth with treacle, honey, tar, or any such viscid fluid. Imagine one end of the trough to give way, the bottom still remaining horizontal, if the friction of the fluid against the bottom be greater than the friction against its own particles, the upper strata will roll over the lower ones, and protrude in a convex slope, which will be propagated backwards towards the other or closed end of the trough. Had the matter been quite fluid the whole would have run out and spread itself on a level; as it is, it assumes precisely the conditions which we suppose to exist in a glacier. The greatest disturbance or maximum separation of the parts takes place at the lower end, and there (the retardation at the sides being proportional to the *absolute* velocity, see page 147,) the separation will be most violent, and the loops on the surface will be most elongated. Near the origin the declivity is less, and the loops are more transverse. This is true of the glacier, (see page 167.)

Fig. 5.

Now, let the trough be a little inclined, so as to aid the gravitating force derived from the mere depth of the fluid. Each particle will be urged on by a force due to the slope, diminished by the resistances opposed to it. The particles near the lower termination of the stream have no resistances, except their attachment to those behind them,— they, therefore, roll straight on; but those in the middle of the glacier will easier raise the weight of a certain superincumbent stratum of ice, than push the entire glacier before them; they may do part of both, but will undoubtedly rise towards the surface, and thus *slide upwards and forwards* over the particles immediately in advance.

Though I am not aware that this form of fluid motion has been pointed out, its existence is scarcely to be doubted from very ordinary mechanical considerations, and several obvious phenomena also indicate it. Thus, such a viscid stream as we have supposed, be it tar, mortar, treacle, glue, plaster of Paris, slag, or cast metal, *invariably* presents *wrinkles*, or curvilinear arrangements of the floating matter, accompanied by a crumpling or inequality of the surface. These inequalities, these lines indicative of motion, which were the very first indications which led me, when studying the " dirt bands," (page 162,) to discover their nature, must be due to *inequality* of motion, and to no other cause. It is vain to look for any original linear or tabular arrangement of the particles when a fluid is poured from a ladle into a sloping channel, and afterwards becomes modified into the curves we have described. In the scum or froth on a sluggish current of water, there was no original arrangement of particles transverse to the stream, which has become *deformed* into the elongated loops in which the bubbles arrange themselves, as it were spontaneously, during their progress. These curves are the direct result of the unequal mutual pressures of the particles; and the whole phenomena, in the case of any of the semifluids I have mentioned, are such as—combined with the evidence which I have given, that the motion of a glacier is actually such as I have described that of a viscid fluid to be—can leave, I think, no reasonable doubt, *that the crevices formed by the forced separation of a half rigid mass, whose parts are compelled to move with different velocities, becoming infiltrated with water, and frozen during winter, produce the bands which we have described.**

---

* The wave-like figures of floating matter on a sluggish surface are not at all to be confounded with the actual direction of motion of the fluid particles. They are curves of *differential velocity* merely, and are always most perceptible near the *sides* of the stream, where the variations of velocity are greatest. A stream, like a mill-race, covered with sawdust, will show these linear markings inclining towards the centre of the stream; but the motion of any floating body, as a bit of cork, is sensibly parallel to the sides. I have proved the same thing by actually performing the experiment suggested in the text, of pouring plaster of Paris and glue into a narrow rectangular box, and sluicing it up by a bit of wood, removeable at pleasure. The surface of the viscid mass was then strewed, whilst level, with a coloured powder, and the sluice withdrawn. The liquid flowed exactly as I have described, and the colouring matter was drawn out into threads, precisely resembling, on a minute scale, in delicacy and continuity, the veined appearance of the glacier surface. The explanation appears to be this :—that the velocity of the central strata tends to pull the lateral strata towards the centre, as well as parallel to the length of the glacier ; this produces a slight lateral as well as longitudinal discontinuity, for the actual motion of the side strata towards the centre is exceedingly small, and (as the phenomena of moraines tell us, which act like the floating cork in the experiment above described) does not sensibly disturb the parallelism of motion of the parts of the ice. In short, the internal movements are of an order so inferior to the general movement of the stream, that they may probably be left out of account in describing that general movement, although by the fissured structure which they induce, they have sufficient evidence of their existence. But if the slope and consequent hydrostatic pressure be great, the movement towards the

I have succeeded in illustrating this theory by constructing models of a viscid material, (a mixture of plaster and glue, which does not *set* readily,) poured down irregular channels, representing Alpine valleys.

---

centre *may*, as supposed in the text, be of an order to modify appreciably the direction of movement of a particle. In an ordinary liquid like water, the direction of the ripple marks, occasioned by the friction of a stream in proceeding from a wider to a narrower channel, is an example of the same thing, namely, lines of maximum mutual friction of the particles against one another. They converge rapidly towards the centre of the stream, whilst the motion of the fluid, indicated by a floating body, deviates but little from the direction of the axis of the channel.

The same view, *mutatis mutandis*, explains the frontal dip and conchoidal form of the bands between the top and bottom of the glacier. There is here also a *drag* acting from the upper to the lower strata, and fissures are produced in consequence of the sluggish lower strata refusing to follow the swifter upper ones. This may also be a quantity of an order so inferior to the actual rate of motion of the ice, as to make it inappreciable by direct experiment.

The experiment, on a model described in this note, is more strictly analogous to the glacier phenomena, than those of a more striking kind described in the text, page 379, where the succession of colours, naturally gives to the mind the impression of a primitive structure near the origin of the glacier, which is mechanically *deformed* into these conchoidal surfaces. They strikingly recall, however, this important fact, that the direction of maximum distension of the particles must be, not parallel to the length of the glacier, but in the direction of the branches of these elongated loops, since their elongation is the simple result of the mechanical tension to which they are subjected ; hence, a motion in *parallel* directions, with *unequal* velocities, of a series of unorganized particles, confusedly arranged, must induce a mutual linear distension in a direction inclined to their real motions ; and this being unequal for adjacent portions, induces the delicately fibrous arrangement of the powder on the surface of the plaster model, and the lamellar arrangement of the ice in the interior of the glacier.

The least distance which can ever exist between a *side* and a *central particle* of a canal-shaped glacier, is half the breadth of the glacier. But the unequal motion of the centre and sides tends continually to separate them wider apart, and to distend the row of particles which connects them. The structural bands are, therefore, perpendicular to the line of greatest tension, and hence crevasses will naturally occur, *crossing the structure at right angles*, which I have found empirically to be the case, (see page 170, and the lines marked *a*, and which represent crevasses in fig. 3, p. 29). In pursuance of this principle the crevasses in an oval glacier are radiating ones ; those of a canal-shaped glacier must be slightly convex upwards, and this is perfectly confirmed by the crevassed appearance which the models described at the commencement of this note present. They are fissured in a direction exactly perpendicular to the striæ of the powdered surface.

In order to trace the motions better, I composed the streams of alternate doses of white and blue fluid, poured in successively. I have had a great number of such experiments made. The results have been preserved, and sections made of them, which I exhibited to the Royal Society of Edinburgh in March 1843, and described in their Proceedings. It may safely be stated, that the results of artificial sections of many of these experimental models, were not to be distinguished from the glacier sections transmitted by me from Geneva six months before, as the results of my observations, which are reprinted from the original woodcuts in the preceding pages, Figs. 1, 4, and 5. I subjoin a figure of one of the experimental models of a viscid fluid, which has been separated from the bed in which it was run, and divided so as to show its various sections.

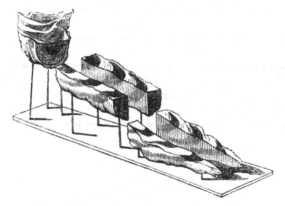

Fig. 6. *View of a model, showing the curves generated (experimentally) by the motion of a viscous fluid.*

It was objected by M. Agassiz* to this theory of the veins, that were it true, so soon as two glaciers united they would each lose their

---

* Proceedings of the Ashmolean Society; *Athenæum Journal*, February 1843. In this communication, M. Agassiz confirms my observation of the "dirt bands;" adopts the name of "annual rings," (*Edin. Phil. Journal*, October 1842,) and endeavours to prove the conformity of their intervals to the actual motion of the Glacier of the Aar, as I had already done on the Mer de Glace. M. Agassiz still insists, that glaciers are *stratified*, (see page 31 of this volume,) and he distinguishes these strata, as he calls the annual rings, from the pro-

individual structure, and have single loops due to the union of their streams, whereas his observations led him to conclude, that the loops of two united glaciers remain distinct. Now, in the first place, I reply, that though the distinct structure of the double stream is maintained for a time, it is always finally *worn out* if the glacier be long enough, and the structure then forms single loops, *cutting at an angle* the medial moraine of the two glaciers, (pages 167, 168 ;) and secondly, I maintain, that this is precisely what a semifluid body might be expected to do. For the structure near the centre is always imperfectly developed, exactly because *there* the differential motion is least; I mean, that there is least discontinuity of parts, because the velocity is nearly the same throughout a considerable space, (page 146 ;) and if two glaciers unite, and move tolerably uniformly together, they will preserve, for a long way, the structure which they had already acquired, before the

---

per veined structure of the ice. Having maintained, in all his earlier writings, that a glacier is horizontally stratified throughout its whole extent, (*Etudes*, page 40,) he now adopts my figure 1 of page 373, for the lower end of his glacier, and connects it with the névé, by a convenient series of interposed strata, first rising, and then falling, as represented in the annexed cut, which is accurately copied from the original in Leonhard and Bronns' Jour-

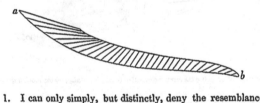

nal, 1843, Heft 1. I can only simply, but distinctly, deny the resemblance to nature of this scheme, and reiterate the observation already several times made in this work, that the structure of a glacier is, and must be formed in the glacier itself, not in the névé, from which it is often separated by an ice fall, *which has ground the integrant parts of the névé to powder*, as in the Glacier of La Brenva, (page 201,) the Glacier of Miage, (page 197,) the Glacier of Taléfre, (page 169,) and of Allalein, (page 352,) with many others. Not to mention the section, page 347, of the Glacier of Macugnaga, where the two structures are seen at once, and perpendicular to each other.

Yet more extraordinary is the assumption made by M. Agassiz, in order to account for this supposed prolongation of the beds of the névé into the inferior glacier. In order to explain the alternate rise and fall of the strata, he affirms, that near the origin of the glacier, the ice, in contact with the bed, moves faster than at the surface, but everywhere else, slower !

new one (representing a single united stream) is superinduced upon it.
Now this, as we have seen in Chapter VIII., is exactly what takes
place at the union of the glaciers of Léchaud and Géant,—of the two
branches of the Glacier of Taléfre, and of the Glacier of La Noire and
Le Géant, all of which, originally double in structure, finally become
single, and cut the separating moraine at an angle. But I appealed
here also to experiment, and found, that by pouring double streams of
viscid plaster down a single channel, the separate forms were *very
slowly* worn out indeed, and perpetuated far beyond the point of union
of the streams. Thus the proposed objection became a strong confirmation of my theory. One of these models, also shown to the Royal
Society, is represented in the annexed figure.

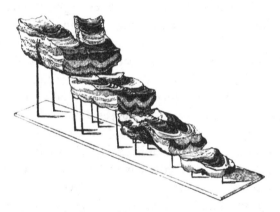

FIG. 7. *View of a model, showing the effect of the union of two streams on the motion of a viscid fluid.*

The illustrations now given will, it is hoped, show that there is a
striking conformity between the *facts of motion* and the *facts of structure*
in a glacier, and that the two, mutually supporting and confirming one
another, lend strong countenance to a theory which includes both. It
would be very easy to enlarge upon and multiply these illustrations
and coincidences, but I am satisfied that I have said enough to put the
intelligent reader in possession of the strong points of the theory, whilst
to many this chapter will appear already too long. A few circum-

stances which have not been here insisted on, appear in the letters to Professor Jameson in the Appendix.

THE idea of comparing a glacier to a river is any thing but new, and I would not be supposed to claim that comparison or analogy as an original one. Something very like the conception of fluid motion seems to have been in the minds of several writers, although I was not aware of it at the time that I made my theory. In particular, M. Rendu, whose mechanical views are in many respects more precise than those of his predecessors or contemporaries, speaks of " glaciers d'écoulement" as distinct from "glaciers reservoirs ;" and in the quotation at the head of this chapter, he evidently contemplates the *possibility* of the mutual pressures of the parts overcoming the rigidity.* He is the only writer of the glacier school who has insisted upon the plasticity of the ice, shewn by moulding itself to the endlessly varying form and section of its bed, and he is also opposed to his leading contemporaries in his conjecture that the centre of the ice-stream would be found to move fastest. But M. Rendu has the candour not to treat his ingenious speculations as leading to any certain result, not being founded on experiments worthy of confidence. " The fact of the motion exists," he says—" the progression of glaciers is demonstrated ; but the manner of it is *entirely unknown*. Perhaps by long observations and well made experiments on ice and snow, we mây be able to apprehend it, *but these first elements are still wanting.*"†

I feel bound also to quote the significant expressions of Captain Hall, pointing to the conception of a semifluid glacier. " When successive

---

* See also page 107 of his work for a comparison between a glacier and a river.

† " Le fait du mouvement existe, la progression des Glaciers est démontrée ; mais le mode est entièrement inconnu. Peut-être avec de longues observations, des expériences bien faites sur la glace et la neige viendra-t-on à bout de le saisir ; mais ces premiers éléments nous manquent encore."—*Theorie des Glaciers*, p. 90.

layers of snow," he says, speaking of the Glacier de Miage, "often several hundreds of feet in thickness, come to be melted by the sun and by the innumerable torrents which are poured upon them from every side, to say nothing of the heavy rains of summer, they form a mass, not liquid indeed, but such as has a tendency to move down the highly inclined faces on which they lie, every part of which is not only well lubricated by running streams resulting from the melting snows on every side, but has been well polished by the friction of ages of antecedent glaciers. Every summer a certain but very slow advance is made by these huge, sluggish, slushy, half-snowy, half-icy accumulations."* It is plain, I think, that the author had an idea that liquid pressure might drag a mass over its rocky bed, which would not move upon it as a solid.

But such speculations could not pass into a theory until supported by the definite facts of which M. Rendu deplores the want. I too, like my predecessors, though independently of them, had compared the movement of glaciers to that of a ductile plastic mass, in 1841, when I spoke of the Glacier of the Rhone as "spreading itself out much as a pailful of thickish mortar would do in like circumstances,"† and again, when I likened the motion of glaciers to that of a great river, or of a lava stream.‡ But I knew very well that such analogies had no claim to found a *theory*. I knew that the *onus* of the proof lay with the theorist,—(1.) To show that (contrary to the then received opinion) the centre of a glacier moves fastest; and, (2.) to prove from direct experiment that the matter of a glacier is plastic on a great scale, a fact which seems so repugnant to first impressions as lately to have been urged in a most respectable quarter,§ as rendering the

---

* "*Patchwork*," vol. i., p. 104, *et seq.* The whole passage, which is too long to quote, gives an admirable picture of the glacier world.

† *Ed. Phil. Journal*, January 1842.

‡ *Edinburgh Review*, April 1842, p. 54. Both these articles were written in 1841.

§ *Bibliothèque Universelle*, January 1843.

doctrine of semifluid motion untenable. No one had a right to maintain the theory of fluid motion as more than a conjecture, until at least these preliminary obstacles were removed by direct observations.

These observations have been made, and the result is the viscous or plastic theory of glaciers, as depending essentially on the three following classes of facts, all of which were ascertained for the first time by observations in 1842, of which the proofs are contained in this work.

1. That the different portions of any transverse section of a glacier move with varying velocities, and fastest in the centre.

2. That those circumstances which increase the *fluidity* of a glacier, —namely, heat and wet,—invariably accelerate its motion.

3. That the structural surfaces occasioned by fissures which have traversed the interior of the ice, are also the surfaces of *maximum tension* in a semisolid or plastic mass, lying in an inclined channel.

There is only one other point to which I would invite attention, and it is this. We have noticed, page 153, the enormous depression which the surface of the ice undergoes during the warmer months of the year. We may be sure that, in some manner or other, this is made up for during winter and spring. I already suggested, in my fourth letter to Professor Jameson, in Appendix, No. II., that this may be partly owing to the dilatation of the ice during winter by the congelation of the water in its fissures, producing, at the same time, " the veined structure." The glacier is very far indeed from being frozen to the bottom in winter, for we have seen that physical principles are opposed to this, as well as the fact that the motion continues during all that period, showing that a great portion of the icy mass is still plastic. It is, however, extremely probable that the congelation extends to a considerable depth, and produces the usual effects of expansion. I think, however, that the explanation, though correct so far as it goes, is inadequate, and that the main cause of the restoration of the surface is the diminished fluidity of the glacier in cold weather, which retards (as we know) the motion of all its parts, but especially of those parts which move most rapidly in summer. The disproportion of velocity throughout the length and breadth of the glacier is therefore less, the ice more pressed together, and less drawn asunder; the crevasses are consolidated, while the increased friction and viscosity causes the whole to swell, and especially the inferior parts, which are the most wasted. Such a hydrostatic pressure, likewise, tending to press the lower layers of ice upwards to the surface, may not be without its influence upon the (so-called) rejection of blocks and sand by the ice, and may even have some connection with the recurrence of the "dirt bands" upon the surface of the glacier. But I forbear to enlarge upon what is only as yet to myself conjectural.

I have no doubt, however, that the convex surface of the glacier, (which resembles that of mercury in a barometer tube,) is due to this hydrostatic pressure acting upwards with most energy near the centre. It is the "renflement" of Rendu, the "surface bombée" of Agassiz. Exactly the contrary is the case in a river, where the centre is always lowest; but that is on account of the extreme fluidity, so that the matter runs off faster than it can be supplied; but in my plaster models, this convexity, with its wrinkles and waves, was perfectly imitated.

In its bearing on the theory of the former extension of the Swiss glaciers, (Chap. III.,) we find, that the doctrine of semifluid motion leads us to this important conclusion,—that as large and deep rivers flow along a far smaller inclination than small and shallow ones, (a circumstance depending mainly upon the weight increasing with the *section*, and the friction, in this particular case, with the *line of contact* with the channel,) the most certain analogy leads us to the same conclusion in the case of glaciers. We cannot, therefore, admit it to be any sufficient argument* against the extension of ancient glaciers to the Jura, for example, that they must have moved with a superficial slope of one degree, or, in some parts, even of a half or a quarter of that amount, whilst in existing glaciers the slope is seldom or never

---

* ELIE DE BEAUMONT, *Annales des Sciences Géologiques, par Rivière,* 1842.

under 3°. The declivity requisite to insure a given velocity, bears a simple proportion to the *dimensions* of a stream. A stream of twice the length, breadth, and depth of another, will flow on a declivity half as great, and one of ten times the dimensions upon 1-10th of the slope.*

Poets and philosophers have delighted to compare the course of human life to that of a river; perhaps a still apter simile might be found in the history of a glacier. Heaven-descended in its origin, it yet takes its mould and conformation from the hidden womb of the mountains which brought it forth. At first soft and ductile, it acquires a character and firmness of its own, as an inevitable destiny urges it on its onward career. Jostled and constrained by the crosses and inequalities of its prescribed path, hedged in by impassable barriers which fix limits to its movements, it yields groaning to its fate, and still travels forward seamed with the scars of many a conflict with opposing obstacles. All this while, although wasting, it is renewed by an unseen power,—it evaporates, but is not consumed. On its surface it bears the spoils which, during the progress of existence it has made its own;—often weighty burdens devoid of beauty or value,—at times precious masses, sparkling with gems or with ore. Having at length attained its greatest width and extension, commanding admiration by its beauty and power, waste predominates over supply, the vital springs begin to fail; it stoops into an attitude of decrepitude;—it drops the burdens, one by one, which it had borne so proudly aloft,—its dissolution is

---

* This results approximately from the formulæ of Dubuat and Eytelwein,—the velocity varies as the square root of the slope, and as the square root of the mean hydraulic depth.

inevitable. But as it is resolved into its elements, it takes all at once, a new, and livelier, and disembarrassed form;—from the wreck of its members it arises, " another, yet the same,"—a noble, full-bodied, arrowy stream, which leaps, rejoicing over the obstacles which before had staid its progress, and hastens through fertile valleys towards a freer existence, and a final union in the ocean with the boundless and the infinite.

Source of the Arveiron.

APPENDIX.

No. I.

ON A REMARKABLE STRUCTURE OBSERVED BY THE AUTHOR IN THE ICE OF GLACIERS.*

" THE object of the present short communication is little more than to announce and describe a peculiarity which the Ice of Glaciers frequently exhibits, interesting in itself as connected with the theory of their formation and propagation, and perhaps having a bearing upon the explanation of some facts long felt by geologists to be perplexing.

" Had I yielded to my own first impulse, this communication would have formed but a part of a much more extensive one, intended to give such an account, as I best might, of the present views entertained respecting the mechanism and conservation of glaciers, and the curious and interesting question of their ancient extension, and perhaps vast geological influence in producing some of the latest evidences of revolution on the surface of the globe. When I considered, however, the great extent which such a communication, to be generally intelligible, must necessarily have—and farther, that a large share of the material must be drawn from the works and the observations of others,—when I recollected, besides, that however earnest and sustained had been my investigation of these curious points, there was still much left obscure or unproved to my own mind ; in short, that the communication I should lay

---

* Read to the Royal Society of Edinburgh on the 6th December 1841, and published in the *Edinburgh New Philosophical Journal* for January 1842.

before the Society could not have that completeness, determination, and origi-
nality, which could properly entitle it to a permanent place in the Transactions of our Body, it seemed to me that the wish which had been expressed
by very many of those to whose judgment I am most willing to defer, that I
should make such a detailed communication, was one with which, in my offi-
cial position as Secretary, and having in some degree the control of the order
and distribution of business, I could not properly comply.

" I do not, however, relinquish the idea of laying before the Society, and even
at considerable length, the conclusions which I may ultimately form respecting the great physical and geological questions now at issue, and the facts and
reasonings upon which these conclusions are founded. The Glacier Theory,
whether it regards the present or past history of those mighty and resistless
vehicles of transport and instruments of degradation, yields to no other physical speculation of the present day in grandeur, importance, interest, and, I
had almost said, novelty. I look forward to the prospect, which I hope may
be realized, of extending much farther, during another summer, my direct
observations and experiments, and in the meantime I desire to prepare myself
for the task, by a thoughtful review of the experience I have already had,
and a close analysis of what has been already argued and written upon the
subject. Should the result be successful, the Society may, a year hence,
expect the communication of it. For the present, I mean to confine myself
to the description of a single fact, which appears generally, if not universally,
to have escaped the notice of former travellers amongst the Glaciers.

" On the 9th of August last (1841) I paid my first visit to the Lower Glacier of the Aar, upon or near which I spent the greater part of three weeks in
company with Professor Agassiz of Neufchatel, and Mr. J. M. Heath of Cambridge. It is surprising how little we see until we are taught to observe. I
had crossed and recrossed many glaciers before, and attended to their phenomena in a general way ; but it was with a new sense of the importance and
difficulty of the investigation of their nature and functions that I found something to remark at every step which had not struck me before; and even in
the course of the walk along *our own* glacier, (as we considered that of the
Aar, when we had taken up our habitation upon it,) we found on its vast
and varied surface something each day which had totally escaped us before. It
was fully three hours' good walking on the ice or moraine from the lower
extremity of the glacier to the huge block of stone, under whose friendly shelter we were to encamp ; and in the course of this walk, (a distance of eight

or nine miles, on a moderate computation, allowing for the roughness of the way,) on the first day, I noticed in some parts of the ice an appearance which I cannot more accurately describe than by calling it a *ribboned structure*, formed by thin and delicate blue and blueish-white bands or strata, which appeared to traverse the ice in a vertical direction, or rather which, by their apposition, formed the entire mass of the ice. The direction of these bands was parallel to the length of the glacier, and, of course, being vertical, they cropped out at the surface, and wherever that surface was intersected and smoothed by superficial water-courses, their structure appeared with the beauty and sharpness of a delicately-veined chalcedony. I was surprised, on remarking it to Mr. Agassiz as a thing which must be familiar to him, to find that he had not distinctly noticed it before—at least if he had, that he had considered it as a superficial phenomenon, wholly unconnected with the general structure of the ice. But we had not completed our walk before my suspicion that it was a permanent and deeply-seated structure was fully confirmed. Not only did we trace it down the walls of the crevasses by which the glacier is intersected, as far as we could distinctly see, but, coming to a great excavation in the ice, at least 20 feet deep, formed by running water, we found the vertical strata or bands perfectly well defined throughout the whole mass of ice to that depth. Where the plane of vertical section was eroded by the action of water, the harder seams of blue ice stood protuberant; whilst the intermediate ones, partaking of a whitish-green colour and granular structure, were washed out. We did not sleep that night until we had traced the structure in all directions, even far above the position of our cabin, and quite from side to side across the spacious Glacier of the Finster Aar.

" During the whole of our subsequent residence amongst the glaciers, the phenomena and causes of this structure occupied our thoughts very frequently. We had much difficulty in arriving at a correct description of the manner of its occurrence, and still more in forming a theory in the least plausible respecting its origin.

" Its importance, however, as an indication of an unknown cause, is very great; not only because all that can illustrate what is so obscure as the manner of glacier formation and movement, is so, but because it is precisely on this very point of " What is the internal structure of the ice of a glacier?" that the question now pending respecting internal dilatation as a force producing progression, mainly hangs. Some consider ice as compact, others as granular; some as crystallized, others as fractured into angular fragments;

some as horizontally stratified, others as homogeneous; some as rigid, others as plastic; some as wasting, others as growing; some as absorbing water, others as only parting with it; and yet no one seems to have observed, or at least observed as an object of study, this pervading slaty or ribboned structure, to be found probably in one part or other of every true glacier.

" With regard to *extent*, this structure was observable on the Lower Glacier of the Aar, from its lower extremity up to the region of the *firn* or *névé*, where, the icy structure ceasing to exist, it could not be looked for; yet even there, where frequent thaws, induced by the neighbourhood of rocks or stones, produced a compacter structure, the veins became apparent. In some parts of the glacier it appears more developed than in others: in the neighbourhood of the *moraines*, and the walls of the glacier, it was most apparent. This would seem to infer a relation to the frequency of thaws and recongelation.

" It penetrates the *thickness* of the glacier to great depths. It is an integral part of its inmost structure. That it could not be the production of a single season I was speedily convinced, by observing, that where old crevasses fissured the glacier transversely, the veined structure not only was reproduced on either side, but frequently with a *shift* or dislocation, or series of parallel fissures, presenting sometimes a series of dislocations advancing in one direction.

" The *course* of the veined structure was, *generally speaking*, on the Glacier of the Aar, strictly parallel with its length, and that with a degree of accuracy which seems extraordinary, if we attribute its production to the remote influence of the retaining walls of the glacier, distant at least half a mile. Near the inferior extremity, where the declivity becomes rapid, the structure varies its position in a manner very difficult to trace satisfactorily. There can be little doubt, however, that the nearly horizontal bands which appear on the steep declivity of the glacier at its lower termination, are nothing else than the outcropping of these bands, which have there totally changed their direction, being *transverse* instead of longitudinal, and leaning forwards in the direction in which the glacier moves at a very considerable angle. The ice in this part of the glacier is distinctly granular, being composed of large fissured morsels, nicely wedged together; and the ribboned structure is greatly obscured. There seems no doubt, however, that the horizontal stratification in the lower part of glaciers, insisted on by several writers, is merely a deception arising from this cause, so familiar to the geologist who gets a section perpendicular to the dip of strata, which, therefore, appear horizontal.

Towards the sides or walls of the glacier, at its lower extremity, the veins have their plane twisted round a vertical axis, having now their dip towards the centre of the glacier, and rising against the walls ; and this inclination sometimes extends nearly to the axis of the glacier, or the medial moraine, where I have observed the veins deviating from the vertical by an angle of about twenty degrees, the bands inclining from the centre, (or rising towards the walls,) as if the pressure arising from the superior elevation of the glacier under the moraine had squeezed them out. The whole phenomenon has a good deal the air of being a structure induced *perpendicularly to the lines of greatest pressure*, though I do not assert that the statement is general. Whilst the glacier is confined between precipitous barriers with a feeble inclination, the structure is longitudinal. As the glacier, by its weight, falls over the lower part of its bed, and moulds itself into the form which the continued action of gravity on its somewhat plastic structure impresses, the longitudinal structure is first annihilated, (for throughout a certain space we could detect no indications of one kind or other,) and the bands then reappear in a transverse direction, as if generated by the downward and forward pressure, which, at the lowest part of the glacier, replaces the tight wedging which higher up it received laterally. It is not easy to convey without a model a clear idea of the forms of surface here intended, and which yet require considerable correction.

" I may mention, however, that the Glacier of the Rhone, which I have carefully examined, presents a structure in conformity with the view thus developed. It will be recollected by all who have seen that magnificent mass, that it pours in colossal fragments over the rocky barrier which separates the Gallenstock from the valley of the Rhone, and having reached the last-named valley, it spreads itself across and along it pretty freely —much as a pailful of thickish mortar would do in like circumstances. The form into which it spreads is rudely represented in the annexed figure. In this particular case, even the strongest partisans of the dilatation theory will hardly deny, that the accumulated ice descending from the glacier

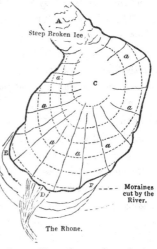

Steep Broken Ice

Moraines cut by the River.

The Rhone.

cataract A would form a centre of pressure at C, and that the lines of equal pressure would be found in the direction of the dotted lines, following nearly the periphery of the glacier. Now, these dotted lines precisely trace out the course of the veined structure alluded to; and, moreover, they bend more and more forwards as we proceed from the centre of pressure C, especially in the direction of D, the line of greatest inclination of the bed, and down which gravity urges the icy mass. The front of the glacier, about E D F, presents the fallacious appearance of horizontal strata, as in the Aar Glacier; but these are found to dip inwards at an angle of 10° or 15°, which angle continually increases as we approach the heart of the glacier, rising to 40°, 50°, 60° and even 70° as we approach C. It cannot be doubted, that these facts are so far favourable to the view which we have taken, although the establishment of it would require far more extensive observation; and in several glaciers which I have visited, the observation of the convolutions of the veined structure is very difficult and obscure. Before quitting the subject, I must add an observation which I made on the Glacier of the Rhone, and which I am pretty confident is well founded. *The lines of fissures, or crevasses, are always perpendicular to the conical surfaces of the veined structure.* These fissures are denoted in the figure by the full lines *a a a.* Perhaps the primary cause of these fissures is, that the pressure of the ice at C forces the glacier to distend itself into continually widening rings, which its rigidity resists, and therefore it becomes traversed by radial crevasses.

" The veined structure itself, I have already said, arises from the alternation of more or less compact bands of ice. The breadth of these varies from a small fraction of an inch to several inches. The more porous of these bands are the likeliest vehicles for the transmission of water from the higher to the lower part of a glacier; and that opinion receives some confirmation from the fact, that, at a certain depth in crevasses, we may see the veined structure marked out and exaggerated by the frozen stalagmite, which is protruded from the section of the more porous layers.

" In conclusion (for the present,) this structure deserves the attention of geologists generally, as showing how the appearance of the most delicate stratification, and of sedimentary deposition, may be produced in homogeneous masses, where nothing of the kind has occurred. For a short time, indeed, I was of opinion that this structure resulted from true stratification; but a closer examination of the mass convinced me that, inexplicable as the fact remains, it must be accounted for in some other way. We have endeavoured to show an empirical connection which appears to exist between the structural planes and the sustaining walls of the glacier, and likewise that the recurrence of congelation and thaw appears to strengthen the formation of the bands. But this cannot be considered as in any degree amounting to an explanation. The analogous difficulty of slaty cleavage in rocks, presents itself as not improbably connected with a similar unknown cause, whose action pervaded the mass of the crystallizing rock undergoing metamorphic change, as this pervades the mass of the crystallizing glacier. In the former case, we have cleavage planes perfectly parallel, almost indefinitely extending with unaltered features over vast surfaces of the most rugged country, changing neither direction, dip, nor interval, with hill or valley, cliff, or scar, and passing alike through strata whose planes of stratification, horizontal, elevated, undulating, or contorted, offer no obstacle or modification to the omnipotent energy which has rearranged every particle in the mass *subsequent* to deposition. The supposition of Professor Sedgwick, who has minutely described and considered this geological puzzle, that " crystalline or polar forces acted on the whole mass simultaneously in giving directions, and with adequate power,"* can hardly be considered as a solution of the difficulty, until it is shown that the forces in question have so acted, and can so act. The experiment is one which the boldest philosopher would be puzzled to repeat in his laboratory ; it probably requires acres for its scope, and years for its accomplishment. May it not be that Nature is performing in her icy domain a repetition of the same mysterious process, and that in another view from the one which has recently been taken, the Theory of Glaciers may lead to the true solution of geological problems ?"†

---

* *Geological Transactions*, Second Series, iii. 477.

† It will be understood that I do not *now* suppose that there is any parallelism between the phenomena of rocky cleavage and the ribboned structure of the ice. On all the other points considered in this paper, my opinions are not only unchanged, but entirely confirmed by additional experience.—J. D. F., June 1843.

No. II.

FOUR LETTERS TO PROFESSOR JAMESON, CONTAINING AN ACCOUNT OF
OBSERVATIONS ON GLACIERS, MADE IN 1842.*

" COURMAYEUR, PIEDMONT, *4th July* 1842.

" MY DEAR SIR,—Knowing that you will be glad to hear of my safe arrival amongst the Alps, and of my farther proceedings, I hasten to give you an
account, in a few words, of what I have as yet done. Finding the season
more than usually advanced, I hastened to reach Chamouni, in order to ascertain whether the Mer de Glace was as yet accessible in all its extent; and I
arrived at the Montanvert on the 24th June, and remained there for a week.
I was fortunate enough to convey all my instruments to their destination,
without, I believe, injury to any one of them. The Mer de Glace, so continually visited by the curious, but so little studied, seemed to me to offer
great advantages for the prosecution of the objects which I proposed to myself. At first sight it appeared to me steeper and more crevassed than I
recollected it to be, and I doubted for a moment whether it was adapted for
my experiments; but that doubt vanished upon closer examination; and in
the course of the single week which I have been able to spend there, being
favoured by most excellent weather, I have obtained results so far definite
and satisfactory, that, imperfect as they necessarily are, and only the commencement of what I expect to accomplish during the remainder of the season,
I will state them shortly.

" You will recollect that, in my lectures on glaciers delivered last December
and January, and afterwards in an article written by me in the Edinburgh
Review, I insisted on the importance of considering the mechanism of glaciers
as a question of pure physics, and of obtaining precise and quantitative mea-

---

* From *The Edinburgh New Philosophical Journal* for October 1842, and January 1843.

sures as the only basis of accurate induction. I pointed out, also, the several experiments of a critical kind which might be made; such, for instance, as the determination of the motion of the ice at different points of its length, in order to distinguish between the theories of De Saussure and De Charpentier; for, if the glacier merely slides, the velocity of all its points ought (in the simplest case) to be the same; if the glacier swells in all its mass, the velocity of the inferior part ought to be greatest. Of course, I do not now advert to the many causes which might accidentally invert this law, and which would require to be fully taken into account; still less do I mean to say that anything I have now to state can be considered as critically decisive between rival theories; but my experiments certainly do show that the kind of precision which I desired to see introduced into reasonings about this subject, is practically attainable, even in a far higher degree than I expected.

" For example :—The motion of glaciers by the measurement of the distance of blocks upon its surface from a fixed point, from one year to another, has marked indubitably the annual progress of the ice. I do not know that any one has attempted to perform the measurement in a manner which could lead to any certain conclusion respecting the motion of the ice at one season compared with another, or from month to month; still less has any one been able to state, *with precision*, whether the glacier moves by starts and irregularly, (as we should certainly expect on the sliding theory,) or uniformly and evenly; and if so, whether it moves only at one part of the twenty-four hours, and stands still during the remainder (as we should expect on the dilatation theory, as commonly expounded.) Now, I have already been able—

" 1st, To show and measure the glacier motion not only from *day to day*, but from *hour to hour*, so that I can tell nearly what o'clock it is by the glacier index. That you may have an idea of the coincidence which these experiments present, I give you the longitudinal motion of a point on the Mer de Glace during four consecutive days.

15.2 inches.   16.3 inches.   17.5 inches.   17.4 inches.

" 2d, This motion, evidently incompatible with sudden starts, takes place in the glacier *as a whole*, undisturbed by the most enormous dislocations of its surface, for these measures were taken where the glacier was excessively crevassed.

" 3d, This motion goes on *day and night*, and if not with absolute uni-

formity, at least without any considerable anomaly.　On the 28th–29th June
the motion

from 6 P.M. to 6 A.M. was 8.0 inches,

...　6 A.M. to 6 P.M. ...　9.5　...

29th–30th,　...　6 P.M. to 6 A.M. ...　8.5　...

...　6 A.M. to 6 P.M. ...　8.9　...

seeming to show a greater motion during the day.

" 4th, In the particular case of the Mer de Glace, the higher part (the
Glacier de Léchaud) moves *slower* than the lower part near the Montanvert
in the proportion of 3 to 5.

" 5th, The central part of the glacier moves faster than the edges in a very
considerable proportion ; quite contrary to the opinion generally entertained.

" There cannot be a doubt of the accuracy of these results *within the limits*
in which the experiment has been made.　They prove how completely problems of a purely physical character admit of accurate investigation ; and
when a larger induction shall have freed the results from local errors, it is
evident that we shall have the solid foundations of a theory.　My wish to see
the total eclipse of the sun on the 8th, has brought me to the south side of the
Alps sooner than I could have wished ; but I have now fixed so many points
on the Mer de Glace, that, on my return thither, I shall be able to obtain
more comprehensive results.　But what is most important in the whole matter is this,—that an observer furnished with the proper instruments and methods may, by spending a few days on a glacier, determine at any particular
season the amount of its motion·at all the essential points, within the limits
which any glacier theory can require.

CHAMOUNI, 10th *August* 1842.

" MY DEAR SIR,—Since I last wrote to you on the 4th July from Courmayeur, I have examined, in detail, the two principal glaciers of the Allée
Blanche ; and having re-crossed the Alps from Courmayeur by the Col du
Géant, where I had the satisfaction of still finding the remains of Saussure's
Cabane of 1788, I have pursued for a fortnight my experiments on the motion of the Mer de Glace.　Being composed, as you know, of several tributaries which are in some degree independent, and presenting also a considerable
variety of surface, this glacier seems as proper as any for detailed experiments,
such as those which I am attempting.　Being about to quit this place on a

tour to Monte-Rosa and the glaciers east of the Great St. Bernard, I wish to explain to you now in what respect my observations differ from those formerly undertaken on the glaciers, and to mention a few results, which, of course, being as yet only partial, ought not to be considered as altogether decisive of the truth or falsehood of any theory; still I believe it will be admitted that the facts established in my last, and which farther experience has confirmed, militate strongly against some of the received opinions as to the cause of glacier motion.

" You are aware, that, in my lectures on glaciers in December and January last, and in an article in the Edinburgh Review for April, I insisted, and so far as I know it was for the first time, on the importance of considering the glacier theory as a branch of mechanical physics, by which I mean that the cause of movement should be ascertained inductively from the observed motion, carefully and *numerically* ascertained at different points. It is because authors have considered the problem as too simple a one to require a systematic analysis, that we find little or nothing done in this respect; and it may be affirmed, without any disrespect to the ingenious persons who have assigned probable causes for the movement of these masses of ice, that their solutions have been, like the astronomical theories of the earlier cosmogonists, based upon somewhat vague analogies with better understood phenomena, as when the analogy of magnetic attractions seemed to offer a parallel to the mechanism of the heavens in the theory of Gilbert, and that of fluid currents gave rise to the Cartesian vortices. The Newtonian theory was based upon its coincidence with the empirical laws of planetary motion. We have as yet no empirical laws of glacier motion, consequently no proper mechanical theory can as yet be adequately tested. I endeavoured to point out in my lectures how a mechanical theory might be deduced from observation, and how these observations might be practically made. I believe that I have also obtained for the first time, the numbers on whose importance I insisted. I am not aware that any one had hitherto proposed to determine the diurnal velocity of a given point of a glacier with reference to three co-ordinates. The analogy with the empirical laws of astronomy is both striking and just; an exact acquaintance with the path described by any molecule of a glacier, will almost as certainly lead to a knowledge of the cause of its motion, as the theory of gravitation sprung from the three laws of Kepler. We have to deal, indeed, with an effect more complex and varied; but the results contained in my last letter, already show how much of numerical precision may be attained. I have already determined the diurnal motion of 10 points of the Mer de Glace with a probable error, not exceeding, I think, a quarter of an inch in any case; and when these observations shall have been pursued, as I expect to do, until the end of September, there will be a tolerable basis for sound speculation.

" In particular, you will recollect that I pointed out last winter two experiments for distinguishing between the prevailing theories of De Saussure and De Charpentier, those of gravitation and of dilatation. One was the exact measurement of a space along the ice to be measured after a certain time, in order to ascertain whether any expansion had occurred. The other was the determination of the linear velocity of the glacier at any point, which, on the theory of Saussure, ought (if the glacier be of nearly uniform section) to be uniform throughout; on the theory of Charpentier it ought to increase from nothing at the upper extremity of the glacier, to a maximum at its lower end. The former experiment had, I have since learned, been suggested by Professor Studer to M. Escher last year, and attempted to be put in practice (though unsuccessfully) by the latter, on the Glacier of Aletsch. Admitting Charpentier's theory, however, this dilatation would be too small to be successfully observed in a moderate time, and with the geometrical methods which the uneven and varying surface of the glacier enables us to employ; I have therefore not attempted it. The other method, in fact, embraces both ends; for if the movement of the glacier in its upper and lower part be determined (by *upper* I mean near its origin), the difference of the motions determines the dilatation or contraction of the intermediate part of the ice, and is liable to none of the great errors arising from the measurement of long distances. The observation, in the simplest and best form which I employ, resembles perfectly that of determining with the transit instrument the progress of a planet.

" I have already said that my later observations confirm those which I previously communicated; any variations, indeed, arise solely from a change of circumstances or season, and not from errors of observation. (1.) The continuous imperceptible motion of the glacier is entirely confirmed; its bearing upon the sliding theory is very obvious. (2.) This motion is not by any means the same, however, from day to day, and from week to week, as indeed already appeared from my first results. (3.) This variation of motion appears to be common to every part of the glacier, as well where compact and completely even, as where most fissured; nor perhaps is the variation of velocity greater in one case than in the other. (4.) From numerous observations, made in all parts of the glacier, it invariably results as before, that the centre moves faster than the sides of the ice-stream. In the lower and faster moving part of the glacier this disproportion is greatest, varying from one-third to one-half of the smaller velocity. Near the origin of the glacier it appears to be one-fourth or one-fifth of the smaller velocity. (5.) The variations of glacier motion affect the central parts most sensibly. (6.) The greatest daily motion which I have observed, nearly opposite the Montan-vert, amounts to 27.1 inches. (7.) I have ascertained the velocity of motion much nearer the origin of the glacier than when I last wrote. This, which would appear to be nearly, if not quite an *experimentum crucis* between the sliding and dilatation theories, does not yield a result so favourable to the latter as I had at first supposed; for though it is undoubtedly true, as stated in my last, that the head of the glacier moves slower than the foot, the middle part moves rather slower than either, owing probably to the greater width and thickening of the ice there. This source of error from the varying section of the glacier I had fully anticipated; but still, when we push the experiment to a limit, and take the velocity very near the origin, the velocity ought to diminish, on the theory of Charpentier, with a rapidity not to be mistaken. Yet very near the head of the Glacier de Léchaud, the diurnal velocity is considerably more than a foot per day. I am far, however, from thinking that I am yet in a position to judge finally of the merits of any theory; my belief is, that both of those cited will as yet require great modification.

" By insisting upon the treatment of the problem as one of pure mechanics, I am far from denying that the kind of investigations to which the glacier theorists have hitherto almost exclusively referred, are also of great value, such as those on the temperature and structure of the ice. The latter, in particular, is a sort of standing evidence of its mechanism, and, rightly understood, must lead to the most important confirmation of any mechanical theory. This you may believe I have made an object of very particular attention. I have now examined so many glaciers as to have a very clear idea of the empirical laws which that structure follows. Lately, I began to perceive a connection between that structure and the facts of motion already cited. If these two classes of facts can be well brought into harmony with one another, we should have a very good chance of consolidating them into something like a theory. In my next letter, I will give you some account, at all events, of my observations on the subject, which are sufficiently definite; and probably also (without considering it as proved), of what seems likely enough to be its true explanation. I go to-morrow to the Great St. Bernard, to meet M. Studer.—Believe me, very sincerely yours,

JAMES D. FORBES."

ZERMATT, North Side of Monte Rosa,
22d August 1842.

" My Dear Sir,—I arrived here two days ago, by a very interesting and unfrequented route. I mentioned in my last, that M. Studer and I had agreed to visit together the valleys eastward of the Great St. Bernard. The Convent was our place of rendezvous, and we afterwards descended to Orsières, and turned into the Valley of Bagnes. Crossing the Alpine chain at the head of the valley, by the Col de Fenêtres, we went down to Valpelline on the Italian side, and ascended that valley quite to its origin. We then crossed to the western branch of the Valley of Erin, by the Col de Collon or Arolla, a very striking glacier pass. Thence M. Studer went to the Val d'Anniviers, and rejoins me here by way of Visp, whilst I ascended the other branch of the Eringer Thal from Evolena by way of the Ferpêcle glacier, and crossed over the mountains to this place, by a pass higher and much longer than the Col du Géant, which presents, certainly, the grandest views I have hitherto met with in the Alps. I must not, however, stop to describe, as my present object is to fulfil the promise in my last respecting the structure of glacier ice.

" The internal veined or ribboned structure presented by all glaciers in a greater or less degree, appears to be the only true essential structure which they possess, and which, you will recollect, I described in a paper printed in your Journal for January last. The existence of granules divided by capillary fissures, as well as of large crevasses, are equally unessential to glacier structure, and subordinate to the other. Whatever other result may flow from the examination of glaciers this summer, by the many persons who are probably at this moment directing their attention to them, this, I am sure, will be admitted, that the veined structure is not peculiar to some glaciers, as some would maintain, nor to some years, as has been alleged by others; but that it is perfectly general and systematic, having one general type or form, which is varied according to external mechanical circumstances. Being then the most essential and intimate part of the glacier formation, as well as one of its most obvious and universal features (especially on those glaciers which are most commonly visited), it is equally singular that it should not have been sooner noticed, or if noticed, never once alluded to by the eminent and ingenious authors who have treated of existing glaciers and their effects.

" With respect to the general type or form of this structure, I am happy to say, that I have found not the slightest reason to modify the description which I have given in the paper above alluded to, of the conformation of the glacier of the Rhone. The description is characteristic, not of that glacier only, but of every other, with certain modifications similar to the variation of the parameter of a curve; variations, therefore, not in kind but in degree. The most beautiful structure I have ever met with is in the glacier of La Brenva in the Allée Blanche, which was one of the earliest I examined this season, and in which I found all that I had seen, though imperfectly, on the Glacier of the Rhone (which it resembles in the circumstances of being derived from an icy cascade, and in having a considerable breadth in proportion to its length), developed in a manner so clear and so geometrically precise, as gave me the most lively satisfaction. I refer to my former paper for the figure and description of that structure ; I have found the same conoidal surfaces, and the same false appearance of horizontal stratification on the terminal face of the glacier, arising from the veins dipping inwards at first at an angle of only 5°, rising to 10°, 20°, up to 60° and 70°, if we follow the medial line of the glacier, or axis parallel to its length. The sides of the glacier, in like manner, have their cleavage planes or veins dipping inwards towards the centre at an angle determined by the declivity of the rock or moraine which supports them, gradually becoming more vertical as the centre of the glacier is approached, where they twist round by degrees, so as to become transverse to its length, and to form part of the system of planes dipping inwards first described. Fig. 1 exhibits a section parallel to the length. Fig. 2, a transverse section.

FIG. 1.   FIG. 2.

" You are already aware that this structure consists in the alternation of more or less perfectly crystallized ice in parallel layers, often thinning out altogether like veins in marble, not unfrequently parallel and uniform like a ribboned calcedony or jasper.

" I will, for brevity, merely state the modifications which this fundamental
type undergoes, bringing together glaciers of all classes, but reserving the detail of examples and proofs, of which my experience has already furnished me
with a great number, to another occasion.   If a glacier lies long and narrow,
as the Lower Aar, or the Mer de Glace of Chamouni, the frontal dip is the
least conspicuous part of the phenomenon ; and if it terminate in an icy cascade, as in the second case, it might escape observation altogether.   The vertical planes parallel to the length, or nearly so, usurp nearly all the breadth
of the glacier, and only in the centre is a narrow space, where not unfrequently the structure appears quite undefined.   I have satisfactorily made
out, however, in every glacier which I have had the means of examining with
that view, that the conoidal structure, however obscured, exists in all parts of
the true glacier, modified, according to its length and breadth, in the manner
which Figs. 3 and 4 indicate.   I need not add, that these rude sketches are

Fig. 3.   Fig. 4.

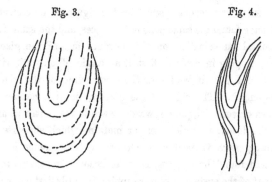

not intended to be considered as rigorously exact, but only to explain generally my meaning.

" There is yet another modification, but only a modification, of the above,
namely, in the case of extremely steep glaciers, but which are coherent, and
not crevassed into pyramids.   There are numberless examples of these in all
the higher valleys of the Alps, which do not descend into the hollows, but festoon the steep sides of snowy mountains.   They are, I believe, what Saussure
called glaciers of the second order, and have no relation to *névés*, so far as I
can attach a meaning to that term.   They are of hard ice, and almost invariably present an appearance of stratification parallel to the soil on which they
rest.   This stratification is only apparent ; the cleavage planes dip forwards

and outwards, instead of dipping inwards, as in the terminal portion of glaciers of less inclination. The surfaces of crystallization have, in this case, absolutely the form of a scallop-shell, the lip or front being always inclined below the horizon. I attach importance to the community of feature in glaciers of every form and inclination, because it indicates that the origin of the structure cannot be unimportant, considering its generality; and, in this particular case of small steep glaciers, it appears, I think, that M. de Charpentier, who has justly denied the stratification of glaciers in general, has wrongly admitted the existence of strata in the case in question, which he regards as formed by the intercalation of mud from the soil in some manner, which, if I recollect rightly, he does not very clearly describe. Now, these seeming strata of mud I have examined in a multitude of cases, and found invariably to result merely from the percolation of dirt from the moraine, sometimes even accompanied by small fragments of rock, into the more spongy and less crystalline veins of the glacier mass which already existed; and the proof is, that, by cutting with a hatchet, we gradually gain the pure ice, equally veined with the exterior, but not discoloured. I may observe, in passing, that the fissures which, in the lower part and near the sides of glaciers, form the granules, about which so much has been written, are stopped by the independent formation of the veins in the ice, which thus demonstrate their prior origin.

" One afternoon I happened to ascend higher than usual above the level of the Mer de Glace, and was struck by the appearance of discoloured bands traversing its surface nearly in the form indicated in fig. 4. These shades, too indistinct to be noticed when near or upon the surface, except upon very careful inspection, are very striking and beautiful when seen at a distance by a light not too strong, as in the afternoon or by moonlight. They are evidently bands of dirt on the surface of the ice, having nearly the form of very elongated parabolas merging in the moraines on either side, widest apart from one another in the centre, and confounded towards the edge. For some time I was at a loss to conceive how these sort of false moraines could spread from side to side of the glacier, but I at length assured myself that it was entirely owing to the structure of the ice, which retains the dirt diffused by avalanches and the weather on those parts which are most porous, whilst the compacter portion is washed clean by the rain, so that these bands are nothing more than visible traces of the direction of the internal icy structure, and of course correspond with what has been already stated as to the forms in which the conoidal surfaces intersect the plane of the glacier. I counted distinctly six-

teen of these bands on the surface of the ice then in view. I afterwards
traced them to the higher part of the ice-field ; and the only distinction which
I there observed was, that the loops of the curves were less acute, or more
nearly circular, fig. 5. All glaciers do not show this external
evidence of their structure equally, as there are some glaciers
which possess the structure itself more developed than others.
The cause of the dazzling whiteness of the Glacier des Bossons
at Chamouni is the comparative absence of these layers of granular and compact ice ; the whole is nearly of uniform consistence, the particles of rock scarcely find a lodgment, the whole
is washed clean by every shower. The superficial bands are
well seen on the Mer de Glace of Chamouni, and, to quote
another example, one of the last I have seen, very admirably
on the Glacier of Ferpêcle, in the Valley of Erin, where I
counted above thirty in view at once.

FIG. 5.

" I am quite persuaded that these bands, and of course the structure which
they represent, have their origin in the movement of the glacier ; and if the
laws of movement, ascertained independently, shall coincide with, or confirm
the phenomena of structure, we shall be better able, from the comparison of
the two classes of facts, to decide upon the *cause* of movement.

" What I have hitherto stated is matter of *fact*. I will state very briefly
what I am disposed to deduce by way of hypothesis.

" It is impossible to consider these structural bands on the surface of the
glacier, in combination with the fact established in my former letters, that the
centre of the glacier moves considerably faster than its edges, without believing that the bands are an indication of the motion, and that the motion gives
rise to the veined structure. These dirt-bands perfectly resemble those of
froth and scum which every one has seen upon the surface of slowly-moving
foul water ; and their figure at once gives the idea of *fluid motion*, freest in the
middle, obstructed by friction towards the sides and bottom. It will be found
that the analogies are entirely favourable : the glacier struggles between a
condition of fluidity and rigidity. It cannot obey the law of semifluid progression (maximum velocity at the centre, which is no hypothesis in the case
of glaciers, but a fact) without a solution of continuity perpendicular to its
sides. If two persons hold a sheet of paper, so as to be tense, by the four
corners, and one moves two adjacent corners, whilst the other two remain at
rest, or move less fast, the tendency will be to tear the paper into shreds

parallel to the motion ; in the glacier, the fissures thus formed are filled with percolated water, which is then frozen. It accords with this view, 1. That the glacier moves fastest in the centre, and that the loop of the curves described coincides (by observation) with the line of swiftest motion. 2. That the bands are least distinct near the centre, for there the difference of velocity of two adjacent stripes parallel to the length of the glacier is nearly nothing ; but near the sides, where the retardation is greatest, it is a maximum. 3. It accords with direct observation, (see my last Letter,) that the *difference* of velocity of the centre and sides is greatest near the lower extremity of the glacier, and that the velocity is more nearly uniform in the higher part; this corresponds to the less elongated form of the loops in the upper part of fig. 5. 4. In the highest part of such glaciers, as the curves become less bent, the structure also vanishes. 5. In the wide saucer-shaped glaciers already spoken of, which descend from mountain-slopes, the velocity being, as in shallow rivers, nearly uniform across their breadth, no vertical structure is developed. On the other hand, the friction of the base determines an apparent stratification, parallel to the slope down which they fall. 6. It also follows immediately, (assuming it as a fact very probable, but still to be proved, that the deepest part of the glacier moves slower than the surface,) that the *frontal dip* of the structural planes of all glaciers diminishes towards their inferior extremity, where it approaches 0, or even inclines outwards, since there the whole pressure of the semifluid mass is ˙unsustained by any barrier, and the velocity varies (probably in a rapid progression) with the distance from the soil; whilst, nearer the origin of the glacier, the frontal dip is great, because the mass of the glacier forms a virtual barrier in advance ; and the structure is comparatively indistinct for the same reason that the transverse structure is indistinct, viz., that the neighbouring horizontal prisms of ice move with nearly a common velocity. 7. Where two glaciers unite, it is a fact, that the structure immediately becomes more developed. This arises from the increased velocity, as well as friction of each, due to lateral compression. 8. The veined structure invariably tends to disappear when a glacier becomes so crevassed as to lose horizontal cohesion, as when it is divided into pyramidal masses. Now, this immediately follows from our theory ; for so soon as lateral cohesion is destroyed, any determinate inequality of motion ceases, each mass moves singly, and the structure disappears very gradually.

" I might add more illustrations ; but let these suffice for the present. It is not difficult to foresee, that, if my view should prove correct, a theory of glaciers may be formed, which, without coinciding either with that of Saussure or Charpentier, shall yet have something in common with both. Whether that of M. Rendu may not avail something, I am unable to say, not yet having been able to procure his work.

" It yet remains to decide, what is the cause of the succession of dirt-bands at considerable distances on the surface of the glacier, indicating the succession of waves of more or less compact ice. In all the glaciers where I have yet distinctly observed them, they appear to follow a regulated order of distances, nearly the same for a considerable space, but closer the farther we ascend the glacier. I cannot help thinking that they are the *true annual rings** of the glacier, which mark its age, like those of a tree, only increasing, instead of diminishing in breadth as the ice grows older, coinciding again with the fact which I formerly established, that the higher part of a glacier moves, generally speaking, more slowly than its lower extremity. The different states of the glacier at different seasons, the presence or absence of snow, or even the simple difference of velocity at different seasons, would be sufficient to account for this alternation of structure. There is no cause so likely to produce it as some *annual change.* I may add, that some observations which I have already made on the distances of these bands, as well as information which I have endeavoured to collect, lead me at least to have some doubt as to the correctness of the opinion generally entertained, that the glaciers are stationary in winter, perhaps even, that there is any very great inequality in their march at different times of the year. I am, my dear Sir, yours very truly,

JAMES D. FORBES."

Professor JAMESON.

GENEVA, 5th October 1842.

" MY DEAR SIR,—Since my last letter from Zermatt, I have had an opportunity of examining the glaciers on different sides of Monte Rosa, particularly those of Lys and Macugnaga, and those near the Valley of Saas; and on my return to Chamouni early in September, I devoted a day to each of the glaciers of Trient and Argentière, before resuming my station at the Montanvert, where I remained until almost the last days of the month.

" What I think it most interesting now to add as supplementary to my former statements, is, not a description of these various glaciers, but with parti-

---

* Originally printed *annular rings* by a typographical error. See editor's note *Ed. Phil. Jour.*, April 1843, p. 382.

cular reference to the Mer de Glace, to mention what the extended period of examination which I have been able to give to it has enabled me to conclude beyond what is contained in my previous letters, respecting the Theory of glacier movement generally. Having accurately observed the condition and motions of this glacier throughout by far the greater part of the season at which it, or indeed any glacier is easily accessible, or sufficiently free from snow for accurate observations,—having also, especially during the month of September, observed it under every circumstance of weather, and a great range of atmospheric temperature, I believe that I have obtained the chief data necessary for basing a theory of its motion, upon sound mechanical principles. The changes which I have witnessed upon its surface, during the period of above three months during which I have studied it, are so great and remarkable, and in some respects so unexpected, as to be of capital importance in any theory which may be proposed.

"I was very greatly struck with the change in the general appearance of the glacier during my absence, from the 10th August to the 10th September. I left it comparatively high and tumid in the centre, at no great depth below the *arrête* of its natural boundary, the moraine by its side; and fissured by crevasses, deep and rather narrow, with well defined vertical walls. On my return, the icy mass had most visibly sunk in its bed, it seemed to me to have a wasted, cadaverous look; the moraines protruded far higher than before from its sides, and the ice itself clinging to the moraine at a considerable height above its general level, was covered by the fallen masses of stone and gravel which had rolled down the inclined plane formed by this central subsidence. The whole resembled somewhat the Wye, or some of those narrow tidal rivers whose muddy banks are left exposed by the retreat of the ocean. That this subsidence was in a good measure occasioned by the melting of the ice in contact with the bottom of the valley in which it lies, and by the falling together of the parts in a soft and yielding state, owing to a complete infiltration of the whole mass with water during the warm season of the year, was proved by a variety of circumstances which I shall not stop to detail. I may mention, however, that the crevasses were wider, but less deep and regular, —excessively degraded on the side to which the mid-day sun had free access and in many places where several crevasses nearly joined, the icy partitions had sunk gradually towards a level, and thus rendered the fissured parts of the glacier more easily traversed than at an earlier part of the season. It is plain, too, that the fact of the more rapid advancement of the centre of the glacier mentioned in my earliest letter, implies a subsidence of that part, and a consequent drain from the lateral ice, to supply the vacuity which it leaves.

"It will at once be understood that the change of which I speak in the external figure of the ice, its crevasses and inequalities, is an effect due to the season, and must be repeated every year. Were the summer considerably prolonged, the annihilation of the glacier would take place from a simple continuation of the process, namely, the increased velocity of the central part, the exaggeration of the crevasses in width, and the falling of their walls, or rather the gradual subsidence of the elevations, softened by the warmth, into the hollows which separate them, whilst the moraine would be left in all its continuity as a witness of the original boundary of the glacier. The ice must possess within itself some reproductive power (if the phrase may be permitted,) to restore it in spring to the level from whence it had descended; and since crevasses thus form, extend, and again vanish,—perhaps in a single season, but certainly in a very few years,—we must consider the glacier as a much more plastic body than it has commonly been imagined.

"I state it, then, as a result of observation the most direct, that, in the early part of summer, the glacier level is highest, and the fissures less numerous. The latter form and widen especially during the months of June and July; and, in the beginning of August, the glacier is most difficult to traverse, (generally speaking,) owing to the multitude and sharpness of these cracks; but later, the prolonged sunshine and autumnal rains not only reduce the ice to water, and thus carry off a part of its surface, but leave the remainder in a softened and plastic state, in which the tendency is to a general subsidence of all the elevations, whilst the prolonged excess of velocity of the central above the lateral parts, causes an increased hollowness and subsidence there, and produces a great fissuring, the lateral ice still clinging to the moraines, which it is compelled gradually to uncover. Before spring, by some process which it remains to explain, the level of the ice is restored, (supposing the glacier not to be permanently wasting.)

"Another mode of considering the successive conditions of a certain portion of the glacier, will lead also to the admission of the ever-varying state of its aggregation and subdivision. In a glacier, like the Mer de Glace of Chamouni, which presents a great many and well marked "accidents" of surface in its different parts, it is yet perfectly well known, that, though continually moving and changing, the distribution of these "accidents" is sensibly invariable. Every year, and year after year, the water courses follow the same lines of direction,—their streams are precipitated into the heart of the glacier by vertical funnels called " moulins," at the very same points ; the fissures, though forming very different angles with the axis or sides of the glacier at different points of its length, opposite the same point are always similarly disposed ;—the same parts of the glacier, relatively to fixed rocks, are every year passable, and the same parts are traversed by innumerable fissures. Yet the solid ice of one year is the fissured ice of the next, and the very ice which this year forms the walls of a " moulin," will next year be some hundred feet farther forward and without perforation, whilst the cascade remains immoveable, or sensibly so, with reference to fixed objects around. All these facts, attested by long and invariable experience, prove that the ice of the glaciers is insensibly and continually moulding itself under the influence of external circumstances, of which the principal, be it remarked, is its own weight affecting its figure, in connection with the surfaces over which it passes, and between which it struggles onwards. It is, in this respect, absolutely comparable to the water of a river, which has here its deep pools, here its constant eddy, continually changing in substance, yet ever the same in form.

" With reference to the yet more essential modifications of *structure*, I mean the veined structure which I formerly described, I showed in my last letter, that it is equally mutable and subjected to the momentary conditions of external restraint, and, that far from being an original structure in the higher part of the glacier, variously modified in its subsequent course, but never annihilated, it owes its existence at any moment to the conditions of varying velocity in different parts of the transverse section of the glacier, and that it is not unfrequently entirely destroyed in one part of the glacier, to be renewed in a totally different direction in another. A molecule of ice is as passive and structureless a unit as a molecule of water, so far as it has not that structure impressed by something external at the time. Like the water in the river, myriads succeed one another, and might be mistaken for the same.

" Few words will suffice to show how intimately what I have stated is connected with the first rudiments of a theory of glacier motion, which I endeavoured to sketch in my last letter, and the truth of which all that I have since seen has tended greatly to confirm. The centre of the glacier stream is urged onwards by pressure from above (how caused we shall immediately consider,) which is there resisted less than at the sides and bottom, owing to the comparative absence of friction. The lateral parts are dragged onwards by the motion of the centre, and move also, but it is quite compatible with this idea of semifluid motion, that the bottom of the glacier should remain frozen to its bed, as some writers have supposed to be the case, though I am far from asserting this to be the fact, or even supposing it probable. Why, then, are the fissures generally *vertical*, and also where a glacier is most regular, simply *transverse*, and not convex towards the lower extremity? The first of these questions had always till lately appeared to me a serious difficulty. The *fact* stated in the second, combined with the positive certainty that the centre of a glacier moves faster than its sides, in the ratio frequently of five to three, shows that an answer *must* be found, and, therefore, that it offers no insurmountable objection. The explanation is to be sought in the continually varying condition of the glacier, the perpetual renewal of the crevasses, the action of water in tending to preserve verticality, and the really small variation of velocity of different parts of the ice towards the centre of a glacier of immense depth. From these circumstances it follows that a crevasse is either renewed or altogether extirpated before its verticality is sensibly effected. For the same reason, a stick several feet long, inserted vertically in the ice, remains sensibly vertical so long as it stands at all, for the velocity of the surface is sensibly the same as that at 10 or 20, or probably even 100 feet deep in most glaciers. It is only near the bottom or bed that the velocity is materially affected, as I have found also, that, in respect to breadth, it is in the immediate neighbourhood of the sides that the velocity diminishes rapidly, and that, for half its breadth in the centre, the velocity does not vary by more than from $\frac{1}{10}$ to $\frac{1}{20}$ of its amount. It is farther worthy of notice, that whenever a glacier is of no great thickness, and, at the same time, highly inclined, that is, in circumstances calculated to produce a great difference between the motions of points of the glacier in a vertical line, there the fissures are not transverse but radiated, as in almost all glaciers of the second order, and, therefore, the fissures are not liable to distortion.

" I might put it rather as a direct result of observation than as a hypothesis, that the motion of a glacier resembles that of a viscid fluid, not being uniform in all parts of the transverse section, but the motion of the parts in contact with the walls being determined mainly by the motion of the centre; but it yet remains to be shown what is the cause of the pressure which conveys the motion, whether it is the mere weight of the semifluid mass, or the dilatation of the head of the glacier pushing onwards. The answer to this question involves the fate of the rival theories of De Saussure and De Charpentier. I still entertain the same difficulties with respect to both, which I have stated in an article in the *Edinburgh Review* ; but these difficulties amount, I think, to a proof of insufficiency, if taken in connection with the observations which I have made this summer. On the one hand, if it were possible that the glacier could slide by the mere action of gravity in a trough inclined only 3, or 4, or 5 degrees, it is probable that one of two things would happen : either it would slide altogether with an accelerated velocity into the valley beneath, or else it would move *by fits and starts,* being stayed by obstacles until these were overcome by the melting of the ice beneath, or by the accumulated weight of snow above and behind. Now, neither of these things happen ; the glacier moves on day and night, or from day to day, with a *continuous* regulated motion, which, I feel certain, could not take place were the sliding theory true.

" But, if possible, still stronger, as well as more multiplied, objections are to be found to the theory of dilatation, and I trust I shall not be accused of levity in thus, as it were, in a few lines, dismissing a theory which has so much *prima facie* plausibility to recommend it, and which has been maintained with so much ingenuity by men such as Scheuchzer, De Charpentier, and Agassiz. It is essential to the aim of this letter, that I state briefly the grounds of the conclusions at which I have arrived, whilst it is equally essential that my observations should be confined within small compass. In another place, I shall give them all the development that may be requisite.

" Summarily, then, (1.) The motion of the glacier, in its several parts, does not appear to follow the law which the dilatation theory would require. It has been shown (*Edinburgh Review,* April 1842, p. 77) that the motion ought to vanish near the origin of the glacier, and increase continually towards its lower extremity. I have found the motion of the higher part of the Mer de Glace to differ sometimes very little from that several leagues farther down ; whilst, in the middle, owing to the expansion of the glacier in breadth, its march was slower than in either of the other parts. (2.) Whilst I admit that the glacier is, during summer, infiltrated with water in all or most of its thickness, (a point on which I had last year great doubts,) I feel quite confident that, during some months of the year during which the glacier is in most rapid motion, no congelation takes place in the mass of the ice beyond a depth of a very few inches, much less during the cold of each night, and least of all, at *all* times, as appears to be now the opinion held upon the subject. Whilst I say that I am confident of this, I will state one proof. Less than ten days since I traversed the Mer de Glace up to the higher part of the Glacier de

Léchaud, while it was covered with snow to a depth of six inches at Montanvert, and three times as much in the higher part. It was snowing at the time, and for a week the glacier had been in the same state nearly, the thermometer having fallen in the meanwhile to 20° Fahr. Yet I had abundant evidence that the effect of the frost had not penetrated farther into the ice than it might be expected to have done into the earth under the same circumstances. All the superficial rills were indeed frozen over; there were no cascades in the "moulins;" all was as still as it could be in mid-winter; yet even on the Glacier de Léchaud, my wooden poles, sunk to a depth of less than a foot in the ice, were quite wet, literally standing in water, and consequently unfrozen to the walls; and in the hollows beneath the stones of the moraines, by breaking the crust of ice, pools of unfrozen water might be found almost on the surface. Is it possible, then, that the mere passing chill of a summer night, or the mere cold of the ice itself at all times, can produce the congelation which has been so much insisted on?

" But, (3.) What was the effect of the congelation, trifling as it was, upon the motion of the glacier? So sharp and sudden a cold succeeding summer weather, must inevitably, it seems to me, were this theory true, have produced an instantaneous acceleration of the mean motion of the glacier. But the contrary was the fact; the diurnal motion fell rather short of its previous value, and so soon as the severe weather was past, and the little congelation which had taken place thawed, and the snow reduced to water, the glacier, saturated in all its pores, resumed its march nearly as in the height of summer.

" (4.) It has been inferred from the dilatation theory, that whilst the surface of the glacier continually wastes, it is at the same time heaved bodily upwards from beneath, so that its absolute level is unchanged. My experiments, as well as the most ordinary observation, (as has been already remarked) disprove this hypothesis. I find that, between the 26th June and the 16th September, the surface of the ice near the side of the Mer de Glace had lowered absolutely TWENTY-FIVE *feet* 1.5 *inches*, and the centre had undoubtedly fallen more. The observation of the waste of the surface by the protrusion of a stick sunk to a determinate depth in a hole, is very inaccurate, and gives results *below* the truth.

" I am perfectly ready to admit, with M. de Charpentier, that the congelation of the infiltrated water of glaciers is an important part of their functions; only, I conceive that it occurs but once a year to any effective extent, instead of daily or continually, as he supposes. Every thing which I have seen on the glacier, during cold weather and when covered with snow, confirms the idea I have always entertained, that the progress of congelation in the mass of the glacier is very similar to that of a mass of moist earth, and that, therefore, the daily variations of temperature can make no sensible impression with respect to the mass of the infiltrated ice. The prolonged cold of winter must, however, produce a very sensible effect; and, considering that the temperature of the mass is never above 32°, it may be expected that the congelation of the water in capillary fissures in ice will, in the course of months of tranquillity, reach a great depth. I apprehend that there is only an annual congelation, and that its effect is not to move the glacier onwards by sliding down its bed —for that the friction of so enormous a body seems evidently to render impossible—but (what Mr. Hopkins has very well shown is the only alternative, and which he has used as an argument against Charpentier's theory) to dilate the ice in the direction of *least resistance*—that is, vertically—and consequently to increase its thickness. The tendency of such a force would, therefore, be to restore, during the winter, the thickness of ice lost during the summer; and in those winters which are less severe, a less depth of ice being frozen, a less expansion would occur, and a permanent diminution of the glacier would result. Nothing can be more certain than the fact, so well stated by Charpentier in his 10th section, that the glacier does not owe its increase to the snow of avalanches, nor indeed to any snow which falls on the greater part of its surface.

" In conclusion, the admission of semifluid motion produced by the weight of the ice itself, appears to explain the chief facts of glacier-movement—viz., (1.) That it is more rapid at the centre than at the sides; (2.) For the most part, most rapid near the lower extremity of glaciers, but varying rather with the transverse section than the length; (3.) That it is more rapid in summer than in winter, in hot than in cold weather, and especially more rapid after rain, and less rapid in sudden frost; (4.) It is farther in conformity with what we know of the plasticity of semisolids generally, especially near their point of fusion. Many examples will occur to every one, of what they have observed of the plasticity of hard bodies—such as sealing-wax, for example— exposed for a long time to a temperature far below their melting heat, and which have moulded themselves to the form of the surfaces on which they rest. (5.) When the ice is very highly fissured, it yields sensibly to the pressure of the hand, having a slight determinate play, like some kinds of limestone, well known for this quality of flexibility. (6.) I have formerly endeavoured to show how such a condition of semirigidity, combined with the determined movements of the glacier, accounts for the remarkably veined structure which pervades it. I am, my dear Sir, yours very truly,

JAMES D. FORBES."

PROFESSOR JAMESON.

No. III.

ON SOLAR RADIATION.

The experiments mentioned in the text, (page 215,) referred to a curious inquiry which has occupied my attention for some years, namely, the loss of force which the sun's rays experience in passing through the earth's atmosphere. It might seem, at first sight, an impossible task to determine the comparative measure of the sun's heat in the state in ·which it arrives at the earth's surface, and that which it would have attained were the atmosphere wholly removed. Some approximation to such a result has, however, been obtained by a very simple, though indirect method. The thickness of air traversed by a sunbeam is, of course, least when the sun is vertical, and greatest when he is near the horizon; at intermediate elevations the heat is intermediate. Now, by comparing the thermometric effect of the sun's rays (which is the object of the actinometer) at several different thicknesses of atmosphere, the *law of extinction* is approximately found, and an inference is made as to what the intensity would be when the thickness of atmosphere is nothing. This inference will be proportionally more accurate as the observations are pushed to a less thickness of interposed air; and I have shewn in the paper already referred to,* that the previous estimates had greatly under-rated the intensity of the unimpaired solar beam, and had also under-rated the absorptive power of the atmosphere, owing to the observations on which they were founded having been made only when the sunbeam had already traversed a great thickness of air, when the law of absorption is very different from the law at small thicknesses.

---

* *Philosophical Transactions for* 1842, page 225.

Now, to obtain observations of solar heat at small thicknesses, we must, in the first place, ascend in the atmosphere, and also use the sun's rays when his elevation is greatest, that is, near the solstice. I mounted the Cramont in hopes of prosecuting these experiments, when the sun had still 21° of northern declination and after having left below me a thickness of 9000 feet of the densest part of the atmosphere. Unfortunately, as we have seen, these delicate experiments were prevented by indifferent weather.

It will probably surprise many persons to be told, that even when the sun's rays shoot vertically through a pure atmosphere, as between the tropics, they lose in their passage (owing to the opacity of the air) very nearly half their intensity.* The intercepted heat goes, of course, to warm the air.

The object of this note is, however, to record a different set of observations performed with an instrument of inferior delicacy to the actinometer, but still capable of yielding very remarkable results,—I mean Leslie's photometer. Its principle may be briefly described as measuring the difference of the heat absorbed by a dark and clear thermometer-ball. It is well known that this instrument gives, on some occasions, results which appear highly paradoxical, but which, if consistent, require to be explained, and ought, therefore, to be distinctly established. My observations with it were directed to two points.

I. To ascertain the effect of the presence of a coating of snow on the ground in magnifying the apparent Solar Radiation. To this effect has been ascribed the extraordinary force of the sun's rays observed in arctic climates, and also some singular variations from one season to another, supposed to depend on the presence of snow on the ground.† Now, the few experiments which I obtained before breaking my instrument last summer (1842) gave me the following most striking results :—

Surrounded by grass, and exposed to direct sunshine, the
photometer indicated, ... 78°
Exposed upon snow instead of grass, it rose to . . 121°

The whiteness of the snow is all-important in this respect; dirty snow produces comparatively little effect, and so does ice. Thus,—

The Photometer, exposed on a dirty part of the Mer de
Glace, stood at . . . . . 70°
Placed upon a neighbouring patch of snow, . . 140°

---

* *Phil. Trans.*, as above.
† *Edin. New Phil. Journal* 1841. A paper by Dr. Richardson, with remarks.

This action fully explains the intensity of the effect of fresh snow upon the eyesight. I have myself found that exposure for several weeks to the moderate glare of sunlight reflected from a glacier surface, produced little effect upon me, whilst I suffered severely from a single day spent amongst the pure snows of the highest summits.

II. On the photometric effect of the diffuse light reflected from the sky, Professor Kämtz first, I believe, announced the startling fact, that *half* the photometric effect on Leslie's instrument is due to the diffuse light of the sky, the other half only being the effect of the direct rays of a bright sun! This singular paradox also manifests itself by the fact, that cloudy weather, if the sun be not itself greatly obscured, apparently increases the effect of solar radiation. Of the truth of both of these facts, I had also, last year, sufficient evidence, of which I shall quote one or two examples.

On the 28th June, 1842, a warm and clear day, at 6000 feet above the sea, at 1h. 40m.,

The photometer, in the sun, placed on snow, stood at . . . 121°

An alpine pole, an inch in diameter, was then stuck into the snow, so as to throw its shadow on the instrument, thus intercepting the *direct* sunbeam only. It fell to . . . . 82°

Leaving for the *direct* effect of the sunlight, ... 39°
the remaining 82° being derived from the reflection of scattered sky-light, and from the snowy surface.

At the same place and time, the photometer, in the sun, surrounded
by grass, stood at . . . . 78°
Shaded by a stick, as before, ... 30°

Direct sun effect,* . . . 48°

Now, if we look to 78° as the total effect of the sun's light and its reflection from the ground, in one of the hottest days of June, in a fine climate and 6000 feet above the sea,—it appears to be inconceivably small, when we know that the same instrument often stands at 120° in moderately fine weather in

---

* The difference of this and the last is probably due in a great measure to a defect in the principle of the instrument, the momentary increment of heat necessary to maintain the temperature 1° above 120°, being greater than what is necessary to maintain it 1° above 77°, for example. There are likewise other sources of error.

Scotland. What is the reason? The sky does not reflect so much light when it is pure as when it is milky, and its surface being immense, compared to the apparent surface of the sun, (100,000 times greater,) a small addition to its reflective power increases the photometric indication more than the thin haze of vapour which produces that effect diminishes the direct sun-light.

Again, on the 29th of June, a clear and warm day, in the same position, at 11, A.M.,—

The photometer standing on snow, ... 114°
Shaded from the *direct* sun, ... 63°
                                                          ——
Direct sun effect, ... 51°

I never observed these effects so strongly as on the 30th June, a day of the most intense solar heat, when, at a height of 7600 feet, the sky exhibited a deep indigo tint, unusual for that moderate elevation. I was engaged, for some hours, in making trigonometrical observations, on an exposed promontory of rock, with scarcely any shelter from the piercing sun-beams. At length I was so exhausted as to be obliged to thrust my head now and then behind a stone for protection and relief. Now, at this time, the photometer, directly exposed on the rock to the sun, stood only at . . . 88°
When shaded from the *direct* sun-beam it fell to . . . . 22°
the smallest result for diffuse atmospheric radiation and reflection from the soil combined which I have witnessed. This, it will be observed, leaves for *direct* sun-heat, ... 66°
which is large compared with the previous results. It is certain from these experiments that the photometric effects thus measured bear no kind of proportion to the physiological effects of direct and reflected heat.

No. IV.

MR. HOPKINS' EXPERIMENT. REFERRED TO, PAGE 364.

Mr. Hopkins of Cambridge, who has recently espoused the Sliding Theory of De Saussure in opposition to De Charpentier's Theory of Dilatation, has illustrated it by an ingenious experiment. He placed a mass of rough ice, confined by a square frame or bottomless box, upon a roughly-chiselled flag-stone, which he then inclined at a small angle, and found that the gradual dissolution of the ice in contact with the stone produced a slow and uniform motion, at a very inconsiderable slope, such as a degree, or even less. As showing that such a sliding motion may take place without acceleration, this experiment is undoubtedly very interesting; but the parallelism to the case of a glacier is so far incomplete, for the reasons which I have stated in the text, and others which might be mentioned, that I cannot suppose it possible that any glacier can move as a *rigid* mass in this manner. At the same time, I readily admit, that the motion of the superficial and central parts having commenced by their own weight, and being rendered possible by their plasticity, the inferior and lateral portions of the glacier may slide over the surface on which they rest somewhat in the manner of the ice in Mr. Hopkins' experiment; although I think it may admit of doubt, whether, even in the least elevated glaciers, the amount of fusion is sufficient to produce a great effect, and whether, in the higher regions, or in glaciers of the second order, it can act sensibly. But Mr. Hopkins' views will no doubt soon receive a farther development, in which these objections may be more or less removed.

POSTSCRIPT to Page 152.

In closing this Volume, I am happy to be able to give the movement of the ice near the Montanvert, as observed by Auguste Balmat, down to the 8th June 1843. From the 4th April to the 8th June, the great stone, D. 7. moved   .   .   .   . 88 feet 1 inch.
Or, daily, ...   16.3 inches.
Connecting the observations of pages 151, 152, with those of the motion of the neighbouring station D 2, in the summer of 1842, (page 139,) we obtain for the total motion of the *lateral* part of the Mer de Glace at the Montanvert—

June 29 to Sept. 28,	.	.	.	132 feet.
Oct. 20 — Dec. 12,	.	.	.	70
Dec. 12 — Feb. 17,	.	.	.	76
Feb. 17 — April 4,	.	.	.	66
April 4 — June 8,	.	.	.	88
Motion in 322 days,	.	.	.	432 feet.
Proportional motion for the whole year,				483 feet.

The movement of the centre is probably, at least, two-fifths greater, corresponding closely with the intervals of the "dirt bands" of the glacier. See page 165.

INDEX.

EDINBURGH : PRINTED BY T. CONSTABLE,
PRINTER TO HER MAJESTY.

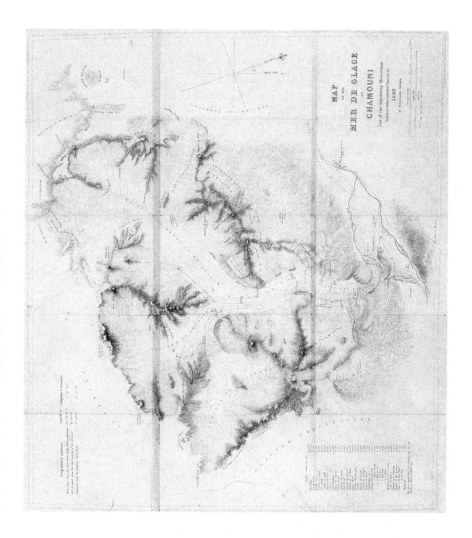

The material originally positioned here is too large for reproduction in this reissue. A PDF can be downloaded from the web address given on page iv of this book, by clicking on 'Resources Available'.

Printed in the United States
By Bookmasters